ABOUT THE IMAGE ON THE FRONT COVER

The picture on the front cover is one of the most celebrated frescos in the world, called "The School of Athens". It was painted by Raphaello Sanzio da Urbino (1483-1520) or Raphael, as he is known to us all. His works were conspicuous for their grace and perfection. He was youngest of the three Italian painters, who dominated the period known as High Renaissance. Pope Julius II had commissioned him to paint some frescos on the walls of the Stanza della Segnatura in the Apostolic Palace at Vatican. The School of Athens was one of them. It was completed between 1511 and 1512.

Even though the black and white rendering of this work, given here, robs from it much of its original beauty, nevertheless the reader will find it still carries with it the impact of Raphael's genius.

On looking at the picture our attention comes to rest on the two central figures absorbed in deep discussion. They seem oblivious of the many figures on either side of them. It is their gaze that holds our attention all the more on the two moving towards us. On our left, the elder figure is Plato, the famous Greek philosopher. The other on right is Aristotle, his renowned pupil, whose theories had been the mainstay of the European civilization for many centuries,

The architecture of the building is imposing enough to make any monarch to be its proud owner. The great arch above the two central figures and the other two behind it adds to its magnificence by giving the fresco a sense of third dimension. Aristotle and Plato appear to be coming towards the steps leading out of the great building, whose walls are embellished with many statues - we can only see two of them that face us. On one side is the statue of Apollo holding his lyre and on the other is Minerva, identified by her shield with the gorgon's head. The space in front of these two Greek deities and the steps leading out of the building are filled with people. They are variously occupied. Some alone in thought or absorbed in their work, while still others in earnest discourse.

When we come to learn their identity; it appears to be an anachronism, because so many great talents from across the ages are gathered together in one place at the same time. However, when our eyes settle on the foreground we come to realize that Raphael reconciles the anachronism through the arch standing on two ornate pillars on either side in the foreground, which actually reveals it to be a stage. We then realize that what we are seeing are only actors depicting the great talents.

Some of the figures in the image that follows are numbered. It will enable the reader to identify them by using the key given below. In the appendix at the back, I have added the details of what they represent and a map, which shows where the ancient philosophers lived and worked.

Key to the some of the identities of the persons depicted in the School of Athens

1: Zeno of Citium 2: Epicurus 3: Federico II of Mantua? 4: Anicius Manlius Severinus Boethius or Anaximander or Empedocles? 5: Averroes 6: Pythagoras 7: Alcibiades or Alexander the Great? 8: Antisthenes or Xenophon? 9: Hypatia (Francesco Maria della Rovere)[9] 10: Aeschines or Xenophon? 11: Parmenides? 12: Socrates 13: Heraclitus (Michelangelo) 14: Plato, (Leonardo da Vinci) 15: Aristotle 16: Diogenes of Sinope? 17: Plotinus? 18: Euclid or Archimedes with students (Bramante)? 19: Zoroaster 20: Ptolemy? R: Apelles (Raphael) 21: Protogenes (Il Sodoma, Perugino or Timoteo Viti).

The names in brackets are those of Raphael's contemporaries. It is believed he used them as models to depict the ancient philosophers.

CONTEMPLATIONS

PART I
THE GATHERING OF KNOWLEDGE

http://en.wikipedia.org/wiki/The_School_of_Athens

THE SCHOOL OF ATHENS by RAPHAEL

AuthorHouse™ UK Ltd.
500 Avebury Boulevard
Central Milton Keynes, MK9 2BE
www.authorhouse.co.uk
Phone: 08001974150

©2010 Arun Bose. All rights reserved.

No part of this book may be reproduced, stored in a retrieval system, or transmitted by any means without the written permission of the author.

First published by AuthorHouse   15/12/2010

ISBN: 978-1-4520-4914-4 (sc)

This book is printed on acid-free paper.

TO
MY WIFE SUDESHNA
FOR HER PATIENCE IN BEARING
WITH ME OVER THE YEARS
AND
OUR DEAR DAUGHTER
SUDAKSHINA

ACKNOWLEDGMENTS

In writing this book, I was aware of most of the things in this book from previous knowledge. Also my father had taught me well. He had a very large collection of books, which ran to over four thousand. Unfortunately due the vicissitudes of life only a few stragglers remain. He presented me with the Encyclopaedia Britannica, when I was nineteen, which has been much used. In this context, I also wish to remember the late Sourangshu Mohon Chowdhury, who was my father's childhood friend and my friend also. He was a frequent visitor to our house. His discussions with my father and myself had taught me much. He was a very erudite person. I also had the opportunity to go to the best schools in India at the time. Even today, after all these years I can vividly recollect many of their lessons, which is the hallmark of a good teacher. I was lucky enough to have been taught biology and physics through practical experiments from the early age of ten. Not many in India had such an opportunity during those days. Today the internet is a wonderful place to pick up knowledge, learn and update knowledge. I have found the Wikipedia especially helpful in this aspect. I have also used it to ascertain many of the facts and figures, which I have I have discussed.

I have used many images from NASA and United States Federal Government and some other official sites. It is wonderful that they provide such a service to the public and I wish to thank them. I am grateful to those individuals who have released their work in the form of images into the Public Domain and others who have allowed their images to be shared by others by through Creative Commons. I have acknowledged each of their contribution under their respective images. I also thank Mr. Peter Ifland and Mr. Victor van Wulfen for giving me permission to use their pictures from their websites. All the coloured images are excellent and some of them even breath-taking. It is my misfortune that I have had to render these images in black and white, because of publishing requirements that were unavoidable.

I wish to thank Mr. Pradip Sett, Dr. Amalendu Chakraborty and Mr. Arun Mullick for going through the manuscript and giving their opinion. Mr. Sett is a friend from my earlier days. I met him first while playing hockey. His life has been colourful. In fact too colourful! I will only recount only one incident. It is said that he is the only person to have travelled motorway in the wrong direction for quite a long distance. One can only imagine the horror of the oncoming traffic! I am assured he was not intoxicated at the time. He went on to become a neurosurgeon and now works in England. He has been kind enough to go through the preliminary manuscript of the whole book, which I had sent him. However, he is not too convinced about my hypothesis about the image of the lens being responsible for changing the structure of the brain, which I have described in the second part. I do not take his comments seriously, because he is merely a neurosurgeon, whereas I have to formulate a theory, which looks at the broader picture. He thinks the chapter on *New Concept of God*, in the third part, is weak. He would also like a few more stories to lighten the reading in parts of the book. I do agree that the facts presented in the book could have been interwoven more closely with stories to make the book even more interesting. As for example Tycho Brahe, Napier and Shakespeare all have a connection.[1] Prospero of Shakespeare's Tempest is drawn from Tycho, who worked his "magic" from an Island (Hven or Ven). The logarithmic tables, which was of great help to the mathematicians, scientists and students before the days of calculators and computers was developed by Napier, after heard about Tycho's methods. This is how it came about.

When Anne of Denmark sailed for Scotland to be united with James VI of Scotland, later to be James I of Great Britain, whom she had married by proxy, there arose a great tempest and the ship had to retreat back to Denmark. On hearing what had occurred; James in a romantic gesture, like a gallant knight and lover of legends, sailed forth to bring home his wife. In Denmark, he and the gentlemen of his court, who accompanied him, went to visit Tycho Brahe's famed state of the art observatory. His personal physician was with him, who came to learn about some of Tycho's method of calculations. Apparently, he had sufficient knowledge to describe his experiences to Napier. Napier was to invent logarithms. This enabled Shakespeare to obtain a plot for his play The Tempest. There is of course another standard version of where he got his idea, but this one is equally plausible. The storm, the lonely prince (Tycho) turned magician (Tycho used to dabble in alchemy), who lived in a lonely island (Hven) with his beautiful daughter (here Anne of Denmark) who married a shipwrecked prince (though the shipwreck was a writers license, but the prince here was James). Thus, today, Tycho Brahe lives on in the image of Prospero. Shakespeare for all his genius depended on his plots from various sources. He also knew which side his toast was buttered. He knew how to keep his royal patron pleased. As further confirmation of this fact, elsewhere he had portrays Macbeth as evil and Duncan as good, because the latter was James' ancestor. Contrary to the impression many have from Shakespeare's play, Macbeth was an able monarch. He ruled Scotland in relative peace for 17 years in those unruly times.

I feel too many stories tend to deflect the mind of the reader from the main theme of any book. So I do not think it is a good idea. Moreover, neither my publishers nor my readers would be too happy about making the already long book any longer.

Amalendu Chakraborty is a very erudite person. He can discuss almost any topic. He would have been considered a sage if he had been born two centuries earlier. He has also been kind enough to go through the manuscript in detail. His opinion is that no one will read the book. Maybe he is right. I will leave it to the reader to decide.

I am grateful to Arun Mullick, who was introduced to me by my late father 56 years ago. He has been a good friend and only person to encourage me throughout the process of writing. The reader must have noted his first name is the same as mine, interestingly our spouses happened to have the same first name also, so we gave our daughters the same name. I give him my thanks for his time, especially to discuss Tycho Brahe's last illness, evolution and clearing my understanding on biological portions.

---

1   Website: http://www.mathpages.com/rr/s8-01/8-01.htm
Kevin Brown: *Reflections on Relativity*, Chapter 8.1 Kepler, Napier, and the Third Law.

I am grateful to Dr. Bhattacharya of the Physics Department, Calcutta for going through the parts dealing with physics. The details of which are given in the introduction.

Mr. Tarun Ghosh came to my help regarding the images. He is a professional artist, working in Calcutta. I believe people from overseas have taken interest in his work. He gave me his time unstintingly. I am also grateful for his advice.

I wish to thank my daughter, Sudakshina, for giving her advice and opinion from time to time by interrupting her busy schedule.

I have had more than my fair share of illness; some of them life-threatening. Therefore before closing I have to thank all those who cared for me during my various illnesses, because without their help I would not have been in a position to have even attempted the book.

At one point while writing the book, I thought I would loose my eye. I had been to many doctors and to premier institutions in India without any result. In fact, my condition had deteriorated by the time I had returned to Calcutta. I wish to thank Dr. Kusal Chowdhuri, who was able to give me back my vision on that and another occasion. I am grateful to Dr Suresh Ramasubban for extending his considerable help to my family and myself during the early part of my present illness. I am also deeply indebted toDr. Utpal Chaudhuri and his friend Dr. Kanjaksha Ghosh for their continuing care of my haematological problem. My thanks go to Dr. Ranjan Rashmi Paul for his delicate handling of my dental problems due to my present condition.

Lastly, I would like to thank my publishers, AuthorHouse for their guidance and help to me as a first time author.

Contents

INTRODUCTION — xiii
   AN ALLEGORY — xiii
   AN ELABORATION — xiv
   HOW AND WHY I CAME TO WRITE THE BOOK — xix
   MY EXPERIENCES IN WRITING THE BOOK — xxvi

PART I THE GATHERING OF KNOWLEDGE

1 A PERSPECTIVE — 3
   INTRODUCTION — 3
   THE SOLAR SYSTEM — 19
   CONCLUSION — 42

2 TERRA MUNDI — 44
   EARLY BELIEFS ABOUT THE EARTH' SHAPE — 44
   THE CUMULATING EVIDENCE — 45
   LURE OF SPICES — 51
   THE FINAL PROOF — 77
   THE ACTUAL SHAPE OF THE EARTH — 81

3 KNOWLEDGE FROM THE HEAVENS — 83
   INTRODUCTION — 83
   SETTING THE SCENE — 86
   EARLIER GLIMPSES OF THE HELIOCENTRIC IDEA — 87
   THE RISE OF GEOCENTRISM — 89
   UNDERSTANDING THE CONTROVERSY IN ITS PROPER PERSPECTIVE — 96

4 THE TRIUMPH OF THE HELIOCENTRIC THEORY — 104
   THE ARCHITECTS OF THE HELIOCENTRIC THEORY — 104
   NICOLAS COPERNICUS — 105
   TYCHO BRAHE — 109

JOHANNES KEPLER	115
THE DFENDERS OF HELIEOCENTRIC THEORY	126
ISLAND UNIVERSES	127
THE FALL OF THE GEOCENTRIC HYPOTHESIS	128
THE FINAL VICTORY OF THE GEOCENTRIC THEORY	128

5 GRAPPLING WITH TIME — 129

AN INTRODUCTION TO THE NATURE OF TIME	129
MAN'S INNATE SENSE OF TIME	131
BIOLOGICAL TIME	132
EXTERNAL TIME AND HOW MAN CAME TO ABIDE BY IT	136
INSTRUMENTS FOR MEASURING TIME	144
THE NEW CHALLENGE	151

6 ARISTOTLE, GALILEO AND NEWTON — 158

ARISTOTLE'S IDEAS	158
FILLING IN THE HISTORICAL BACKGROUND	159
GALILEO	164
NEWTON	171

7 UNION OF SCIENCE WITH PHILOSOPHY - A BIOGRAPHY OF EINSTEIN — 192

INTRODUCTION	192
EINSTEIN	193

BOOKS AND INTERNET SITES CONSULTED — 224
APPENDIX — 225
REFERENCES — 229
INDEX — 231
About the Author — 247

INTRODUCTION

AN ALLEGORY

It was through an error of assumption man committed his original sin, which was the sin of pride and not because he had partaken of the fruit from the Tree of Knowledge. He presumed that he owned the garden in which he lived, along with everything in it; that his home was at the centre of the universe; and finally that he was the ultimate in creation. Had he the sense to have eaten from the Tree of Knowledge, his knowledge would have been complete! He would not have had to earn this knowledge the hard way and know that his assumptions were not true.

Then there was this wise serpent, who lived in the Garden of Ignorance with man. It was called so by the other creatures that lived there. Not because they themselves were ignorant in any way, but due to man's ignorance in claiming the garden to be his own. However, the other creatures were tolerant of man and did not mind. Man had arrived there only recently, so he had not had the time to learn their ways yet. Our friend the serpent was a nice creature. He generally minded his own business, always ready to help and give good advice; to which incidentally no one listened.

One warm afternoon, when the man was resting in the garden and deep slumber had not quite overtaken him yet. The serpent thought, it would be the best time to put an idea in his ear. He was wise this serpent, he knew that when fully awake man would throw up tantrums. It always hurt his ego to learn from others. Consequently, man could be very unreceptive to any new ideas, which were proposed to him when fully awake. So it came about that the serpent whispered softly in man's ear and told him all was not quite as it appeared to him! In his rapid eye movement sleep, the man thought he had dreamt that the serpent enjoined him to look for certain discrepancies in the natural phenomena, which he saw around him. If he looked hard enough, they would give him the clues to the right order of things. In the end it would help him to understand his place in relation to the scheme of the universe and eventually through this he would come to know more about himself. The good serpent also indicated that the answers lay in knowing about the very earth on which "his" garden grew and in the skies above him. This knowledge would set him on the right path that would ultimately lead to wisdom, a thing that all wish to possess in the end.

When the man woke up his mind was troubled by what he first thought was only a dream. Anything that troubled him made him feel insecure and consequently it upset him. This feeling of insecurity should not surprise anyone, as because being the first of his kind he did not have a mother. Later, more he thought about his dream the more he suspected it was the doing of the serpent, but he could not prove it. So he cursed the serpent and tried to go on his own way. But whatever he did from that day onwards, the feeling of unease caused by the serpent's words would not go away. The serpent was indeed wise. He knew how to get man's ear and would set him going on the right path eventually.

Now man began to mediate on the significance of his dream. He was no fool. Only he had not had the time to mature yet. You will remember he was the last to arrive, so everyone had pampered him. On looking around, he glimpsed that there were indeed some discrepancies between his ideas about the world and what he actually observed; just as the good serpent had indicated in his dream. He was now faced with the dilemma of how to reconcile these discrepancies. Trying to think of it all, he soon realized, once he started there would be no respite until he discovered the answers; otherwise it would be like an unfinished story whose end would be tantalizingly beyond reach. He concluded that it was time to go on a quest. He again cursed the serpent (it was his favorite thing to do whenever he was annoyed) for upsetting the comfort of his life, because at the back of his mind he knew well that this road to knowledge would only be toil and sweat. Never-the-less he set out on his quest, as he knew it was the only way he could regain his peace of mind once more. Realizing by this time that this was all the serpent's doing, he became the serpent's life-long enemy.

It would be after many, many ages when his children's children and their children's children would gather the knowledge to be able to turn it into wisdom. Thus through the serpent's advice his descendants would one day learn how to live in peace respecting all other life forms on this planet. Men would then once more be reconciled to the serpents again. Though that day is yet to come, it can only be hoped for man's sake it will not be too far off.

AN ELABORATION

In a sense, this allegory embodies the gist of what I have to say and therefore represents the essence of this book. I have therefore used it to introduce the book.

It appears to be loosely based on the story of the Garden of Eden. Though this is not the case, the garden here is only relevant in that it provides a stage for the story. It is only the presence of the serpent, which draws a parallel with the Garden of Eden. I could have chosen any other garden for the story with, say, a wise owl instead of a serpent. I chose the serpent over all the other creatures, not for the reasons in the Bible, but because today we know, the serpent is a much-misunderstood creature. Only some of them are poisonous. Their poison is not intended for us. Nature has dictated their bite should be poisonous, so that they may paralyze their prey, as because they do not have any hands. Just put yourselves in the serpents' shoes. Oops! Sorry, I forgot; my apologies to the serpents. Imagine you are trying to eat food, which was struggling to get away and your hands and legs were tied. The chances are high that the prey would escape and you would go hungry. So many serpents have got out of the predicament by evolving poison. The poison that happens to be their saliva is much superior to our own as far as digestion is concerned. For them this higher quality is needed to digest fur, tooth, claws, bones and all. In this, their anatomy gives them no choice, because they have to swallow everything.

I felt, here I could contribute my mite to correcting the distorted view we have of this much-maligned creature. After all, it is through evolution they have given up their legs to scramble on their ribs, while we strut about like lords with our heads in the air and with little regard for those creatures that live on the ground. Most of them scurry past when we pass by, without us noticing them. Only a small number amongst them will strike and then only if our actions happen to appear threatening to them. If it comes to a tally, we kill more serpents then they kill men. They do so only in self-defense, while we do it out of fear and vengeance born from our ignorance.

Regarding man in the allegory, though some may not agree, I hope I have represented him more or less accurately. It is here that the allegory's resemblance with the story in the Genesis ends. There is no God in the story, because the presence of God in the conventional sense is not necessary to the point I am trying to make. There is no Eve, that much-misused woman, whom God created and put beside man as a complement to the whole. It was men, who forgot that nature had destined her to nurture their kind and instead they subjugated her for their own interests. I will concede, some may contend that it was women who subjugated men. To many this has been a source of endless argument. I will not further the point, except to say it was men in general, who subjugated women and it was only women in particular, who were able to subjugated men. Lastly, just in case the reader wonders what happened to the Tree of Knowledge, I left it out of the story. Knowledge, unfortunately, does not grow on trees; it has to be earned. The allegory ends by man setting out on the quest for knowledge. What then is this knowledge, the man in the allegory went out to seek?

What is outside the universe we may never know! Our concern is with the universe. All knowledge we may possibly hope to attain lies within its compass. At first sight, this knowledge would seem too vast to acquire. Luckily, the principles of physics on which the universe happens are the same in all corners of the universe. With this, the volume of knowledge becomes much reduced. Yet this knowledge, which we speak of, is still very vast. Many may doubt as to whether man will ever be able to attain it all. Nevertheless, I will attempt to explain what this knowledge is about, which the man in the allegory went out to seek.

In essence all knowledge is interrelated and therefore to be taken as one. Nonetheless, it is convenient to treat it in separate parts. On one side, there is knowledge about the universe. On the other, there is knowledge about us. The knowledge of the universe has two aspects. First, there are the fundamental constituents of the universe, which are energy, matter, space and time. Second, we should appreciate; the universe is a dynamic entity as a result of energy and matter interacting in the background of space and time to produce events and events bring change. Thus events form the unit of change. The universe is continually changing with time. It is like a never-ending story, which is ever unfolding. We call this saga – *evolution*!

Today we have earned this knowledge through science and come to know our cosmic origin. Truly, we are the children of stars that have died in the distant past of our Universe's history Everything we see around us is star stuff. Thus, everything shares a common origin with us. Even though today we have attained this understanding, yet we are slowly distancing ourselves from the web of life, because of our misuse of technology and increasing population. It is only when if we come face to face with the spiritual aspect of our lives, in a broader sense and not in the narrow religious sense, only then we can integrate ourselves with the rest that share the planet with us. Otherwise, we will alienate ourselves to such an extent that we will no longer be a part of the web of life. The choice is stark; we are either within this web or out of it. If we stray away then we must exit ourselves. In plain terms, it implies our extinction!

So how do the other forms of knowledge, which we learn at schools and universities, fit into what I have said until now? Physics is the mother of all sciences. Physics started as the study of energy and this continued until the beginning of the last century. It then came to include the study of matter below the atomic level and rightly so because now properties of matter becomes dualistic and starts to show properties of energy as well. Chemistry was initially the study of matter. Today all chemistry can be explained by physics. The laws of physics form the basis for all nature. It lies at the root of all other branches of science such as astronomy, chemistry, biology, geology, geography etc. Mathematics is the language of physics and therefore of nature. Knowing mathematics gives us a better understanding of our universe. In language, literature, poetry and art we see our expression flower. History records all. The subjects like trade, manufacturing, economics, accountancy, law, forensics, engineering, medicine and the like; we use these to work out our lives. Culture, manners and politeness make our lives seamless. One cannot list everything, but briefly, this orients us to the different branches of knowledge that we come across. In what follows, is the story of that quest the man in the allegory took upon himself. It represents the spirit of man by which he distinguishes himself from all other creatures.

Nonetheless, like the man's initial reluctance there is the vast majority who do not want to know. That is also in our nature. However, the time of ignorance is past. The future of the planet is at stake! It is time to shake ourselves from the torpor of our existence and shed light on our ignorance. It is to this end the book has been written.

The book consists of three parts. Each part is a series of essays on various topics, which I wrote for different reasons, under different circumstances of my life and spanning across several years. It has been like a spiritual quest for me to find the answers to the many questions, which came to my mind from time to time. Finally, the answers fell into place. Though the various chapters may be read independently, hopefully the reader will find that I have knitted them together coherently to give a fuller sense, if the three parts are read as a whole. Its ultimate aim is to bring about a broader awareness to the reader, before discussing the problems faced by the planet due to man's presence. In the end, I have discussed some solutions.

The first part describes how man acquired knowledge. I have named it *"The Gathering of Knowledge"*. The second, I have called *"The ingredients of the Universe and Evolution"*, because it describes the ingredients of the universe and how they evolved matter and life. The last part is introspection about us. It places the age-old concepts that we have about religion, God, good and evil, and our morality on an objective basis, so that we may approach the problems we have created for the planet rationally and without bias. It is only after being armed with all this knowledge, the reader may go on to the last chapter, which offers some solutions to the problems man has created for the planet. This part is named, *"The Spirit of Man"* in the hope our spirit may rise to the occasion to meet the challenge. I consider this last part the more important part of the book. Since the book was born out of my contemplations, so I have named it *"Contemplations"*.

Today man has disturbed the ecology and polluted the planet to such an extent that it is fast developing into a crisis. If things continue in this fashion, then it will soon reach to a point of no return. All forms of life then would be affected severely, except probably the lowest forms. They are the only ones who can survive such a calamity. We have already caused much damage and are about to inflict more. Any effort that has been made to correct the error of our ways has not had any significant effect till now. There have been conventions, resolutions and laws by various governments, but to no real effect. This is because such efforts are never under a central control, whose authority extends across all the nations of this planet. Neither have we conceived of any world authority, whose diktat may not be crossed without impunity.

Thus the Japanese still continue to consume whales with relish. They have no qualms about wiping out whole species by sending them down their gullets. The Chinese still believe tiger bones, teeth and claws have some miraculous properties to enhance their jaded sex lives. Like contended cows chewing cud in the summer shade, the Indians watch passively, oblivious to extinction of protected species taking place right under their very noses. We do not hear of any Arabs contributing towards the environmental cause, some may even doubt whether they understand such things at all. The African countries are struggling to maintain their own stability. Famine, war and strife are their inheritance, bequeathed to them by their many colonists. Today, most of the inhabitants there are fighting for their survival, so that they are unable to give any heed to the environment. The poachers thus have a field day, whilst the wildlife continues to suffer. The only efforts being initiated are by people from other parts of the world, with some local help. It is heartening to know these crusaders are the descendents of the erstwhile colonists. Without their initiative, the battle would have been already be lost. Many South American countries are busy cutting down their rain forest, oblivious of all pleas made by others. The Brazilians are prominent amongst them. They have made it clear to the world that it is nobody's business as to what they do with "their" trees, as if a tree can belong to somebody! What they have not realized is that the world is dependent on rain forests. There is a saying in Bengali, which roughly translates to "even the mad know when to draw the line, where their life is concerned".

The developed countries are occupied with their own majesty. Their governments' thoughts are far removed from what happens to a tree in some distant country, to a bird in some tropical forest, a turtle on a distant beach or a whale in a far off sea! After all, they do not want to interfere with the internal affairs of any country. Of course, when it comes to their own interests their ideas of noninterference vanishes into thin air. They invade countries with evil regimes and depose them. In one instance, a leader, who after being propped up by them once, was hanged by proxy. In their misguided evangelism, they create untold horror. They are more interested in selling arms and controlling oil rich countries by whatever means available, even if it means sacrificing the lives of their own youths for no just cause. While they shout their lungs out against terrorism, at the same time, they not only countenance countries that nurture terrorism, but also give them support through superabundant aid, while their own poor go neglected. They are also quite happy to suffer military juntas, who keep their democratically leaders under house arrest or those who subjugate whole nations on some flimsy pretext.

Almost anybody can be a politician today. Therefore, mostly governments are formed by the less than mediocre. Most politicians neither have the right education nor breadth of vision needed for the position they occupy. They seek power and very often money also, because without either of these they would not amount to anything. When they are about to be elected, they are too absorbed in counting how many votes they are going to get. The price to them does not matter. Once elected, they know fully well that those who have put them in their seats will call back their debts. The latter care more about their interests rather than think about the environment. Money is their god. Thus, the groups and individuals who point out their concerns for the planet get a short shrift from their respective governments, except probably only to the extent, which suites the leaders politically. They dare not cross the line for they have to appease the people who are their true masters. Even the rich and powerful have their masters. It is only rarely we hear about them. All this makes one wonder if democracy works at all! If all this seems bitter to some, it is because it comes from a bitter heart.

Thus the concern for the welfare of the planet and through it the responsibility of our fellow creatures and plant life have come to rest not on the governments, but on the shoulders of the common man. As a testimony to what I am saying, much of the good that is being done for the planet today is happening through individual effort and by non-government organizations, rather than by any government. It is only through widespread realization amongst the common folk that will turn the tide. This has been shown from experience in the Himalayan foothills, where individuals from surrounding villages turned out and literally embraced trees to prevent contractors from cutting them down with police support. There can be no doubt that the contractors, police and politicians were hand in glove. Nothing happens in India without money changing hands. Astoundingly, even in a country like India, the poor villagers without any significant outside support claimed victory! Such sporadic efforts will hardly make any difference to the politicians, but this goes to show collectively the common people can do good for the environment. Like once the will of the monarchs were taken away from them by the people. Today, if the will of the politicians works against the good of the planet, it must be broken by the common man. Once these politicians realize, if their policies do not have respect for the environment, they will be voted out of power, it is only then they will be made to mend their ways. Thus, the average person needs to be educated to such a level, so that they are not only able to understand the planets needs, but distinguish between what is right and what is wrong for the planet. In this way only, they can decide what changes can be brought about peacefully in an acceptable way.

Most people do not see the true nature of the predicament that the planet faces today. To comprehend what all this is about, it is not enough for people to know a few words like "pollution", "environment", "ecology", "nature", "habitat", "global warming", "green house effect" and such others. We must have a deeper understanding of their implications and learn about the connection that lies between them. Such an understanding can come only through broader knowledge. This form of education, unfortunately, is not imparted through the normal course of our studies. Instead, in many parts of the world, they still have religious classes, social study classes and some are even given political lessons at school in the hope of elevating their minds, so they may become model citizens. The futility of all this will not be lost on the sensible. In contrast, today people are left to their own devices to find out by themselves, as to how they may become responsible citizens of the planet. This, as we shall come to see, rests on a clear understanding of the evolutionary process, the great cycles of nature and how our actions affect them.

HOW AND WHY I CAME TO WRITE THE BOOK

In writing this book, I have drawn on a spiritual Odyssey I undertook a long time ago. In fact, on looking back, it stems from my childhood days. Here the term "spiritual" does not have the religious connotations the word may imply, when used in its more usual sense. Thus in this context it has nothing to do with something that stems from the divine. Rather, it comes from a profound feeling in the mind, which arises from a deeper and wider perception of any matter that forms the subject of serious contemplation. So that when the understanding finally comes, it transcends our everyday perception of the point in question. Together with this revelation comes a feeling, which seems to elevate us temporarily and thereby in some measure change our view of the matter concerned. Thereby brings us a step closer to our understanding of the whole. However, at the same time, our wisdom tells us we can never achieve the sum total!

One may point out revelations are known to occur in religious experience also. This is quite true. In fact, from the psychological point of view, both these experiences may well be rooted in the same origins inside our brain. Nevertheless, there is a distinction between the two. In a scientific quest, such feelings are restricted to revelation of knowledge gained through reason; at the end, it brings with it the sort of feeling, like lightness of heart or a great burden falling off from the shoulders or simply the feeling relief of having completed a job to one's satisfaction. These sentiments may be combined with excitement or even a very small touch of ecstasy, like in the story of Archimedes streaking across the streets crying, "Eureka! Eureka!" This, however, never takes the intense form, as sometimes seen in divine ecstasy, such as trance like conditions or fainting, which may be a part of religious experiences we read about or even hear of sometimes. Neither form of such revelation can be conveyed to others in their totality. Therefore, others cannot vouch for them. Others can know of such an experience, only if they have been through it themselves. Nonetheless, such experiences do exist. The word "spiritual" in this book, however, will not be used in a religious sense.

To the pragmatic and rational mind, initiation of such spiritual quests can only arise from profound questions that come up in the mind from time to time during our lives. It is only when the desire to pursue such ideas takes root and starts to grow inside our mind and finally pervade our mind. We are then compelled, both by thought and by action, to try to bring it to its proper conclusion. In the process, we forgo many of our other needs. Thus, it should be obvious, that to conduct such a search is beyond the ability of any normal child. I hope I was a normal child. Nonetheless, as far as I can tell my spiritual Odyssey started then, though I was not aware of it in such a context at the time.

It all began with a simple question I had asked at the age of five. Even at this tender age, the answer I had received had not been to my satisfaction. Even though I was unable to find an answer then, the urge however remained and from time to time, over the years, it surfaced in my mind. It was this and other questions, which would ultimately take me on my spiritual journey. Today in the winter of my life I can say, I was able to obtain some of the answers I was seeking. As a result my mind, which had been much troubled then, is much more peaceful now.

This is how it all came to happen. I often visited my maternal grandmother during the weekends. There I had great freedom, one because the house was large and the other because I was away from my mother's over strict eyes. At the time, my grandmother used to stay with her father, because my grandfather had died. I remember my great grandfather at his morning prayers. He sat in the lotus position on a deerskin spread across a blue and white porcelain mosaic floor. To me he looked like some sage of antiquity with a long flowing white beard, whom the years had treated kindly. In front of him, there were the images of the various gods in a polished wooden case with a glass face, which was placed in an alcove built into the wall. Those were the days before the age of concrete and houses had thick walls. They kept the house cool in the summer and warm in the winter. As he prayed, the smell of burning incense filled the room. The early morning light would stream into the room, through the high windows, filling the room with radiance as it passed through the incense smoke. I sat beside him and watched him perform his rituals. His great age and venerable appearance commanded my awe and respect. Even at that age through my social training, I knew it was not considered polite in our society to question elders, least of all someone of his years. However, I always wondered how and when the gods partook of the sweets, which were offered to them. One day, I asked my grandmother. She answered the gods would come in their own time to have them. I was not satisfied with the answer; because I knew these offerings, known as "prasad", would be eventually distributed to the members of the family and I would get a share. Yet I could not disbelieve my grandmother also, because of my close attachment and love I had for her. However, I kept my own thoughts to myself.

Nevertheless, I wanted to know if the gods would come. I knew mortals could not look on the gods, but I could see the sweets and I was curious to see if any portion from the sweets disappeared. After his prayers, my great grandfather always left the room closing the door behind him to go to the baithak-khana. The word the "baithak-khana" comes from two words, "baithak" here means a meeting or discussion and the word "khana" means a portion of a building. The baithak-khana was a room, which was situated in the outer portion of the house, where the senior male members of the family spent part of their day. In those days, people had much time to spend and time for them rolled much more slowly, than it does for us today. It served as a sitting room cum drawing room, where guests and acquaintances could be received. I usually did not follow him there in the mornings, because I had much play to catch up with. I came around time-to-time and peaked through the shutters to see if any part of the offerings were missing. After few days of such inspection, I was disappointed. Soon I realized the gods did not come to take what was offered by mortals. This was the time when the first seeds of doubt formed in my mind regarding the existence of gods. At this stage, all this lay dormant in my mind. Later this question happened to be the beginning of my odyssey, which I have spoken of earlier.

When a question remains unanswered or if the answer is not to one's satisfaction, it always leaves a sense of uneasiness in the mind. It is like an unfulfilled longing for an answer that needs to be satiated. All this lay latent for many years until I was in my late teens. As my rationality developed, I could not reconcile my belief in God with my reason. The profound belief in God, which I had in my early life, soon gave way to doubts as I matured. The more I thought on the question of God's existence, the less I could find a rational explanation. Therefore, I turned skeptic and then gradually as the years went by I became an agnostic. In the mean time, my varied interests had taken me to read various subjects. In the process, I tried to understand the inner meaning, which lay behind the superficial. Though I was not always successful, at least the spirit was there. These forays into the various topics seemed unrelated at first. Nevertheless, in the attic of my mind I had collected many ideas over the years. Eventually with time, I was able to see a connection between them. By inserting the missing links, I was able to bring these ideas into a coherent state. It now allowed me to see things in their proper perspective, which now began to make some sense to me.

Initially when the urge came, I wrote down my thoughts as essays. These essays on various topics were written in the spirit of Thomas Carlyle's phrase "An essay is a revere, a frame of the mind, where a man says to himself 'Says I to myself, says I'". *A New Concept of God*, which forms the essay number 19 of this book, was the first essay to be written in the answer to the question I had asked as a child. In doing so, I had turned a full circle from believing in God, then turning agnostic and believing once again. This time, however, as the reader will find out the belief was based on reason. One should not be put off by the title and assume I am another religious crank. If one cares to read what I have to say, though one may not agree with it, they will still find it based on reason. More importantly, it always gives the correct interpretation with respect to any question relating to God. Predictability is one of the hallmarks of science. If on reading it or even before reading it the religious fear their citadel has been threatened, then they are advised not to read this chapter or even the book. All I can tell them is each person has a right to his or her own interpretation of God. Whether this interpretation is right or wrong, it is open to discussion through reason.

Then I wrote a second essay to follow, which was essentially a criticism of this new concept. It has been placed in the same essay as the "A New Concept of God". I always feel it is best to do one's own audit. I believe it should be done as thoroughly as possible, before facing criticism from others. Not because I am adverse to criticism, but I would not like to be in a position, where I cannot answer any rational question put to me. Otherwise, it may make me look foolish. There is a difference in being a fool and being foolish. Though I may be a fool, I would not wish to make a fool of myself. I would advise the reader to go through my criticism, before advancing any opinion. I will however look forward to any reasoned discussion on the matter.

After this essay, I felt the need to explain man's relationship with God and how the concept of God evolved. So appropriately, this essay bears the title "Evolution of the Concept of God". It forms the 18th essay.

Whatever be the concept of God in our minds, the issues of "good and evil" have always been traditionally associated with it. There are two aspects to the idea of good and evil. One is the social aspect. The second is the broader aspect, which is closely bound to nature. It is this latter aspect, which should be our first concern here. I have shown here before man came, the concept of good or evil did not exist in the natural world. Evil only came with man's widespread use of technology. In its wake, it has brought pollution and disruption to the environment. This has happened to such an extent, today it affects the great cycles of nature and impinges on the very process of evolution itself. This discussion on *Good and Evil* forms the subject matter of essay number 20. We will find it brings us out from the world of ideas and concepts to the real world, where the issues involved are real! Thus the effects of evil are also real and therefore they demand real solutions.

All this brings us to the social aspect of good and evil. We are taught to distinguish between good and evil very early in our lives. In this, we are urged by our parents, taught at school, lectured by our religious preachers and expected to comply by society and follow the path of what is defined as good. Paradoxically, as we grow up, the reality of our experience is exactly the opposite of what we have learnt or what was expected of us. We see people deviate from the path prescribed, whenever and wherever their interests are at stake. The more powerful, a person or an institution, the more they try to get their own way. This affects others. Transgression into other peoples' domain is a social problem. However, if one transgresses with respect to nature, it is an environmental issue. This then becomes a concern for all who share this planet with us. Unfortunately, other living organisms of this world cannot show their concerns. If a convention of all living creatures were to be called, there would be no representation on behalf of the other species, only man would be represented. Any matter for discussion would thus always be unopposed and any decision would always be taken in favour of man by default. This one sided approach is where the conflict between man and nature arises. If however the animals did happen to attend such a convention and even if they managed to have lobbies, they would not be able to muster the power of money to buy the necessary votes.

The social aspect of good and evil in turn is closely linked with the question of truth and lies. So the subject of "truth and lies" is discussed together with good and evil. The reader will find that in discussing good and evil I have asked the reader to look to nature for cues as to what is good and what is evil to keep us oriented to the right path. However, when we come to truth and lies, we should not look to the ways of the living world. The living world as we shall see is a web of subterfuge, lies and deception in the game of eating and being eaten. It results from the struggle for survival and is inbuilt into the evolutionary process. It is the world of the "selfish genes". We should bear in mind that man is also a part of this process and remember, like all others, we have also inherited the selfish nature from our genes. Where the welfare of planet is concerned, it is by our wisdom earned through knowledge we may hold back the selfish nature of our genes in rein. Only in this way can the planet be saved.

Before man's appearance on the planet, evolution had inbuilt checks, so no one species could cause untold havoc to the environment. However with man came technology. A point came when he made extensive use of technology to turn things in his favour. In doing so, he has overturned nature's checks. Thus his numbers grow more and more. In this the selfish genes control him and the overall aim of evolution of life becomes self-defeating, if this purpose of the genes continues to have hold on man. The problem of overpopulation and misuse of technology brings us back to the great question of our environment. It is here man must transcend the purpose of his genes and not take more than his due from the environment. It is only by taking this broader view that we can achieve what is good for the environment. To do so, we must think the truth, say the truth and above all act the truth. All viewed not from our point, but in the overall interest of the planet! I felt this was a message I had to convey and so the idea of the book was born, but I also realized it needed further elaboration.

To do any good for the planet, we have to know ourselves and one of the things we must know is what distinguishes us from the other species and why we behave differently from them. When we look at nature, at first we may find, the other species appear to be vastly different from us. At the same time, if we trace any of our individual attributes, we will find it arises from a common basis. Therefore, the distinction between us is actually very small; only the way in which these attributes manifest themselves are different. It had been my interest at, one time, to find out what really distinguishes us from the others. I have made this the subject of the 21st essay, which forms the penultimate chapter of this book. The discussion reveals, only man can go in search of his beliefs, which goes beyond the needs of procreation and survival. What drives us constitutes the *Spirit of Man*. It this, that has led man to seek and find. As a result, much good has come of it, but at the same time, much that is bad has accompanied it. Today the bad is overtaking the good. Yet it is only the spirit of man that can bring him back on the right path.

Today the time has come to repay all our debts to nature for the extravagances of our past. The final essay, in the last part, outlines the problems we face today. In doing so, I have ventured to give a few suggestions. Some might not like what I have said. Nevertheless, the truth is we have not achieved much in what we have done till now and also will not be able to achieve much in what we are about to do, if we follow the current trend. Whenever we have tried to grapple with any serious problem regarding this planet, we have stepped back from our goal. This always happens, because we find the consequences of the actions we have to take, in order to obtain any solution, must necessarily go against human interests. This brings us face to face with the dilemma of dealing with the moral aspect of any solution we have to undertake. They can only be solved through knowledge, wisdom, reason and compassion. To all this we must add spirituality and understanding through which man can gain the strength of mind and transcend his interests to accept such changes. It is then only, we can look for a practical solution.

Today the world has become economically divided, between those who have and those who have not. For one-half life has become narrow and they view the world through tunnel vision. Their lives have become cocooned by technology. To them the planet's future is only an academic problem. They believe there will always be new technology to solve any problem that may arise in the future. Therefore, they think the planet's interests are not their concern. Thus through technology they continue to squeeze the planet like a lemon to get the last drop of juice out from it. The other half, which form the vast multitudes, are not so fortunate. In their desperation to survive, the question of the planet's predicament does not come within the ambit of their lives. They destroy the environment, so that they may survive. To change either mindset is a Herculean task. First, the lots of the have-nots must be elevated, after that the requisite knowledge must be imparted to both the groups. It must also be seen to that this knowledge is absorbed at an individual level. It is only through such knowledge will we be able to correct the imbalance of our past actions and at the same time commit ourselves to preventing further damage to the environment. I am not suggesting the creation of an ideal society. That would be foolish, as has been repeatedly shown in the past. Remember, no man is born equal, because their genes are unequal and no man can avail of equal opportunities even if given them, because such is not in their nature. What we can do is to see that justice is done by simply ensuring that the wrongs are righted and that too swiftly. What people resent is the injustice and not so much their inequality. It is with justice in our lives; people will be in the position to make the sacrifices required of them and pay the penalties in order to rectify the ills we have wrought on the planet until now.

It is in our individual hands that the future of this planet rests. Thus each and every one of us must have a clear understanding of the problems. To reach such a point in our understanding one cannot escape the fact that we need the requisite knowledge, which implies a broader understanding of evolution. It is only with this knowledge we can focus our efforts. In the past, there was little to teach and so much was retained, today there is much to teach and therefore little is retained. Moreover, as the world advances, we are inundated with more and more facts, so it is not humanly possible to keep up with it all. Those who receive education in a way have become specialists, but at the same time by keeping to their own sphere, they become ignorant of other disciplines. I have therefore added a first and second part to the final part, which contains my message.

In writing the book, I have chosen the narrative form, which tells the history of man's struggle and how he came to acquire this knowledge. Unfortunately, this has taken up bulk of the book. This could not be helped as I felt this understanding is crucial to the point I wish to put across in the third and final part. The first two parts reveal an aspect of human endeavor, which can never be projected through textbooks. It is only in this way we can get a better understanding not only of the various aspects of science, but also come to know about the people who were responsible for the knowledge that we posses today. It also gives us an idea of their times and the circumstances under which they struggled to achieve what they did. However, when writing about science in a narrative form, we sometimes fail to convey the essence of the implication of what was achieved through it later. This is not an omission. In trying to keep with the chronological order, we can only convey things form the view of the people who lived then and not as we view things today. In fact, it is only with hindsight that we can get the wider picture. I have therefore tried to correct them as the stories unfold.

The first part of the book opens with an allegory. It embodies the gist of what I have to say. Thus it represents the essence of the book. I thought this would be a good start. It is followed by *A Perspective,* which gives a description of the universe and our solar system. This can be taken as a prelude, which orients the reader to our address in the order of the cosmos. After all, this is where we exist. The next essay, *Terra Mundi,* tells the story of how we learnt the Earth was round. In the two other essays – *Knowledge from the Heavens* and *Triumph of the Heliocentric Theory* that follow, we meet the people who cleared our misconception of the place of our planet in the solar system. We then go on to the next essay *Grappling with Time.* It describes our relation with time and tells us how we came to measure it. In the next essay, we meet *Aristotle, Galileo and Newton,* who clarified our ideas on motion. The essay thus bears their name. Their work would eventually change our ideas on space and time. So the seventh essay is named *Einstein,* who gave us space-time. The first part ends here.

Then the second part contains ten essays about the ingredients of the universe and shows how matter evolved to create the chromosomes. It opens with the eight and ninth essay, which discusses *How Man Unraveled Matter Part I* and *II.* The next essay, the tenth, is about *Thoughts on Space and Time,* which form the other two ingredients of our universe. In the part on space, I have only discussed how inertia may happen. Here I must warn the reader this is a concept of mine and *not* the opinion of physicists. In the same essay, I have tried to explain the nature of time. I hope the complexity of time is presented in a way, which will enable the reader to understand the concept of time more lucidly. I have refrained from discussing the nature of energy, because though we know its various manifestations, we do not know what it really is. I have only pointed out; in creating matter evolution has used the different aspects of energy to build each step in the hierarchy of matter.

The story then moves on to the physicists, who explained to us the nature of *Evolution of Matter*. In here are important ideas, which go to show that matter interacts in presence of energy in the background of space and time to create events. In understanding this we come to recognize that an event is the quantum of evolution. The next two essays are devoted to the story of how we came to understand evolution of life. The twelfth is about *Evolution of Life - Pre-Darwinian Period* and the thirteenth on *Evolution of Life - Natural Selection.* As we gain an understanding of evolution, we come to recognize that our bodies are based on the laws of physics. A point many of the earlier biologists had missed. There is nothing quintessential or sacrosanct about life, like all matter it is based on the laws of physics. What is important to realize, it can only flourish under the right circumstances. I have shown that these combinations are rare in my first essay, "A Perspective". Therefore we should treat life with respect, not only our own, but all life. We must appreciate we stand on a twig that is the result of billions of years of evolution, so if we cut off the main branch we will definitely be consigned to the bin of extinction. This brings us back to what I wanted to say in the beginning.

These two essays are followed by *Mendel and the Hunt for the Genes,* which form the subject matter of the fourteenth essay. The next appropriately discusses the *Chromosomes*. I consider it to be the highest point to which matter has evolved. Life and living forms, including us, are the product of chromosomes. This constitutes the fifteenth essay. The sixteenth essay, *The Wonder that is Our Body*, describes the how are bodies are built according to physical laws. The second part ends with the seventeenth essay. Here I have suggested our chromosomes act by interpreting physical laws to build our bodies, which is the reason why the evolution of life is possible. This last is my idea and does not conform with the view of biologists. Nevertheless I feel the idea is worth considering, because without physical laws nothing can happen.

MY EXPERIENCES IN WRITING THE BOOK

Before ending, I would like to say a few words about my experiences in writing this book, which I feel compelled to record. Before I was struck down with near fatal illness in the year 1991, I thought it was important for man to realize that the key to understanding the universe lay in physics. However, since that time my ideas have gone through a profound change. After many days I am glad to say I recovered from this near fatal illness through my wife's care and my daughter's support. After recovery for a few weeks my mind was very clear and I was fully alert when I woke up, which is unusual for me. I was free to concentrate on various topics, I had been thinking of earlier. At this stage, I also found I was able to think more intensely than usual. It was at this time I found many of the answers I was seeking.

The weeks that followed my recovery were very productive. At this period in my life, I found some of the answers to many of the questions, which had nagged my mind earlier. Questions such as "Why should the speed of light be constant?"; "How bodies travel through space?", which is basically a concept of inertia; there also other concepts like "the case for a second event horizon inside black holes" and a "new concept of God". All these answers may have been possible, because during the brief period of my convalescence I was able to free my mind from the concerns related to earning my bread and the responsibilities to my family.

Today, if asked, I would say the most important thing for every human being is to have an understanding of evolution in the background of physical laws, rather than physics only. It is through this we will find the philosophy in physics, which opens the doorway for us to appreciate the beauty in evolution. I have mentioned this in the last chapter, where I have said our new education should be around this central theme.

Lastly to attempt the task of writing a book, I felt would be beyond my meager capacity. I looked around for help. I went to discuss certain questions on time, space and some other points on physics at a prominent academic institution in Calcutta. I met with a cool reception. On reading some points I had written on the topics, which I wanted to discuss with them, they responded to my queries by asking me whether I was writing a book on astrology. My readers will find after reading, what I have written, there has nothing remotely connected with astrology. One physics professor there asked me what life was. I told them briefly that it was based on the laws of physics. The puzzling look he gave me made me feel that he thought I was pulling his leg. After some more discussion, it seemed to me, they were trying to avoid answering valid questions, which I had put to them. There was a retired professor sitting next to me, who had come to speak to the head of the department about processing papers for his pension. He was listening to my conversation and remarked that Einstein's "twin paradox" was not actually true, but only used to explain certain points regarding relativity. There was the head of the department of physics sitting at the table, but he did not make any comment. I knew then it was time for me to leave. Still I had hopes. I contacted three others who were professors of physics, but none of them discussed the questions I had asked them. Instead, they asked me questions. One asked me whether I thought atoms were conscious. I answered we humans are aware of our world through the medium of energy and it is because of it, we are able to interact with our environment. This is what we may term as being conscious. We are an evolved state of matter, so like matter consciousness has evolved also. We may not recognize it in the atom, but in a way they are also "aware" of each other through the medium of energy, otherwise they could not have reacted to form compounds. We may term this as "primitive awareness", which possibly lies at the root of consciousness. This is not some metaphysical idea. In reading the chapter on evolution, one will appreciate that evolution does not create anything new; it serves old wine in new bottles. It is all there for us to see; only we do not always recognize what we see. The second asked me to explain why the left half of the brain deals with the right half of the body and the right half with the left. I was aware his question was not asked in the right spirit; because he must have known very well that we do not have an answer to this problem yet. Nevertheless, in my opinion it happened to be a rational question. So in response to this challenge I have given a hypothesis in the chapter which deals with the *How Physical Laws can Imprint on the Body*. I hope he will find it satisfactory, if he ever happens to read it. The third was noncommittal about everything I discussed. The thing I found strange was that people were not prepared to discuss rational questions, just because the questions happened to be outside the syllabus they taught. It seemed to me they felt there could be no knowledge outside their books. There was, however, one exception, Dr. Gour Bhattacharya of the Physics Department of Presidency College, Calcutta. He was kind enough to explain the mathematics behind the interpretation of the third law of planetary motion. He wrote for me a short synopsis on Einstein's contribution to general relativity; on which I have based this topic. I also have to thank him for going through my ideas that I have mentioned earlier, which are given in the second part of the book.

From time to time, I had tried to talk to people belonging to different disciplines, but it all proved to be of no great help. I found it rather strange that a person should be turned away when looking for answers to rational questions. It is a sad reflection, not only on these people, but also on the society, which they come from. My impression was that they could not tell me any more than what I could find in the books in our house or in the internet. I gave up trying. I did not now know what to do with my ideas. I realized that the central idea of control of population was the most important message that had to be put across along with some solutions. I concluded the only way I could present these ideas was by writing a book. I therefore set out to write.

I have kept the language simple as possible from the scientific point of view and tried to explain as I have gone along. The specialist in a particular field may find the approach rather elementary where their subject is concerned. They should remember this book has been written for a wider audience and not just for them. The critics should also bear in mind everyone is not well versed in the English language. There are those amongst the readers who have learnt English not through Shakespeare, Shaw and Shelley, but through other ways. The latter are at a disadvantage with classical names so I have taken the trouble to explain the references through stories, just as I have done with scientific references, which I felt was only appropriate. I have given dates so that the reader may correlate them with respect to other world events. I have also given maps of places wherever possible. This will help the readers to keep their chronological and geographical bearings. I feel this is very important to one's understanding.

I hope that in the end, after reading all three parts, the reader will find that it will change their perception of our world in some small way, so that they will see themselves as being responsible members of the planet and not just citizens of their country or adherents of their religious beliefs only.

PART I

THE GATHERING OF KNOWLEDGE

by

ARUN BOSE

Credit: Courtesy NASA/ESA/S. Beckwith(STScI) and the HUDF team.
Website: http://www.nasa.gov/multimedia/imagegallery/image_feature_142.html
1.2 The Hubble telescope digs deep into space and sees a field of galaxies. [Public Domain]
Each point of light seen in the picture is a galaxy in its own right.

1 A PERSPECTIVE

A man who has lost his address is a lost man,
A man who has lost his identity is a lost soul.

INTRODUCTION

We live in a universe so vast; its magnitude is beyond our comprehension. Today we know on good scientific grounds the universe is finite. Nevertheless, no matter how far we let our imagination go, in reality it is much bigger than we can ever conceive. A poet or a philosopher, if asked, would describe it as being "infinite in its finiteness, whose edge is intangible and its centre lost in its immensity".

PERSPECTIVE OF OUR UNIVERSE

Our home in this universe lies in a star system, which is known as the Milky Way Galaxy.

Credit: Courtesy NASA, ESA, and the Hubble Heritage (STScI/AURA)-ESA/ Hubble Collaboration {Acknowledgment: L. Jenkins (GSFC/University of Leicester)}.
Website: http://apod.nasa.gov/apod/ap070418.html

1.1 NGC1672, a barred spiral galaxy, shaped very much like our Milky Way Galaxy. [Public Domain]

It is less than a speck in the vastness of the cosmos, where it is lost amongst billions and billions of other galaxies. Each galaxy in turn is made up of millions and millions of stars of which our Sun is but one. Many have planets. Our Earth is one such planet, which orbits such a star. We call it the Sun or Sol. Some planets have satellites; our Moon is an example. To all this we must take into account the distances separating the satellites from planets; the great distances of planets from their parent star; the even greater distances that divide the stars from one another; and finally, the immense space that lies between the galaxies themselves.

THE LIGHT YEAR

Cosmic distances are so great that we cannot have any idea of their measure by scales we use on the Earth. To get some idea of the magnitude of the universe we require a much larger scale: A scale suitable for astronomical work. There are many such scales used in astronomy. To keep matters simple we will only use one such scale for now - the light year. It is the measure of the distance that light takes to travel through vacuum in one year. We know in one second light travels 186,000miles or 300,000 kilometers, thus in one year light will travel 5,865,696,000,000 miles (Five trillion, eight hundred sixty-five billion, six hundred ninety-six million miles) or 9,460,800,000,000 kilometers (nine trillion, four hundred sixty billion, eight hundred million kilometers). Such is the magnitude of the scale by which we measure the cosmos.

Box 1.1 SCALES USED FOR DISTANCES IN ASTRONOMY

Light year is the distance covered by light in one year.
= 5, 878, 630, 000, 000 miles
= 9, 460, 730, 472, 580 kilometers

Astronomical Unit (AU) is the mean distance of the
Earth from the Sun.
= 93, 000, 000 miles
= 148, 800, 000 kilometers

Parsec is the distance from the Sun to an astronomical object, which has a parallax angle of one arc second.
= 19, 174, 000, 000, 000 miles
= 30, 857, 000, 000, 000 kilometers

Box 1.1 Scales used for astronomical measurements: The light year, astronomical unit (AU) and parsec.

Having had a flavour of the extent of the light year, we are now ready to consider some of the facts concerning the dimension of various objects within the universe. We may then go on to obtain some idea of the distances, which separate these bodies from each other. This will then give us a somewhat better appreciation of the size of our universe. However, before that I have included some measurements of objects in the solar system in terms of light years. This will highlight the magnitude of the light year and at the same time show it is too big for our solar system. This is why we have to use a smaller scale where the solar system is concerned.

THE SIZE OF THE UNIVERSE

The Earth's diameter is eight thousand miles; it would take light 0.043 part of a second to cross this distance. In terms of light years it would be, as small as, 0.000000001364. The Moon is 240,000 miles away from us, which comes to 0.00000004092 light years. The Sun is a little over eight light minutes away from the Earth, which is to say 0.00001585 light years. Pluto, when it is at its furthest from the Sun, has a distance of 4,566,000,000 miles. It takes 2.8 days only for light to span this distance; making it equivalent to 0.0007784 light years.

It is outside our solar system the light year comes into its own. Our closest star, the Proxima Centauri is 4¼ light years away. The Milky Way Galaxy to which we belong is 100,000 light years across. The stars in the night sky, which we see with our naked eyes, are all a part of our star system. We cannot see the stars in other galaxies with the unaided eye. This should not surprise us, as most of these galaxies are too far away for us to even see, let alone seeing any stars that lie within them. However, we are lucky enough to be able to view some of the larger bodies that lie outside our own galaxy with our naked eyes. They are the galaxies and other objects, which lie in the neighborhood of the Milky Way Galaxy. The Greater and Lesser Magellanic Clouds are two such examples. They are immense clouds of hydrogen, known as gaseous nebulae, which appear like a veil, in the reflected light of stars that exist within them.

Credit: Image courtesy of Victor van Wulfen, used with permission – www.clearskies.eu

1.3 Greater and Lesser Magellanic Clouds with a meteor chancing between them. [By written permission]

The Magellanic Clouds are visible on the horizon from just north of 10° latitude north of the equator, though they are better viewed from the southern hemisphere. Portuguese sailors, who ventured south, along the west coast of Africa, must have been aware of these cloud-like objects long before Magellan. This made these sailors the first Europeans to know about them! Their existence became widely known in Europe only after Magellan's expedition of circumnavigation of the globe. His voyage took him far south of the equator. He noted, unlike other clouds, they did not drift or change shape and like the stars in the night sky, they did not change respective position throughout his long voyage. These objects were named in honour of Magellan, because he was the first person to record these observations correctly. One must not however be tempted to think this was something new the Europeans had discovered. People living in the southern hemisphere had been aware of these unchanging so-called clouds, ever since the time they made their home in lands south of the equator.

These clouds are nebulae or irregular galaxies. They represent unformed galaxies. The Greater and the Lesser Magellanic Clouds lie some 160,000 light years and 200,000 light years respectively from us. Along with some other galaxies, including our own, they belong to what we call the *local group*. The Andromeda Galaxy is another member of this group. It is a spiral galaxy very much like our own and similar in size. It is the most distant object in the sky, which is to be seen with the naked eye, lying at a distance of 2,500,000 light years! As we go beyond our local group, even the light year pales into insignificance.

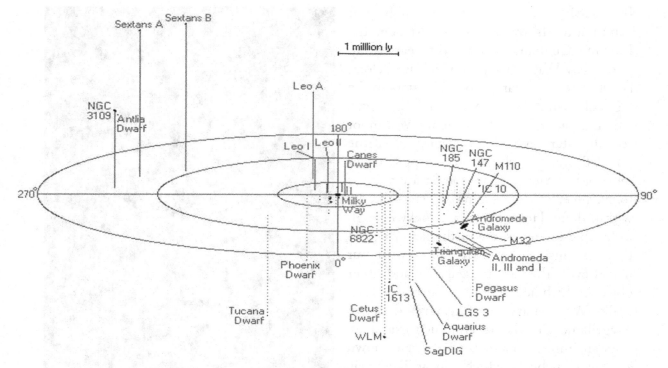

Credit: Courtesy Richard Powell from his website "An Atlas of The Universe".
Website: http://www.atlasoftheuniverse.com/localgr.html

1.4 Our local group lies within 5 million light years of the Milky Way Galaxy. [CC-BY-SA-2.5]

Outside our local group, there lie other "local groups", which together with our own local group form a *cluster*. Our cluster is about 10,000,000 light years across. Many such clusters form a *supercluster*. Lastly, many, many, many superclusters go to make up our universe. Scientists have estimated the *visible universe* has a radius of some 14,000,000,000 light years. As we progress, through the use of better and better instruments, the frontiers are being pushed further back.

Such vast distances are beyond the understanding of the human mind. We may imagine, we may calculate, we may give analogies and after that we may think we understand the size of our universe, but the truth is we shall never be able to comprehend its awesome magnitude.

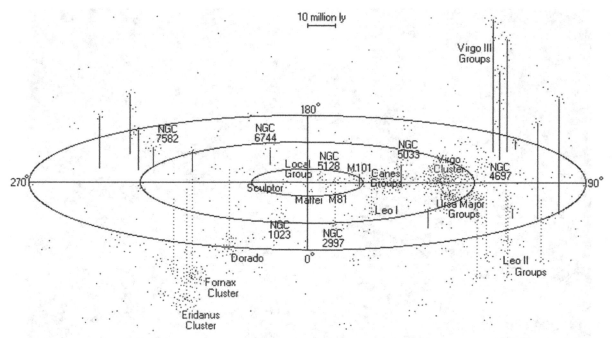

Credit: Courtesy Richard Powell from his website "An Atlas of The Universe".
Website: http://www.atlasoftheuniverse.com/virgo.html

1.5 Our supercluster - the Virgo supercluster, which lies within 100 million light years radius of our Solar system. [CC-BY-SA-2.5]

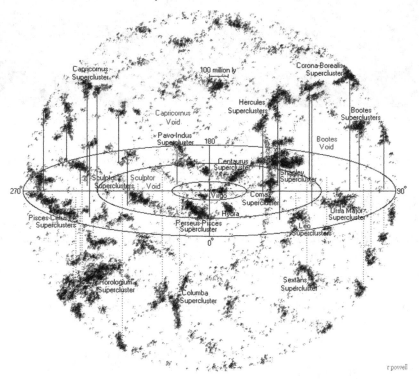

Credit: Courtesy Richard Powell from his website "An Atlas of the Universe".
Website: http://www.atlasoftheuniverse.com/superc.html

1.6 The universe within 1 billion light years radius - our neighbouring superclusters. [CC-BY-SA-2.5]

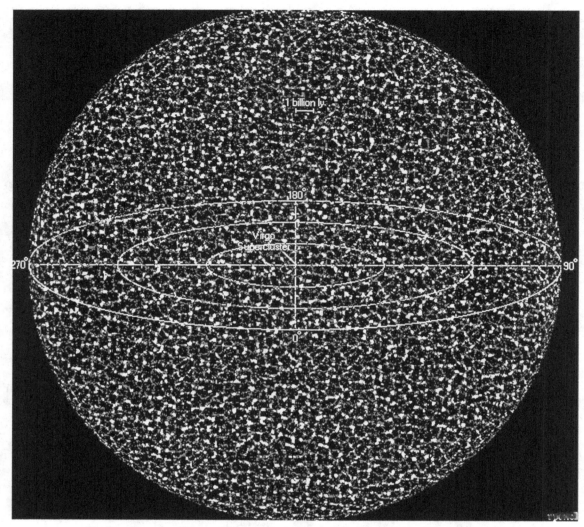

Credit: Courtesy Richard Powell from his website "An Atlas of The Universe".
Website: http://www.atlasoftheuniverse.com/universe.html

1.7 The visible universe within 14 billion light years radius. Each white area is a supercluster. [CC-BY-SA-2.5]

EXPANDING UNIVERSE

The universe is not a static entity. The superclusters, the clusters, the galaxies, the stars, the planets and their satellites are all rotating about their own axes; and in all probability the space in the universe does the same also! At the same time, the universe is expanding, so the galaxies are moving away from each other at great speeds. The space, which houses the universe, is also expanding; its edge is receding from us at the speed of light. Even if we did happen to stand near the very edge, we would never be able to experience this ever-receding boundary of the universe to know its nature.

As far as we are concerned, the edge is intangible. What of the center? From the point of view of man, who lives on a small planet, travelling around an ordinary star, which is one amongst the billions of other stars in our galaxy and in its turn the latter lost amongst the innumerable galaxies that go up to make the universe, the word centre becomes meaningless. The poet and the philosopher are right after all!

By now, some of you have been wondering, where all this energy comes from to bring about the continuous expansion of the universe and cause everything in it to turn. This enormous energy, which beggars all description, arose from the effects of the *big bang* – an event, which brought the known universe into existence. We shall deal with the big bang when we discuss the evolution of matter in the second part of the book. It would suffice to say for now, everything in the primordial universe was energy and the big bang was the first event that led to the creation of time, space and matter, which in course of events evolved into the universe we know today. Though the energy released from the big bang was a one-time occurrence, its lasting effect on both matter and space is a result of the inertia factor. In this connection, we must remember space is considered as rarified matter. Galileo was the first to realize the property of inertia in matter, which Newton was to embody in his first law. It states - any body (read as mass/matter) will continue to be in a state of rest or of uniform motion unless impressed upon by an external force. Therefore, it was the initial effect of the big bang, which was responsible for imparting motion to all heavenly bodies and to ever expanding space, which is rarified matter.

There is of course another factor – gravity. This is but a secondary factor produced as a result presence of matter. It contributes to motion of bodies as matter in the universe evolves. It is a force, which tends to pull things together and thereby imparts motion.

KNOWING THE ADDRESS OF OUR HOME IN THE UNIVERSE

Not to know your address in the order of the universe may not be a crime. It may well have been without this understanding, we would have evolved technologically in other ways, just as the Mayans did, by bypassing the knowledge of the wheel. The Mayans were a civilization who knew about the wheel as evident from the toys they made, but had not added this idea to their practical lives. Thus without the application of circular motion they did not have the wheel and axle. Therefore, they could never graduate to the idea of pulleys, gears, screws, cams, flywheel or propellers. All of which work on the same basic principle, which opens up a whole array of technological developments that resulted from them. On the other hand, their civilization by 1000 AD had been able to advance in many other ways. In their time, they had a calendar, which was one of the most advanced in the world. Their cities had buildings that equaled the pyramids of Egypt, if not in stature at least in grandeur. Their roads had no comparison since the Roman times. Their technique of bringing water to their cities indicated they had a good idea of fluid dynamics. Nevertheless their civilization failed to reach the heights one would have expected from them, because of the absence of the wheel. If we on our part had not attempted to know the true state of affairs regarding the place of our planet's location in the universe and continued to think the Earth was the centre of all things, as we did at one time, then like the Mayans, we too could have developed technologically without this knowledge.

In such an event, if we happened to encounter an alien civilization from some distant galaxy and they found out we thought the Earth was at the centre of the universe. Then they would certainly have taken the same view of us as we do of the Mayans today. It would indeed seem strange to our visiting aliens that we lacked an understanding, which we should have acquired in the normal processes of scientific evolution: A civilization otherwise so advanced in other respects should have the idea that the Earth lay at the centre of the universe.

CLUES TO OUR PLACE IN THE UNIVERSE

If one wants to have an idea of the universe then one has to first look up at the skies. Though man has been doing this since he first walked the Earth, it is only within the last three millennia he has taken to observing the skies in earnest, which has resulted in understanding our place in the order of the universe today. Thus, astronomy can lay claim to being the oldest science. Amongst the first discoveries made through such early observations was the recognition planets behaved differently from stars. The knowledge about their paths would eventually lead us to the realization our planet was not at the centre of our universe. We must remember that all this knowledge was obtained through naked eye observations.

It was only during the last four centuries, people have been observing the skies with telescopes. Recently over the last few decades astronomers have included instruments, which use other wavelengths beyond the visible spectrum of light, such as radio waves, infrared rays, microwaves and x-rays. Their search, especially in the microwave region of the spectrum, has led to the knowledge of how the universe came into being. In order to avoid atmospheric distortions and pollution telescopes have been sent to space, so we might get a better view of very distant objects. Planetary probes have opened up new vistas in our knowledge about the planets and their moons. Today through a deeper understanding about the stars themselves, we have come to learn our relationship with the cosmos and the fact all life has evolved from stars that have died long ago.

THE MILKY WAY

It is only about two hundred years ago, we came to know our solar system was a part of a star system. This started with William Fredrick Herschel (1738-1822), who built up a picture of our galaxy through observations with his 48-inch (120 cm.) telescope. Thus historically this recognition was quite recent compared to the many things we had learnt about the skies over the ages. Nevertheless, we will discuss the Milky Way Galaxy first, so we may continue to keep our order in the description of the universe.

There is a distinction between the *Milky Way* and the Milky Way Galaxy. The Milky Way is a concentration of stars spread out like a path across the night sky. It presents us with one of the most beautiful spectacle on a clear moonless night. Today after spending a lifetime on this planet, most people hardly ever bother to look up at the sky. Thus, many of us who live in big cities have only heard about the Milky Way, but never seen its full splendor.

Credit: Courtesy Dan Duriscoe, U.S. National Park Service, U.S. Department of the Interior. Website: http://apod.nasa.gov/apod/ap070508.html

1.8 The Milky Way. A panoramic view of the skies over Death Valley, California, USA. (A composite of many pictures) [Public Domain]

Ancient Greeks had a story of how the Milky Way was created. Zeus, the king of the gods, had committed an indiscretion with Alcmene, wife of Amphitryon. Out of this union, Heracles was born. Not unnaturally Hera, the queen of the gods and the wife of Zeus, was incensed. Unable to vent her anger against her all-powerful husband, she directed her ire on the infant Heracles. However, Zeus got wind of Hera's intentions. He sent Pallas Athena, the goddess of wisdom, to protect Heracles. She stole the baby Heracles and laid him beside a path in the woods where the queen of gods would be passing. Hera set out to accomplish her purpose, but on her way seeing what appeared to be a forsaken baby her maternal instinct was aroused. Not knowing Heracles' identity, she took him to her breast to nurse him. Heracles with his characteristic vigor, which would mark his life, bit the goddess' nipple instead of sucking on it. The divine milk spurted across the sky creating the Milky Way.

Thus the term "Milky Way" should not be confused with the Milky Way Galaxy. It should be considered as a landmark in the sky. It represents one of the spiral arms of our galaxy, known as the Orion arm. The shape of our galaxy is that of a barred spiral galaxy, but from where we are placed on the galactic plane, we cannot see its full shape. Neither can we see the centre, because of the dust and debris that obstructs our line of sight. If it were otherwise, we would be presented with a surreal view of the centre with its globular collection of stars, instead of just a part of an arm only. The bright centre would appear like a glowing orb with its bright light bedimming the stars clustering around it. The display of stars further out, even though somewhat subdued, would look like fairy lights of different hues all flickering in the distance. Some would be bluish, others yellowish, a few orange coloured and many reddish. The colours represent the different stages in the life of the stars. Had we been a little higher or lower from the galactic plain, we would have seen the spiral arms emerging from this centre. All the stars together would then light up our sky as on a moonlit night.

CONSTELLATIONS

It was observed, stars were "fixed", which meant they did not change their position in relation to each other even as they moved across the celestial sphere. Ancient cultures used their imagination to trace out various shapes joining the stars with imaginary lines to give them form; just as today's children would join dots, in order to make pictures in puzzle books. These arbitrary shapes traced out by the imagination of those early civilizations were known as constellations or assemblage of stars. Not only did it help to pass time by giving shapes to them, but it also helped to identify the stars quite easily. This proved to be a very convenient way to search for any star by first identifying the constellation to which it belonged. It helped in finding the direction when travelling through deserts or over expanses of water, where there were no other landmarks other than the stars above. Also as the constellations passed across the heavens, the month or the seasons could be gauged at a time when there were no calendars. Each civilization gave their own shapes to the constellations and wove their own stories around them. The ancients told these stories around their fires in the evening, bringing them to life. The names the Greeks gave to these constellations, including those of the zodiac, have been retained in modern astronomy with which most of us are familiar today. This is not only in honour of their contribution to early astronomy, but more importantly due to the profound influence the ancient Greeks had on human thought and culture over the ages. Names like Andromeda, Hercules, Leo, Orion, Pegasus, Gemini, Pleiades and others should be familiar even to those who do not know astronomy, but have read poetry and literature or heard about Greek mythology.

Each constellation seen in the northern hemisphere has a story to tell, like Orion the hunter with his dogs, Canis Major and Canis Minor. Orion is one of the easiest constellations to identify in the night sky. He is seen holding his head high, wearing a belt with a sword and walking across the sky with his dogs to hunt. There is a story in Greek mythology as to how Orion came to be amongst the stars. Diana the goddess of the moon and chase, sister of Apollo, fell in love with Orion the son of Neptune, god of the oceans and the waters of the world. One day Orion was wading through the sea with only his head showing. From high in the heavens, his head seemed like a dark speck on the sea. Apollo was not happy about his sister's love for Orion. He asked his sister to show him her skill in archery by shooting at the dark speck on the sea. Diana shot with a fatal aim. On seeing his floating body, she was overwhelmed with grief. The goddess raised his body to the heavens and placed it amongst the stars, where it can be seen even to this day.

Source: Orion constellation in Uranographia (1690) by Johannes Hevelius.
Website: http://en.wikipedia.org/wiki/Orion_(mythology)

1.9 Image of Orion the Hunter superimposed on the constellation of that name. [Public Domain]

Today, to us, the story may seem rather prosaic and appears somewhat forced, in order to explain the shape of the constellation of Orion. However, these stories nearly always have some deeper meaning, which the ancients wanted to convey. In this instance, Diana apart from being the goddess of the moon was goddess of the woods and hunt. She also happened to be a fertility goddess, who belonged to an earlier period. The story comes down to us from that former age. Through the death of Orion, Apollo ensured she did not pass into anyone's sole possession. As a fertility goddess she could not belong to anyone in particular, she had to be there for all nature. Other cultures over the world also have stories to tell about the constellations; all have some deeper symbolism like the one I have just related.

Constellations of the southern hemisphere were not visible to the Greeks, so many of them do not bear names from Greek mythology. Their names were given at a time when the great explorations took place in the seventeenth and eighteenth centuries. This was a time when winds of change were blowing across Europe. It was also an era, when science had caught the peoples' imagination. Thus, many of them were given names of common scientific apparatuses, though in a Latinized form. Latin was still the medium of instruction at the time. Today, if we consult charts of the southern sky we will see constellations, some of them still bear names such as Microscopium (Microscope), Telescopium (Telescope), Antila (Air Pump), Circinus (Pair of compasses) and Horologium (Clock).

The stars of any particular constellation have the shape they have, only because the stars in them incidentally happen to lie that way in the line of sight, as seen from the Earth. If seen by a space traveller from the Moon their shapes would be still recognizable, because the Moon is not far away from the Earth. Further, away, as we receded from the Earth their shapes would start to change slightly. By the time the cosmonaut reached beyond the solar system, many of the shapes would change appreciably. If he or she travelled further on, say to Sirius, which is 8.6 light years away, he would not be able to recognize many of them.

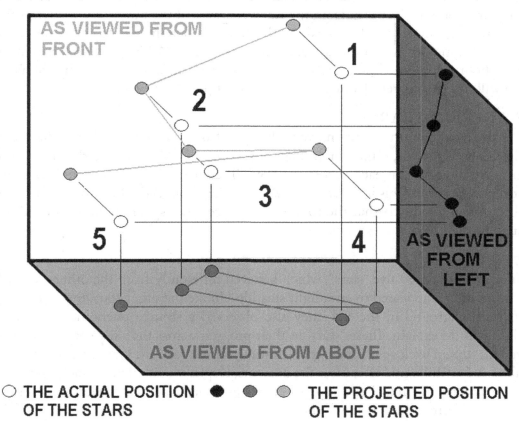

Drawing by Tarun Ghosh.

1.10 Changing pattern of a constellation as viewed from three different sides.

PROPER MOTION OF THE STARS

Apart from the apparent diurnal motion of the stars, which we see due to the Earth's rotation around its axis, all the stars like any other body in the universe have their own motion. This motion is known as its *proper motion*. The term "proper" has become archaic, but appropriately its use is still retained in astronomical science. The only other place I have come across this word "proper", in relation to motion or movement outside astronomy, is in Dumas' *The Viscount de Bragelonne*, which is a sequel to The *Three Musketeers*, where Charles II explains to General Monk that not he, but D'Artagnan had taken the initiative in his capture. He goes on to say, "—and observe, M. Monk, I do not say this to excuse myself,—for M. D'Artagnan," continued the King, "has gone into England on his own proper movement, without interest, without orders, without hope, like a true gentleman as he is, to render a service to an unfortunate King, and to add to the illustrious actions of an existence, already well filled, one fine action more." On hearing this eulogy, we must record that our hero did cough to keep his countenance.

Over the millennia, due to their proper motion the stars will change their positions with respect to each other. The time over which such change would be apparent will depend on their distances from the Earth and their relative speeds. Some stars may move faster and others slower. Thus with time, the constellations will no longer have the configuration we see today. Later in the chapter and elsewhere, we will often use the term *fixed star*. This does not imply the stars are fixed in any way, but just that they do not appear to move in our day-to-day experience as we observe the skies. The reason for this is the great speeds at which they move is not apparent when seen from the Earth, because of their immense distances from us. At such great distances, they do not exhibit the parallax phenomenon to any significant effect. The relative motion of the Sun is about 12 miles/second or 19 kilometers/sec. compared to the neighbouring stars in the galactic arm. However, even at this speed the Sun's relative motion will not be apparent from Sirius over a short period of time.

THE SIDEREAL YEAR

Even as the fixed pattern of stars move each night, from east to west, due to the rotation of the Earth about its axis, a man or woman observing the skies will see the stars set a little earlier every morning, on each successive day. Till after one year they would come back to their original position. This is the *sidereal year*. It comes from the Latin word "sidus" meaning star. The reason for this is that the rotation of the celestial sphere is Sun centered and not centered on the Earth.

THE PLANETS

The ancients also noted five "stars", which behaved differently from the others. These "stars" did not flicker, but appeared to emit light steadily. Unlike the other stars, they moved against the stellar background in the sky. They also observed a strange phenomenon, which they found difficult to explain. These "stars" in their progress across the skies some times stopped, looped and turned back again. They then continued their forward journey across the heavens as before. A few millennia later after the ancients, Kepler would picturesquely compare their path to a pretzel. A pretzel is a type of crisp knot-shaped biscuit flavored with salt, used especially by Germans as relish with beer.

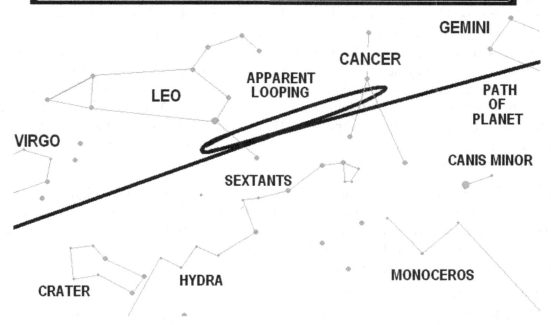

1.11 Retrograde motion or the looping movement of a planet across the background of fixed stars. Redrawn by author form a website, unfortunately I have lost the address. Refer to Fig. 3.9 to see what actually happens.

To the ancients from many civilizations these five "stars" did not fit in with the scheme of the heavens. The Greeks called them *planets*, this was only appropriate, because the word "planet" meant "wanderer" in their language. They named the planets after their gods. However it is their Roman names Mercury, Venus, Mars, Jupiter and Saturn that have come down to us today. Galileo was the first to turn his telescope to look beyond the Moon and seriously explore the planets. In doing so, he observed Jupiter's four larger moons.

Now with the improvements in the telescope observations people were able to define the skies better and better. In time astronomers came to know there were three more planets beyond Saturn, namely Uranus, Neptune and Pluto. Uranus was discovered in 1781 by William Fredrick Herschel. Though it had been observed on 17 occasions before, it was thought to be a star. Neptune's discovery had a more chequered history. At the same time, it was a triumph of the state of the art accuracy, which astronomical science had attained at the period through Kepler's and Newton's work. Long before Neptune's discovery, astronomers had noted Uranus did not follow its predicted path. This was to an extent that went beyond legitimate errors, which might be produced through attempts at accurate observations. Thus the mystery came to be a growing interest amongst astronomers of the day. Many theories came up. Some thought these perturbations could be due to Saturn's presence. However, this would only be possible if of Saturn's mass was much more than its real value. Since this could not be, the theory had to be abandoned. Thus this could only be explained by some undiscovered body beyond Uranus. This led Rev. T. J. Hussey in 1834 to surmise this was the case. John Couch Adams (1819-1892), while still a student at St. John's College, Cambridge, worked out the predicted path of this new planet mathematically. He sent his results to the Astronomer Royal, Sir Biddel George Airy (1801-1892) in 1843. He showed the path of Uranus corresponded to the influence of a planet, whose radius lay at twice the mean distance of Uranus. This would closely match with Bode's Law. Airy was however cautious and procrastinated over the matter. This failure to act on the information in time by Airy caused Adams to loose his true place in the contribution of this discovery. In the mean time, Urbain Jean Joseph Le Verrier (1811-1877) of France computed the position of this unknown planet in 1846 and sent his results to the Berlin Observatory. The new planet was located within few days by Johann Gottfried Galle (1812-1910) and Heinrich d'Arrest (1822-1875). It was named Neptune. In 1930, Clyde Tombaugh (1906-1997) discovered Pluto. Thus bringing the number to nine; including our Earth. However, today, Pluto has now been relegated to the status of a dwarf planet.

Except for the two inner most planets, all the others have moons. The original number of moons is increasing as planetary probes are revealing more and more moons. At present, we cannot be certain of their exact number. Jupiter was known to have 12 moons in the middle of the last century; some of their names I remember memorizing as a boy. On the last count, their number has increased to sixty-three in the year 2007.

OUR SUN'S LOCALITY IN THE GALAXY

The Milky Way Galaxy is 100,000 light years across along its equatorial plane and some 1,000 light years along its axis of rotation. Astronomers describe its shape as a barred spiral, which means its arms originate from two short bars that come out of opposite sides from its centre. Its shape is thus not a true spiral. It has at least 200 billion stars; some have estimated it may be double this number. Our parent star, the Sun, is only one amongst them. It lies close to the main galactic plane and about 26,000 light years away from the galactic center. The centre of the galaxy lies towards the star-clouds in the direction of the constellation Sagittarius.

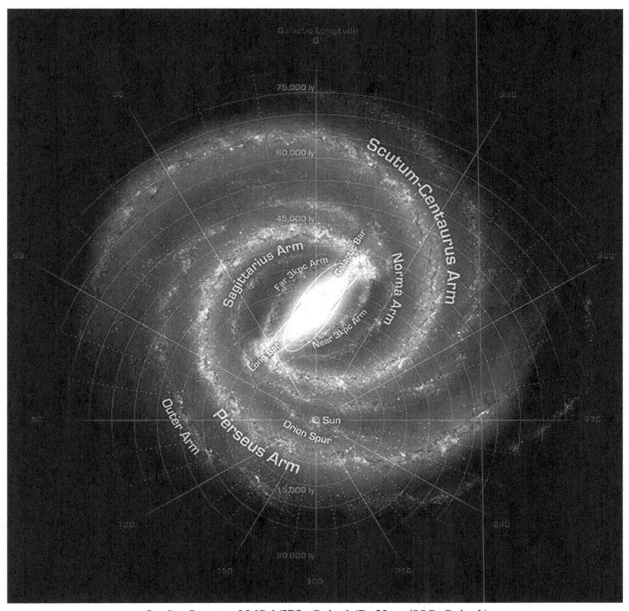

Credit: Courtesy NASA/JPL-Caltech/R. Hurt (SSC-Caltech)
Website: http://www.spitzer.caltech.edu/Media/releases/ssc2008-10/ssc2008-10b.shtml

1.12 The Sun's position in the Orion Arm of the Milky Way Galaxy. [Public Domain]

It may seem to us, the galaxy with its huge number of stars is overcrowded. In reality, the stars have great distances between them, because the galaxy is so huge. Otherwise, their gravity would have torn each other apart. Distances of the stars, which lie within a radius of ten light years of the Sun will give us an idea of the scanty distribution of the stars in the spiral arms. Their density is however greater towards the central globular portion, which hides the denser galactic core.

Our nearest neighbor, the Proxima Centauri or Alpha Centauri C it is classed as a *red dwarf*. A red dwarf is a star, which is much smaller than the Sun and produces much less energy.

Their surface temperatures are relatively cooler than the Sun, so they look red. The Proxima Centauri is a part of the Alpha Centauri system. It reaches closest to us as it circles the Alpha Centauri A and Alpha Centauri B (a paired system) at a distance of 0.2 light years from them. The whole Alpha Centauri system is 4.25 light years away from the Sun. Amongst the other stars, which lie within a radius of ten light years there are three red dwarfs, the Bernard's star at 6 light years, Wolf 359 at 7.8 light years and Lande 21185 at 8.3 light years. The brightest star in the night sky, Sirius (Canis Major) is 8.6 light-years with a mass twice that of the Sun and is circled by Sirius B, a white dwarf. A *white dwarf* is a star, which eventually burns its fuel and becomes the size of the Earth, but its mass is near to that of the Sun. Therefore these stars are very massive, even though they are quit small. The initial whiteness of theirs surface indicates they are very, very hot, but eventually with time they will give off their heat and cool down to become a glowing red star. In spite of their red colour, they are still known as white dwarfs. Their red colour should not to be confused with that of red dwarfs, which are never as massive nor do their lives follow a similar course. There are two other red dwarfs Luyten 726 and 728 at 8.7 light years. They form a binary system. Lastly, there is Ross 154, which is also a red dwarf at 9.7 light years. There are no Sun-like stars within this ten light year radius of the Sun. Tau Ceti (τ Ceti), which resembles the Sun, with 80% of its mass and 60% of its luminosity, lies almost 12 light years away.

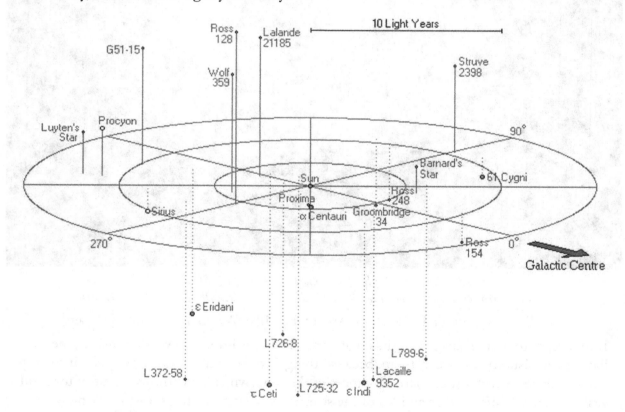

Credit: Courtesy Richard Powell from his website "An Atlas of The Universe".
Website: http://www.atlasoftheuniverse.com/12lys.html

1.13 Stars within a 12.5 light year radius of the Sun. (CC-BY-SA-2.5)

THE SOLAR SYSTEM

The solar system is situated in a spur, which comes out of the inner edge of the Orion spur (Orion arm or Orion-Cygnus arm) of the Milky Way Galaxy. It is also called the *local spur*; it lies between the Perseus arm on the outside and the Sagittarius arm on the inside.

The solar system is ensconced in a "dense cloud" of *interstellar medium* known as *interstellar cloud* or *Local Fluff*. Within the Local Fluff there lies a less dense region, whose density is much less than the average interstellar medium and known as a *bubble*. A bubble in the interstellar medium is a rarified region many light years across, created by the past explosion of a huge star many times the size of the Sun. This stellar explosion pushes the interstellar medium away, creating a relatively sparse region with a denser "cloud" surrounding it. This is the origin of the Local Fluff and the *Local Bubble*.

Our Local Bubble, through which the solar system is now passing, is thought to be about 300 light years across and shaped somewhat like an hourglass. In terms of cosmic time, the bubble is only a temporary phenomenon. Its boundary with the Solar system is not sharply demarcated. The latter's effective limit stops where the Sun's gravitational sphere of influence ends for all practical purposes.

Credit: Courtesy Wikipedia User: "Geni".
Website: http://en.wikipedia.org/wiki/File:Local_bubble.jpg

1.14 An artist's conception of the Local Bubble. [Public Domain]

The interstellar medium is made up of an extremely sparse matter, which occupies the space between the stars. Here an atom of hydrogen, which forms the most abundant material in space, occupies about 2cm³ of space. Thus, we use the term "dense" in a relative sense. The interstellar medium consists of a mixture of ions, atoms, molecules and dust, which is bathed by cosmic rays and the galactic magnetic field. Most of it, about 99% is gas (90% is hydrogen and 9% is helium) and the rest, about 1%, is dust.

All the celestial bodies bound by the Sun's gravity go to make up the Solar system along with the Sun itself. They include the planets, dwarf planets, asteroids, the Kuiper belt objects, the scattered disc objects, comets, meteors and dust. It includes the hypothetical Oort cloud as well, which is thought to exist far beyond the scattered disc objects. Though there is no proof that the Oort cloud actually exists, its presence is hypothesized, as there must be some source of the long-range comets, which appear in our skies from time to time.

Due to its own proper movement the Sun hurtles around the centre of the galaxy at 138 miles/sec or 221 kilometers/sec. This speed in much higher than its relative motion, which is the apparent speed we observe with respect to the other stars in the spiral arm. We will realize how high this speed is, if we compare it with the speed of an aeroplane flying at the speed of sound. At this speed, the plane would be going at 0.2 miles/sec or 0.32 kilometers/sec. This is 690 times slower than the proper motion of the Sun. Even then, it takes 225 million years for the Sun to make one revolution around the center of the galaxy; thus completing one *galactic year* for the Sun. Until now it has completed a little over 22 revolutions and has that many to go before it reaches the winter of its life, when it will no longer be the life giving star we know today. Mankind, however, will no longer be there to witness the event.

As children, we were taught the Sun had nine planets, Mercury, Venus, Earth, Mars, Jupiter, Saturn, Uranus, Neptune and Pluto. Today our conception of a planet has changed. Pluto is no longer considered as a planet. This was indeed long over due since the earlier definitions such as "wanderer" or "a body that goes around the Sun" was unsatisfactory, because other bodies in the solar system such as asteroids, comets, debris and dust also go around the Sun. They are also seen to wander against the fixed starry background when observed through a telescope. But common sense tells us that none of these bodies may be classed as planets. The astronomers got together on 24th August 2006 and for the first time decided to create a definition of a planet. Today, a planet is defined as a body, which orbits the Sun having enough mass to assume a spherical shape and clear its immediate vicinity of any other smaller objects, except those that revolve around it. Therefore today, Pluto has now been relegated to the group of bodies, which are known as *dwarf planets*, along with Ceres, an asteroid and Eris, a Kuiper belt object.

Till now when dealing with cosmic distances we had taken the light year to be our scale, but when we come to the solar system we will find the light year becomes too unwieldy, so we have to use a smaller scale. Even though miles and kilometers make some sense, the astronomers have chosen the *Astronomical Unit* (AU) for this purpose. The average distance of the Earth from the Sun, 93,000,000 miles or 148,800,000 kilometers, is taken as one AU. It is much smaller than the light year. One light year is equivalent to 63,240 AU.

THE SUN

The Sun is a yellow star. It is medium sized and middle aged. In terms of Earth years it is about 5 billion years into its life and probably has about another 5 billion years to go before turning into a red giant, then a nova and finally ending its life as a white dwarf. It is 865,000 miles in diameter, which makes its diameter about 109 times the diameter of our planet. Its volume is 1,300,000 times of the Earth. In comparison to the Earth, it is 332,946 times more massive. The Sun makes up 99.8% of the total mass of the solar system. It is mainly composed of hydrogen, which forms 73.46% of its mass. Other constituent elements in terms of mass make up are helium 24.85%, oxygen 0.77%, carbon 0.29%, iron 0.16%, sulphur, neon 0.12%, nitrogen 0.09%, silicon 0.07% and magnesium 0.05%. Traces of other elements, 0.02% in all, make up the rest.

The Sun's huge size creates a tremendous gravitational pull on its constituent substances causing titanic pressures to build up in its core. Here the density is 150 times of water, whose density is taken to be one. Compared to these figures, hydrogen has the least density at 0.0899 and osmium has the highest density at 22.6 amongst the elements. Iron has a density of 7.86, lead 11.34, mercury 13.6, gold 19.5 and platinum 21.45. This high density at the Sun's core, along with extremely high temperatures (14,000,000°C), causes the nuclei of hydrogen atoms to fuse to produce the next heavier element - helium. In the process there is some excess energy left over. It is this energy, which comes out in the form of radiation, including light and heat. The Sun's energy comes from its hydrogen content. It is the cleanest form of energy created by nature, which man is trying to emulate without success.

The Sun converts a mind boggling 4,000,000 tons of hydrogen to helium every second and in the process, it produces energy as we have seen. However, the rate of the Sun's energy production becomes dwarfed, if we compare it with the rate of energy produced by the human body. The human body produces energy at the rate of 1.2 watts/kg of matter. In comparison, the Sun produces only 6μ watts/kg of matter, which is an incredibly small amount, compared to the human body, by a factor of 200,000 times. Conversely, if we tried to emulate the Sun it would require a 1gigawatt fusion power plant supplied with 166.6 billion tons of hydrogen in the plasma state occupying one cubic mile! Since fusion depends upon density and temperature, such reactors would require much higher temperatures in comparison to the Sun's core. Our present technology is incapable of providing this amount of energy.[1] However, it is only by such low rate of conversion of matter into energy can the life of our parent star be prolonged. This is a key factor, which has allowed evolution of life to progress over the ages. It will continue to do so, until long after man has disappeared from this planet.

The light energy comes out in minute quanta known as *photons*. The photon travels at the speed of light and from this one is tempted to assume, the photon created at the centre of the Sun reaches its surface in 2.3 seconds. This is the time it would take the light to cover this distance. However, this does not happen. The photon after being formed in the core of the Sun cannot come out immediately. It encounters the gauntlet of densely packed atoms inside the solar substances. At every twist and turn it bumps into an atom, the photon is absorbed and it is re-emitted again in a random direction. Thus, in effect the photon bounced back and forth like a tennis ball. In this manner it has to travel through the various zones inside the Sun. At first, it has to come out of the *core* and then successively passes through the *radiation zone*, *convection* zone and the *photosphere*. It is eventually freed at the surface and travels through the *chromosphere* into space. It has been calculated a photon thus produces can take anywhere between about 10,000 to 170,000 years, before it finally reaches the surface and radiates out into space. Once in space, the photon becomes free to travel at 186,000 miles per second and some of them reach the Earth after 8.3 minutes.

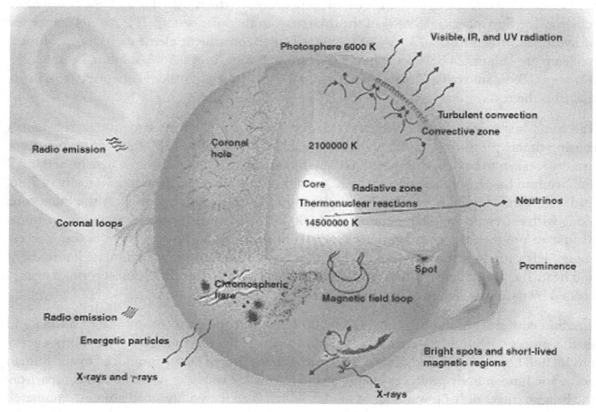

Credit: Courtesy NASA – Goddard Space Flight Center - The Imagine Team: Project leader: Dr. Jim Lochner; Curator: Meredith Gibb; Responsible NASA Official: Phil Newman.
Website: http://en.wikipedia.org/wiki/File:Sun_parts_big.jpg
1.15 Across section of a Sun-like star. [Public Domain]

Apart from the photons, *cosmic rays* and neutrinos there are other particles known as plasma come out. The latter consists of a stream of charged electrons and protons, ejected from the Sun and represents the *solar wind*. This solar wind travels away from the Sun in all directions and reaches out 95 AU into space. Here it slows down on meeting the interstellar medium. At the interphase, there is a zone of turbulence; just like when two moving bodies of water meet. This turbulence encloses the solar system like a sheath. This is the *heliosheath*. Beyond this, the solar wind finally dies down and ultimately blends with the interstellar medium. This boundary is known as the *heliopause*.

MERCURY

Mercury is the smallest of the planets. It has only 5.5% of the Earth's mass and its diameter is 3100 miles. It is also the nearest to the Sun, orbiting at an average distance of 0.38AU. Its orbit is quite eccentric, bringing it to about 29,000,000 miles near the Sun at *perihelion* and 43,000,000 miles at *aphelion*. The word perihelion comes from word "peri-", implying being close and "ap-" signifying away, whereas, the word "-helion" is derived from Hēlios, the name of the Greek Sun god. He was one of the Titans or the elder gods in Greek mythology. It was much later that Apollo came to be identified with the Sun. Even though Mercury is quite small, it receives a great amount of light due to its proximity to the Sun. This light is again reflected back from its surface. This makes Mercury's brightness rival that of Sirius, the brightest amongst the stars as seen from the Earth. Yet for all its brightness, Mercury is difficult to spot.

The reason for this paradox arises due to its close proximity to the Sun. As the Sun makes its daily journey across the heavens, Mercury "hugs" the Sun as it orbits close to it and disappears with the Sun over the horizon in the evening. Thus, it cannot be seen at night like the other planets. Neither do we see Mercury during the day, because like any other planet or star, the brilliance of the Sun masks its brightness. Therefore, the planet is only visible briefly in the diminutive light of early morning and in the fading light of early evening, when the Sun is very low on the horizon or it is just below it. It was for this reason for a long time the ancients thought they were seeing two different bodies. Accordingly, they named them the Morning Star and the Evening Star. Thus due to its elusiveness, Mercury was rightly named after the messenger of the Roman gods. It was said the god so quick, the eye could hardly detect him.

Its closeness to the Sun makes it a dry place. Unlike the other inner planets, the surface of the planet is not its crust, but the exposed mantle, which covers its core. Loss of its crust probably resulted from the constant effect of erosion due to the solar wind, which has been bombarding it over the aeons. Apart from crater marks, which are the result of meteoritic strikes, Mercury's surface is wrinkled. Even though there is no evidence of any geological activity on the planet now, these wrinkles are likely to be an indication of past activity within the planet. Beneath this mantle lies an iron core. The planets only "atmosphere" consists of particles blasted off its surface by the solar wind.

Its axial inclination is very slight and the planet rotates on its axis much more slowly than the Earth. In terms of our Earth days, one Mercurian day is equal to about 58½ Earth days and a Mercurian year equals about 88 Earth days. The planets axial rotation and its orbital period are "locked" in a ratio of 2:3. This means the planet makes two rotations about its axis every three Mercurian years.

VENUS

Apart from the Sun and Moon, Venus is the brightest celestial body. It is 0.7AU or about 67,000,000 miles from the Sun. It is almost the size of the Earth, being 7,700 miles in diameter. It mass is 80% of the Earth. Probes to the planet show, like Earth it has a substantial iron core.

In the past, due to the resemblance in size, people thought of Venus as a sister planet to the Earth. Therefore, some astronomers assumed its history must have taken a similar course. In spite of all conjectures, Venus remained shrouded in mystery. No surface features were discernable during the era of telescopic observations, as it always remains covered by dense clouds. The mystery under the clouds only helped to spur on our imagination. Consequently, many readily agreed with the Swedish scientist and Nobel laureate Svante Arrhenius (1859-1927), when he proposed the idea that life was present on Venus. Arrhenius speculated like Earth, Venus had land and plenty of water where life had evolved and was thriving. He thought conditions there were somewhat like those present during the Carboniferous era on Earth. He went on to assume, primitive forests flourished and animal life had evolved there up to the reptilian stage. However these dreams were shattered when planetary probes reached Venus. The true picture of this siren that had fascinated man for so long was now exposed! Probes revealed the harshest of environments. There is no water and the surface strewn with rocks. Temperature averages over 400°C. This is hot enough to melt lead easily. An atmosphere of carbon dioxide clothes the planet and its density is 900 times that of Earth's. Consequently, the pressure exerted on the surface is about 900 kg/cm2. Compared to this the Earth's atmospheric pressure exerts only 1 kg/cm2 at sea level. All this makes Venus a forbidding place for any form of life.

In spite of our disappointment, it is through planetary probes, we have come to know more about this planet. Its axis appears to be tilted about 3° to the vertical. This is deceptive. The true tilt is 177°, which means the planet is rotating on its "head". Its north pole is pointing south. This fact can be deduced from its retrograde rotation, as compared to the other planets. The planet rotates about its axis once in 243 Earth days, but the year on Venus is equal to 224 Earth days. This makes the day on Venus longer than its year; but to an observer on the planet it still would appear that the year has two Venusian days. The reason for this phenomenon is due to the fact the planet, in contrast to the other seven planets, orbits around the Sun in a very slow retrograde fashion. The day therefore catches up to the year. If a hypothetical observer on Venus could see through its permanent dense clouds, he or she would have the novel experience of seeing the Sun rising in the west and setting in the east twice a year! Each full day on Venus is equal to about 118 Earth days. There is equal distribution between daylight and the darkness of night of 59 Earth days each.

Being an *inferior planet* like Mercury, which means it lies between the Sun and the Earth, it shows phases like the Moon when observed from the Earth through a small telescope or even with binoculars. Some claim the keen sighted can make out these phases through the naked eye. I do not know whether it is true or not. This may be difficult, even though at 24,000,000 miles Venus comes close to us by planetary standards. This distance is closer than Mars can ever come to the Earth by 10,000,000 miles. It is, however, doubtful that even at this distance, whether it would be possible to observe these phases of Venus by the unaided eye. However to the meticulous observer, the planet would seem to brighten and

Drawing by Tarun Ghosh.
1.16 Phases of an inferior planet.

dim as it cyclically passes through its phases. The probes to Venus have detected geological activity. Surprisingly in spite of its iron core the planet does not seem to have any magnetic field. Its core is covered with a silicate mantle.

Like Mercury, Venus does not possess any satellite.

EARTH

The Earth by the very definition of AU is 1AU away from the Sun. Its distance varies from 94,537,000 miles at aphelion to 91,377,000 miles at perihelion, the mean distance being 93,000,000 miles. It is the largest of the inner planets with an equatorial diameter of 7927 miles. The axis is tilted at 23½° and the planet's rotation period of 24 hours 3 minutes and 56.555seconds. It has an iron core with an appreciable magnetic field. There is a mantle covered by a crust. Geological activity is present, as we all know, sometimes to our cost. Both land and water covers its surface. The atmosphere is substantial containing nitrogen, oxygen and a little argon and less carbon dioxide. The Earth is distinctive amongst the planets in that it contains life.

Even today, we do not know how life started, but there is evidence it began some time before 3.5 million years. It is fairly certain that life has been able to get a foothold, because of our planet's position from the Sun, its size and certain other factors, which allowed conditions suitable for life. Its distance from the Sun is such that the temperatures are equitable. These temperatures allowed water to be present in a liquid form in most places all round the year. The evolving atmosphere of the Earth was another crucial factor. The very earliest atmosphere in the Earth's nascent stage probably consisted mainly of hydrogen and some helium. Later these gases became replaced by a mixture of sulphur dioxide, hydrogen sulphide, carbon dioxide, nitrogen, methane and ammonia, all as a result of incessant volcanic activity and lightening due to violent storms that raged continuously. If any significant amount of hydrogen or helium remained from the beginning, then they must have been lost a long time ago. Being lighter gases they would have reached the upper limits of the atmosphere from where they would then pass into space. This happens, because at the extreme upper limits the atmosphere becomes so rarified, the speed of the motion of the atoms these gases exceeds the Earth's escape velocity. Thus, the Earth's gravity would not have been strong enough to hold them.

Today it may seem to us the composition of the early atmosphere was inhospitable to any form of life. We think so, because we would not be able to survive in it; but life is not about us only or what we see around us. The first organisms to appear were microorganisms. They were able to thrive in this environment. The reason for this was that carbon compounds form the basis of all living bodies. The forces that drive life depends the ability to fix carbon from the environment to build bodies. To do this the organisms has to have the ability to extract energy their environment to utilize the available carbon through a complex chemical process known as metabolism for growth, sustenance and repair of their bodies; also for multiplication. Evolution is not particular as to where the organism obtains either the carbon for its body or the energy to drive its metabolism.

The fundamental tenet remains the same no matter how an organism has evolved. Today this process of extraction for growth, sustenance, repair and replication is done by the plants from carbon dioxide and water in presence of chlorophyll and sunlight. The sunlight supplies the energy, the chlorophyll facilitates the process and the carbon dioxide the carbon. Again animals, including us, get the carbon and energy from food, either by eating plants or eating other animals, which eat plants. The earliest living forms had neither opportunity. They could not depend not on the sunlight for their energy, as sunlight was not sufficient at that period. This was for two reasons. Firstly, the atmosphere in the early history of the planet contained dust and debris due to volcanic activity. Thus sunlight did not penetrate well. Secondly, it is known now that the Sun, during the early stage of its life, was much dimmer than it is today. Thus it was not conducive to the development of photosynthesis as a way of life. Neither did they have the ability to devour their fellow species as food. So, living organisms at that period had to extract energy from inorganic chemicals around them to drive their metabolism. They had evolved in a way that were able to break down the gases like sulphur dioxide, hydrogen sulphide and ammonia which constituted the primitive atmosphere, into carbon, sulphur, hydrogen, nitrogen and oxygen. Even today there are such microorganisms present on our planet. They then used hydrogen or oxygen to drive their metabolism. The utilization of oxygen resulted in the production of carbon dioxide by the organism. This now passed into the atmosphere. Thus over the millennia the composition of the atmosphere began to change, carbon dioxide gradually came to occupy a more and more significant proportion of the Earth's atmosphere, and the sulphur dioxide and the other gases gradually disappeared. This reduction in levels of sulphur dioxide was also in part due to decrease in volcanic activity on the planet. The atmosphere cleared and at the same time the Sun had become brighter also, like we know it to be today

This allowed evolution changed its strategy. A new class of organism evolved, known as cyanobacteria (blue-green bacteria or blue-green algae). They were a type of bacteria, which can still be found across the world today. They obtained their energy from sunlight. These organisms lived in water. Nothing lived on land at the time. They distinguished themselves from their predecessors by creating the chlorophyll molecule. This substance gives the green colour to leaves. The introduction of chlorophyll was an important landmark in evolution, because with chlorophyll they could turn water and carbon dioxide into sugar in presence of sunlight. Through this processes the Sun's energy became "locked" in the chemical bonds of the sugar molecule. Oxygen was released in the process. When these organisms broke down the sugar, because of its low metabolic needs, it only utilized a small part of this oxygen. The rest of it passed out into the atmosphere. This would ultimately produce an oxygen rich atmosphere over the ages. There was more oxygen now. This allowed an ozone layer to build up in the upper atmosphere. This would form a shield against the harmful portion of ultraviolet rays and prevent them from reaching the Earth's surface. In the mean time in the presence of relative abundance of carbon dioxide, which is highly dissolvable in water, these organisms were able to flourish in the waters of the Earth. They gradually became the dominant form of life. Some of their ancestors have left their imprint in the form of fossils, which are thought to be 2.8 – 3.5 billion years old. Their legacy the chlorophyll molecule was picked up by a new form of organism – the plants.

At first, the plants were unicellular and kept themselves to the waters of the Earth. From the unicellular organisms, they evolved slowly into small multicellular organisms. However, they did not achieve any more. They did not need to develop any further, because they floated about in the water and obtained their necessary nutrients from their surrounding environment. To do this they did not need any roots. Today's seaweeds are their descendants. They belong to a class of plants known as algae, which still to this day have not evolved any roots. The variety that grow in the seas stop themselves from being washed away by the tide by anchoring themselves to the bottom by a devise known as *holdfast*, which should not to be mistaken for roots.

At this point, the land was inhospitable. There was no soil on land - only *regolith*. Regolith is the result of the debris from volcanic activity and the effects of erosion on rocks. To create soil, it needs organic substances, like dead plants, mixing with the inorganic regolith. At this stage there were no plants or any other life forms on land to supply such organic material. The situation would soon be remedied. The plants moved out of the water to invade land. At first they made their home along the moist areas on the water's edges. As these plants died, they mixed with the regolith and created more and more soil, which enabled them to move further inland. Starting as rootless plants, like the algae, they slowly evolved rudimentary roots like mosses, liverworts and ferns. In time, these species with rudimentary roots evolved further. Now through the process of propagation, such as wind and water, the plants were able to march inland and thrive. Larger plants like, the modern clubmosses and horsetails now evolved and flourished. At one period in the history of this planet, their relatives grew to gigantic proportions and dominated the scene. In turn, they gave way to the gymnosperms, which are true seed bearing plants, such as cycads and the conifers. In time came they were superseded by the angiosperms, which are the monocotyledonous and the dicotyledonous plants. The former represented by the grasses and palms, while the latter by the flowering plants, which flourish in the tropical and temperate forests of our world. Thus at one stage in their evolution the plants ruled the world.

Now with plants, the atmosphere contained more oxygen. It was able to pave the way for a completely different form of life - the animals. This form of life had a completely different way of acquiring energy from the environment, which was through the process of eating. Thus they utilized much larger amounts of oxygen due to their higher metabolic demands arising from movement and other bodily functions. They were no longer fixed to the ground like the plants. However, they were fated to be entirely dependent on the plants for their living, either directly or indirectly. It is obvious animals evolved in response to plants producing too much oxygen. Though nearly all organisms need oxygen to survive, it must be remembered, breathing high levels of oxygen over prolonged periods can poison the metabolism. This is why, even today, at its present level it is important to have nitrogen in our atmosphere. It dilutes the oxygen we breathe. It is quite probable evolution in trying to protect plant metabolism, responded by introducing animals to curb further increase of oxygen in the Earth's atmosphere. Now with both animals and plants, the atmosphere was able to reach a balance between the proportions of carbon dioxide and oxygen. This had been more or less stable over the ages, until man came and disturbed it.

The relationship of life and the changing atmosphere may seem a somewhat strange idea to many. It is however equally true, in the absence of life, the composition of the Earth's atmosphere would not be what it is today! It is important to realize the other planets do not have an atmosphere like ours, because there is no life there. The atmosphere, like all else, has to evolve also!

If life is to develop, it is not just simply a question of Sun supplying energy for all living creatures. The Sun is important for other concerns of life as well. The various chemical reactions in the metabolic process of any living body cannot go on indefinitely. They require rest. It is important to have both day and night. Thus for any planet to harbour life, it would not only be required to rotate around its axis, but this rotation should not take too long or be too short. It appears something near a 24 hours rotation period is reasonable for reasons discussed later.

The rotation causes forces to develop in the atmosphere and cause it to move in certain directions. This lays down the basic pattern of atmospheric movement on which other patterns are imposed. Under such conditions if the rotation is too quick, then violent weather conditions would ensue. If, however, the rotation is too slow or all together absent. The wind patterns, like we see on Earth today, would not develop or at best be poorly developed. They would not then be able to take the clouds too far into the land for it to be watered. This atmospheric movement is further aided by convection currents due to differential warming of the oceans and landmasses by the heat derived from the Sun. Thus, an equitable distribution of land and water is important for life to evolve also.

Together with the tilt of its axis, the revolution of a planet around its parent star is also significant for life. They cause the phenomenon of the seasons and seasons are linked to the cycles of life. Their right combination is a very important factor. Without an adequate tilt in the axis, there would be no seasons and weather would be monotonous. Thus planets, with no tilt like Jupiter or hardly any tilt as with Venus, would have no seasons. On the other hand this would be equally true for planets where the tilt is extreme as in case of Uranus, which actually "rolls" along its orbit as it goes around the Sun. In doing this it always presents one pole to the Sun.

To get some idea of the significance of the tilt of the axis in relation to life, let us imagine what would happen, if hypothetically, the Earth suddenly lost its tilt. This would result in the absence of seasons. The weather would lose its regularity. Consequently, the cycles of birth and death of plants and their seasonal fruiting would become asynchronous. As a result, the availability of food in many places would become much less at any given time of the year for the plant eaters. This would be compounded in times of stress, such as drought. It would further add to the dwindling food supplies. This would create stiffer competition for the meager food available under such circumstances. Thus with food in short supply, herbivores would look for plants they would not normally eat. In this way, the herbivores in order to survive would loose their pattern of foraging habits. It would then lead to a free for all situations. Therefore, the chance of any particular plaint specie being wiped out would be greater. This would lead to further food shortages for herbivores.

Also in such a scenario the seasonal calving, when the calves are born within a few days of each other, would be disrupted. This synchronous calving is very important, because it gives protection from predators through numbers, where some can always chance to survive and grow up to continue their species. However, without seasons, calves would be born at any time. Predators would have a field day chasing after a few vulnerable calves all the year round. Gradually over the years, the number of herbivores growing up to replenish their species would dwindle. Eventually, with time, the herbivores would be wiped out. When this happened, the carnivores would starve and die. Higher forms of life would disappear from the planet!

Of course, some might also argue there may be planets in the universe where there may be no seasons and yet life could still have evolved in them. This may well be true. They should however still try to remember, without a proper tilt in the axis, life could never have evolved to the stage we see on Earth.

The period of revolution of a planet around its parent star is also important. It connects the year with the length of each season. This in turn is related to various biological cycles, importantly growth and aging. Thus a quick succession of seasons would be like the cruel practice of artificially doubling or tripling egg production in poultry batteries, where the light is turned off and on alternately to simulate day and night, two or three times every 24 hours. Though the egg production is increased, the hens die early. If on the other hand seasons were too prolonged, then the biological cycles would be stretched to their limits and many forms of life would not be able to complete the cycles of life before the year was over. Also many young would have to face long winters and in the colder climates they would not live to see the next season.

The size of the planet is important. If the size were too small then the gravity would not be enough to hold on to its atmosphere. It would leak out in time. Even a planet the size of Mars, as we shall see, will not be able to hold on to its atmosphere long enough to allow life to evolve. It is probable all planets may have started their lives with abundant hydrogen. So if a planet were too large, then its huge gravity would hold on to its original hydrogen. Depending on the temperature, its surface would be made of frozen hydrogen or liquid oceans of hydrogen would cover the surface. In addition, the atmosphere would contain too much hydrogen. This would not be congenial for life. It appears the size must be just right to allow the hydrogen to leak out, while retaining the other gases like nitrogen, oxygen and carbon dioxide, which are heavier. Apart from such considerations, the different conditions of gravity are also relevant to how living organisms evolve. Under extreme gravity, muscles and bones would have to be truly massive for any movement was to take place. It would not be worthwhile for evolution to build such forms, because to supply the energy for their movements in order to obtain food, their energy account would be overdrawn. Evolution always acquiesces to an energy audit. Therefore such higher forms of life would never appear on such planets even though other conditions allowed for it. On a more frivolous note, if the gravity were not adequate, living organisms would fly off the planet.

No one knows what should be the minimum or the maximum of such things as the axis tilt, the period of rotation of a planet and the period of revolution around the parent star or even it's mass to make it suitable for life. We do not have any data on such factors until now, because we have not found life on other planets yet. What we do know is they are equally important to life like the other factors such as liquid water, protection from radiation by having a magnetic field and an ozone layer acting a shield against ultraviolet rays. To all this we must take into account the type and age of the of the planet's parent star and other local cosmological conditions such as gravitational effects due to binary star systems, exploding supernova, radiation from nearby pulsars, and such like. Out of the countless planets that must exist in this universe, such criteria as these drastically narrow down the number of eligible planets, which may harbour life. Thus, astronomers before announcing they have found an Earth like planet orbiting a distant star should ascertain all these factors before pronouncing there may be life on such a body. Be that as it may, life cannot be unique to the Earth. Amongst the trillions and trillions of stars in the universe there must be many suitable stars that have suitable planets, which have evolved life. The only question is, in what direction and to what level?

The Earth has only one satellite, the Moon. Its distance from the Earth is 225,740 miles (363,300 km) at its *perigee* and 251,970 miles (405,500 km) at its *apogee*.[2] We have earlier seen the meaning of "ap-"and "peri-"in relation to aphelion and perihelion. The term "-ge" is derived from the name of the Greek mother goddess, Ge. She represented "mother earth". Not the planet Earth, but the agricultural soil, which gave food. Ge was the wife of Uranus, god of the sky. Out of there, union was born the Titans, the elder gods of Greek mythology. The orbital eccentricity of the Moon is 0.0549. When seen from the Earth, its angular diameter varies from 33' 31" at perigee and 29' 22" at apogee. Though the difference is not very apparent, it can be measured easily. The Moon has a diameter of 2158 miles or 3453 km. Its mass is 1/80 of the Earth. It has no atmosphere and no geological activity. The same face always remains turned to us. It is only recently water has been found in the Moon.

The Earth and the Moon forms a *binary system*. To understand this concept we consider two bodies of equal mass. One of them cannot move around the other without the other also moving. Both must revolve around a common centre of gravity between their masses. This is because they behave as if their mass was concentrated on this point, which will lie between the centres of the two bodies. Their common centre of gravity is known as the *barycentre* (barus = heavy in Greek). When one body is more massive, then this barycentre will shift towards it. If we kept the mass of the smaller body the same and hypothetically, we were able to go on increasing the mass of the other, then the barycentre would go on shifting towards the more massive body. A point in time would come, when the centre of gravity would come to reside inside the more massive body. This happens in case of the Earth-Moon system. Their common centre of gravity lies within the Earth, as the Earth is so much more massive than the Moon. Though we may think so, both the Earth and the Moon move about together. With the result, on the ultimate analysis the Moon actually does not go around the Earth in circles - it traces a sinusoidal curve around the Earth's, in the latter's progress around the Sun.

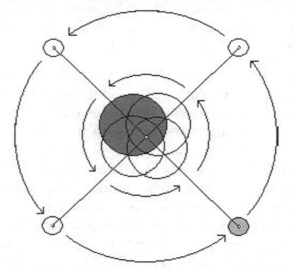

1.17 Earth-Moon system and its barycentre

The Moon is also significant for life in many ways. Like the seasons, the cyclical waxing and waning has some effect on many life forms. The moon is responsible for the tides, which has influence on many aspects of marine life.

MARS

Mars also known as the Red Planet. It has a distinct red tinge, which is easy to see with the eye. The colour red has a powerful symbolism for man. It has always been associated with the colour of blood. So the planet is aptly named after the Roman god of war.

Mars is 141,600,000 miles away from the Sun, which is about 1.5AU. It has 11% the Earth's mass and an equatorial diameter of 4219 miles. It has some similarities to our Earth. Mars' axial rotation period is close to that of the Earth, being 24 hours 37 minutes and 23 seconds. Its maximum temperature reaches 22° C (71.6°F) and the minimum temperature falls to -70° C (-94°F). The similarities end here. The planet has two polar caps like Earth, but these are made of frozen carbon dioxide and not water. There is no geological activity on Mars, at least not at present. However, Mars' surface features still show many features, which stand as testimony to past geological activity. There is the great rift the valley, Valles Marineris and the many volcanoes, the largest of them being Mount Olympus.

The planet has an atmosphere, which is composed mainly of carbon dioxide. If there were other gases like those on Earth then they must have leaked out into space long ago due to its low gravitational field. Carbon dioxide also may be gradually leaking out, but some of it maybe replaced from the frozen reservoirs in the planets poles. Thus, Mars may have started with a denser atmosphere, but the present atmosphere is quite thin. Winds on Mars can reach very high speeds, creating dust storms. They may cause streaking of dust along the surface of the planet. This may be the explanation of the phenomenon of "canals" that we see from Earthbound telescopes. However, this is only a conjecture. The true cause of this phenomenon is not known.

There is no surface water on the planet today, but we can see dried up riverbeds. This is a good indication there must have been rains on the planet in the distant past and atmospheric conditions may have been similar to Earth. When the Martian atmosphere leaked out, the atmospheric pressure started falling. The rate of conversion of liquid water into water vapor increased. In the long history of the planet, there was enough time for the water to have been lost in this way. Without water, we can expect no signs of life. Whether subterranean water exists in this planet with primitive life is yet to be ascertained.

Mars has two diminutive satellites, Deimos and Phobos, which are likely to be captured asteroids.

At one time, there was speculation as to whether there was intelligent life on there, because of the markings that were seen through terrestrial telescopes. Even when I was a boy, before the days of space probes, I had read books, which discussed whether the existence of canals was real. The history of the idea of canals on Mars goes back to the nineteenth century telescopic observations of the planet. It was in 1877, Giovanni Schiaparelli (1835-1910) announced he had observed an intricate network of "canali" on the surface of Mars. Now the word "canali" was an Italian word meaning groves or channels. Such is human nature, instead of looking up the translation; it was much easier to associate the Italian word "canali" with the word "canal" in English, which happened to sound similar. Canals on Earth as opposed to rivers carried water through human intent. Thus, their presence on Mars implied intelligent life. It was assumed therefore the Martians who lived there must necessarily be intelligent. This sentiment fired the imagination of many astronomers. The idea spread in the English speaking nations and from them to others, even to the Italians.

One of the people who believed in intelligent life on Mars was Percival Lowell (1855-1916). He was a rich man, who had more than a passing interest in astronomy. During his life time he made many contributions to astronomical science, amongst them was his part in the discovery of Pluto. After Schiaparelli retired, due to failing eyesight, it was Lowell, who took up the task of observing Mars. In order to have the best observation facilities, free from atmospheric pollution, he built a 24-inch refracting telescope on a hill at Flagstaff, Arizona. He appropriately named the hill, Mars Hill. Observation of Mars and mapping the canals became an obsession with him for the rest of his life. He firmly believed the climate on Mars was dry and some ancient civilization lived there. Thus in order to distribute the sparse water available on the planet, the Martians had designed a network of canals to supply their people with this life giving liquid. However romantic this idea may have been, sadly it did not turn out to be true. The Martian probes have proved there are no canals on Mars and certainly no intelligent life.

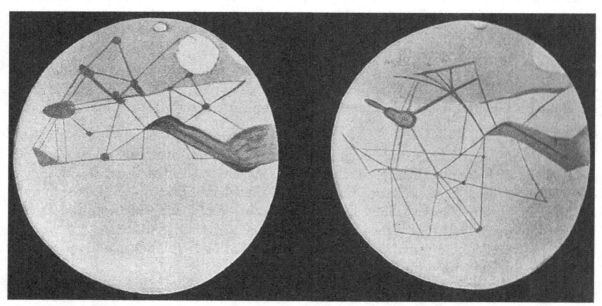

Source: From Yakov Perelman's - "Distant of Miry". St. Petersburg, the printing house Of Soykina), 1914.
Website: http://en.wikipedia.org/wiki/File:Lowell_Mars_channels.jpg
1.18 Percival Lowell's drawing of Martian canals. [Public Domain]

THE ASTEROID BELT
Johann Daniel Titus von Wittenburg (1729-1796) found a mathematical relationship between the distances of the various planets from the Sun, which was popularized by Johann Elert Bode (1747-1826) and became known as Bode's law. The gap between the two planets, Mars and Jupiter, appeared to be inconsistent with the distances between the other then known

planets from the Sun. The astronomers were already aware of this gap for quite some time. The question in astronomers' minds was whether there was any other planet between Mars and Jupiter. All these ideas had occurred in the century before the discovery of the first object between Mars and Jupiter. In the mean time, with all this in mind, Baron Franz Xaver von Zach (1754-1832) got together a group of astronomers to hunt for the hypothetical planet predicted by von Wittenburg. Lady Luck while looking upon Baron von Zach's elaborate preparations must have been mildly amused. Such was the travesty of fate; even before the Baron's team got started, the discovery was already made! It was on the first night of 1801, by chance Giuseppe Piazzi (1746-1826) of Palermo in Sicily made the discovery. He was making a routine inspection of the skies while trying to compile a star catalogue. Piazzi detected an object between Mars and Jupiter. Naturally, he thought it might prove to be the planet, which was expected to be there in that region. He named it Ceres after the ancient Roman corn-goddess. His discovery lent further support to the earlier work of von Wittenburg. The astronomers now came to believe it might be the much sought after planet between Mars and Jupiter.

Ceres' position was confirmed to be in the right place, where the elusive planet was meant to be. Its mean distance was 257,000,000 miles or 411,200,000 kilometres from the Sun. However, being only a little less than 625 miles or about 1000 kilometres, it was considered too small to be a planet. Baron von Zach team of planet-hunters now began looking for the elusive planet with renewed hope. Instead of a planet they found other bodies like Ceres in this region, but all smaller. Their search between 1802 and 1808 yielded Pallas in 1802, Juno in 1804 and Vesta in 1807. Disappointed at not finding what they were seeking he desisted and disbanded the team in 1808. In 1845, a German astronomer named Hencke detected the next asteroid. He called it Astræa. By end of the century, 26 other such bodies had been identified. Today it is estimated there are over 40,000 asteroids in this belt, apart from the dust and debris to be found in this region.

The asteroid belt lies between 2.3 to 3.3 AU from the Sun. It resembles the Kuiper belt, which lies beyond Neptune. Like the Kuiper belt, it contains bodies that range from the size of dust particles to bodies the size of a few hundred miles. Except for Ceres, Pallas and Vesta, which are more than 300 miles, the sizes of the asteroids vary from a few hundred yards to a few miles only. Thus apart from the larger ones that are spherical, the rest are irregular shaped rocks containing some metallic elements - predominantly iron.

Those bodies over a kilometer may reach into thousands. Yet their total mass is only 1/1000 of the Earth's mass. Such a small amount of matter spread over a vast area of space indicates the sparseness of the asteroid belt. This is reflected in the fact planetary probes pass this belt without any record of collisions happening. Even though the Asteroids travel along a belt, they are not evenly distributed. In the belt, there are gaps, so certain regions are free of these bodies, thus forming separate divisions within the belt. These gaps in the belt are known as Kirkwood gaps; named in honour of Daniel Kirkwood (1814-1895), American astronomer and mathematician.

Ceres at 2.7AU distance from the Sun is the largest object in the asteroid belt. It is the only body, which merits the name of a dwarf planet amongst the asteroids. It is too small to posses any atmosphere. It has a rocky composition and possesses some metallic iron.

Some asteroids have irregular orbits. A few even come close to Earth. Some in fact too close. These latter are known as rouge asteroids. In the year 1937 Hermes, an asteroid, whose size is about a mile came as close as 485,000 miles to the Earth, which makes it twice the distance of the Moon. In astronomical terms this is remarkably near. If such a body were to come any closer, then it would be caught up in the Earth's gravitational influence and cause it to strike the Earth. As it neared, the Earth such a body would be broken up by the Earth's gravitational field even before it reaches the Earth's atmosphere. On entering the atmosphere, the smaller fragments would burn up and the rest would strike the Earth's surface. If they landed in the ocean then the result could be a tsunami. If they hit land, there would be impact craters and huge amounts of dust and debris would be thrown up. The larger debris would fall back to the Earth quickly, while the smaller particles and dust would remain suspended in the atmosphere. Depending upon the size and quantity of these objects, it may be days or may be months or even years, before we would see the light of the Sun again. Such an asteroid strike was probably responsible for the extinction of the dinosaurs 65 million years ago. Some asteroids have highly eccentric orbits. Hidalgo's path crosses Jupiter and reaches very near to Neptune's orbit, while Adonis and Apollo's reaches between Venus and Mercury. Eros an irregular body whose long axis is 15½ miles comes as close as 15 million miles to the Earth. Others have orbits, which reach just beyond Mars. At the other extreme, there are those that have strayed from the main asteroid belt, probably displaced by gravitational effects of Jupiter, sometime in the distant past. Like the Trojans, they have left their fellows completely. These latter are a group of asteroids, which follow Jupiter's orbit at a separation of about 60° from the giant, so there is no chance of them colliding. Some asteroids have highly inclined orbits, like Icarus whose inclination lies at 23° to the orbital plain of the other planets. At perihelion, it comes within 19 million miles of the Sun, which is closer than Mercury.

The exact origins of the asteroids remain uncertain. Either they had failed to form a planet or they may be the result of a planet breaking up. The cause of either of these scenarios would point to the gravitational effects of Jupiter. Today their total mass is estimated to be about 0.1% that of the Earth. This does not add up to what we would call a planet. On the other hand, if a planet had been actually torn apart by the gravitational effects of Jupiter, it is quite reasonable to think that most of the parts may have been lost after such a cataclysmic event. Much of the resultant fragments would have been pulled into Jupiter by its gravity, while other fragments would be scattered into eccentric orbits. Most of these would then be lost over the ages by being drawn out of their orbits by the other planets or by the Sun as evidenced by many of their erratic paths today. So it is likely many of them were displaced altogether from the solar system in this manner. Only a small residue was left to settle into their present orbit, forming the present asteroid belt. However all this is in the realms of conjecture. Until we have more evidence, we cannot pronounce any firm judgment on the matter.

THE JOVIAN PLANETS

As we go further away from the Sun, we come to the Jovian planets - Jupiter, Saturn, Neptune and Uranus. They make up 99% of the total mass of bodies, which orbit the Sun. These planets are also known as "gas giants", so called because much of their size is contributed by gases, mainly hydrogen. At their cores there is metallic or liquid hydrogen with helium next in abundance. Their atmosphere contains hydrogen, helium and ices. However, the relative percentages of water, methane and ammonia in their ices may vary. Jupiter and Saturn appear to have more hydrogen and helium than the other two.

JUPITER

Jupiter is the largest planet in the solar system. It is 88700 miles in diameter at its equator. The mean distance of Jupiter from the Sun is 484,300,000 miles or just over 5AU from the Sun. It is a remarkably bright object in the night sky, though it is not as bright as Venus. Its axial inclination is 3°1'. The Jupiter year is equal to 11.86 Earth years. Its proper motion along its orbit averages to about 8.1 miles/second. Its axial rotation is under 10 hours. Its volume is 1312 times and mass is 318 times of the Earth. It is so massive that its mass comprises two and a half times the mass of all the rest of the other planets put together. Due to its immense size, some have even considered it a failed star. Its internal heat is 30,000°C. Even though it comes nowhere near the Sun's core temperature, it is still quite high. However, this temperature is not enough to cause nuclear fusion of hydrogen within, which would have elevated the king of planets to the status of a star.

The planet has a strong magnetic field. Jupiter has as a comparatively stable core of iron and silicates over which lies an inner mantle of liquid hydrogen. Over it, there is an outer mantle of molecular hydrogen. The atmosphere is 600 miles thick and consists of ices, such as water droplets, ammonium hydrosulfide, ammonia and gaseous hydrogen. In the upper reaches of its atmosphere, the temperatures range around -130°C.

Through the telescope, bands of clouds can be seen. They form a distinctive feature of the planet, like so many furrows befitting the brow of a monarch. Not surprisingly, Jupiter has been named after the king of the Roman gods due to its size. The Giant Red Spot is another characteristic feature on its surface. This feature created by a storm, which has been raging for many centuries. Its size is about 25,000 miles by 7000 miles, which is an area equivalent to the whole of the Earth's surface. These characteristic features are however not permanent. They are merely turbulences created by the weather, which results from the internal heat generated by the planet. This is in contrast to the Earth's weather created by the Sun's heat.

Credit: NASA/JPL/USGS
Website: http://photojournal.jpl.nasa.gov/catalog/PIA00343

1.19 Jupiter. [Public Domain]

Jupiter has sixty-three known satellites. Ganymede, Callisto, Io and Europa are the four larger ones. The former was a beautiful youth, who was the cupbearer of the gods. The rest were maidens loved by Jupiter. They are known as the Galilean moons. These moons also produce heat from their interior and Io has been shown to have volcanic activity, as recorded by planetary probes.

SATURN

Saturn is the second largest planet. Seen through the naked eye it is has a slightly yellowish tinge, which imparts a jaundiced look. It is not very bright, because of its great distance from the Sun. Viewed through a telescope the most characteristic feature is its rings.

When Galileo first looked at Jupiter and observed its four larger moons in 1609, he also turned his telescope on Saturn. He saw the planet's rings. Contrary to popular belief, through his small telescope, he did not recognize them as such. In fact, he thought Saturn consisted of three bodies. The central body being Saturn itself and the broad surface of the ring visible on two sides of the planet appeared to him to be two smaller bodies on either side. When he looked at Saturn again in 1613 he found the "two bodies" he had seen earlier had vanished.

Credit: NASA and The Hubble Heritage Team (STScI /AURA: Acknowledgment: R.G. French (Wellesley College), J. Cuzzi (NASA/Ames), L. Dones (SwRI), and J. Lissauer (NASA/Ames)
Website: http://www.hubblesite.org/gallery/album/solar_system/pr2001015a/

1.20 Showing Saturn with its rings at various tilts, as seen between 1996-2000. [Public Domain]

It was left to the Dutch astronomer Christiaan Huygens to recognize the true nature of the rings of Saturn in 1655, but he did not announce his discovery immediately. He wanted to make sure. He, therefore, wanted more time. However, he became concerned that someone else might preempt him in the mean time. So he published the following cryptogram in 1659 – "aaaaaaa ccccc d eeeee g hh iiiiiii llll mm nnnnnnnnnn oooo pp q rr s ttttt uuuuu". If these letters were rearranged to make sense it would read "Annulo cingitur, tenui, plano, nusquam chohaerente, ad eclipticam inclinnato" (It is girdled by a thin flat ring, nowhere touching,

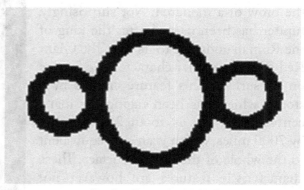

Credit: Courtesy NASA.
Website: http://huygensgcms.gsfc.nasa.gov/Shistory.htm

1.21 Galileo's drawings of Saturn, 1610. [Public Domain]

inclined to the elliptic').[3] Interesting! I wonder if the journals of today would accept such an article. When he finally published, human nature being what it is, the other astronomers of his day did not believe him. Thus his finding was met with considerable opposition, until his work was reconfirmed in 1665 by Robert Hooke (1635-1703) and Giovanni Domenic Cassini (1625-1712).

Galileo had failed to observe Saturn's rings in 1613, because they presented their edge to the Earth at the time. Saturn's rings are aligned with its equatorial plane. Due of the tilt of the Saturn's axis the ring presents different views at different times, when seen from the Earth. It presents its edge once every 13¾ years. This edge of the ring cannot be seen even when using a moderate size telescope of today. This is not very surprising, since it is less than 10 miles thick. When the rings present either their superior or inferior surface, they become apparent with a much smaller instrument. This was the case when Galileo viewed it for the first time. To him it appeared as "three bodies", because his telescope had optical aberrations due to imperfection of his handmade lens [See Fig. 6.9].

Saturn is situated at 9.5AU, which makes the mean distance 932,600,000 miles from the Sun. The volume of this planet is 763 times of the Earth. It is also 95 times as massive compared to our planet. However, its density is only 0.7 times of water. Its orbital inclination is 2° 49'. Its velocity is 6 miles/second as it goes around its orbit. Saturn's one revolution takes 29.46 Earth years. Its diameter at its equator is 75,100 miles. Its axis is inclined at 26.7° is a little more than the Earth's. Its axial rotation of 10 hours 12 minutes compares with that of Jupiter. The surface temperature is around -180°C.

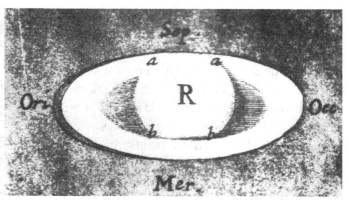

Source: Philosophical Transactions (Royal Society publication) 1666 (volume 1).
Website: http://en.wikipedia.org/wiki/File:Saturn_Robert_Hooke_1666.jpg

1.22 A drawing of Saturn by Robert Hooke. [Public Domain]

R indicates Saturn's globe.

The two letters "a" indicate the expected overlap of the globe on the rings. Hooke's shading on the rings to the right of the right-hand "a" only (i.e. not at the left-hand "a"), indicates the shadow of the globe on the rings.

The two letters "b" indicate the expected overlap of the rings on the globe.

Hooke's shading on the globe between the right-hand and left-hand "b"s could indicate the shadow of the rings on the globe, or the faint C-ring faintly blocking part of the globe.

The abbreviations marked around Saturn indicate directions: Sep. for Septentrional or northern; Mer. for Meridional; Occ. for Occidental or western; and Ori. for Oriental or eastern.

Saturn has fifty-seven known satellites. Amongst the satellites, there is Titan, which is the largest. It is also the biggest satellite in the solar system. In has a diameter of 3600 miles, thus making it larger than the smallest planet, Mercury. It has an appreciable atmosphere and like Triton, one of Neptune's satellites, it shows geological activity.

URANUS

Uranus lies at a distance of 19.6AU or 1,783,000,000 miles from the Sun. It is therefore too far to be seen by the naked eye, which is why its existence was not known to the ancient astronomers. Its equatorial diameter is 32000 miles. It is 14 times the Earth's mass and its volume is 50 times. Uranus' density is 1.65 times of water. Its axis tilt is unique amongst the planets in that it lies at an extreme 98°. Though the planet is featureless when seen through earthbound telescopes, nevertheless this tilt can be deduced from plane of the satellites orbiting it. The planet's rotation period is not exactly known yet, but lies some where between 17 and 23 hours. The year on Uranus is equivalent to 84.01 Earth years and its mean orbital velocity is 4.2 miles/second. Its surface temperature is about 190°C below zero.

Uranus has 27 moons, the larger five being Titania, Oberon, Umbriel, Ariel and Miranda.

NEPTUNE

Neptune is 30AU or 2,768,000,000 miles from the Sun. It is only slightly smaller than Uranus, its diameter being 30,800 miles at the equator. It is 17 times the Earth's mass and with a volume of 61 times of the Earth and a density of twice that of water. Neptune's period of revolution around the Sun is 164.79 Earth years with a mean orbital velocity of 3.4 miles/second. Its orbital inclination is 1° 46'. The axial inclination of the planet is 29° and an axial rotation of 14 hours. Its surface temperature is about -220°C.

Credit: Courtesy NASA and Erich Karkoshka, Univ. of Arizona.
Website: http://hubblesite.org/gallery/album/solar_system/uranus/pr2004005a/

1.23 Tilt of Uranus as indicated by its moons' orbits. [Public Domain]

It has 13 satellites, Triton being the largest. It is of interest to note it shows geological activity. Like Jupiter, Neptune is accompanied by "Trojan bodies". They travel with Neptune in 1:1 resonance. They are known as Neptunian Trojans. Their origins, however, may not be the same as Jupiter's Trojans, since they may have been recruited from the Kuiper belt.

THE TRANS-NEPTUNIAN REGION

As we move away from the Sun, the next locality in the solar system is the trans-Neptunian region. It is made up of two parts, the Kuiper belt on the inside and the "scattered disc" on the outside. The inner edge of the scattered disc overlaps the outer part of the Kuiper belt. This overlap is believed to be due to objects from the Kuiper belt being ejected by the gravitational effects of Neptune to form the scattered disc. This planet is believed to have migrated outwards some time during the early history of the solar system. This caused gravitational perturbations in that distant region and may be why the objects in the scattered disc have erratic trajectories, with some of their orbits highly inclined to the planetary plane; reminding us of Icarus' orbit. A few amongst these are so extreme, being almost at right angles to the other bodies, which move around the Sun. Thus, the name scattered disc.

KUIPER BELT

The Kuiper belt, like the Asteroid belt, is a world consisting of small bodies. However, unlike the asteroids they have what the astronomers' term as ices on them, meaning frozen water, ammonia and methane. They result from the extreme cold, because of their great distance from the Sun. While most are quite small, being about 50 kilometers in diameter or less. Some of them like Quaoar, Varuna and Orcus may reach nearly one-fifth the size of the Earth. All of them have now been reclassified as dwarf planets. Their orbits are at an incline to the ecliptic. Many of them have satellites.

It is believed the short period comets arise from this region. The Kuiper belt extends approximately between 30AU and 50AU from the Sun. This makes Pluto's orbit fall within the Kuiper belt, because at perihelion, it is nearly 29.7AU from the Sun and at aphelion it is 49.5AU away. Its orbit is inclined at 17° to the ecliptic. All these characteristics along with its size make it fit in well with an object belonging to the Kuiper belt. This clearly shows why Pluto does not agree with our concept of a planet any longer. Having relegated Pluto to its proper status it is also now understood, Charon, which was earlier considered to be Pluto's moon, forms a binary system with Pluto. They orbit under a common barycentre. They have two moons, Nix and Hydra circling them both.

SCATTERED DISC
The scattered disc as we have already mentioned lies beyond the Kuiper belt. Most of the scattered disc objects have aphelia going out as far as 150AU from the Sun and their perihelia lying within the Kuiper belt. Thus, both the regions overlap.

Eris is the largest body detected in the scattered disc. It is slightly larger than Pluto with a diameter of 1500 miles. It was this body, which gave rise to the controversy of "What is a planet?" It has a highly eccentric orbit with its perihelion at about 38AU and aphelion at nearly 98AU, its orbit markedly inclined to the ecliptic. It also has a moon, Dysnomia.

COMETS
Andreas Celichius was a Lutheran superintendent of Magdeburg. His position as superintendent was equivalent to a bishop. This is what he had to say about comets. In a book he wrote in 1578 he explained, "the thick smoke of human sins, rising every day, every hour, every moment, full of stench and horror before the face of God, and becoming gradually so thick as to form a comet, with a curled and plated tresses, which at last is kindled by the hot and fiery anger of the Supreme Heavenly Judge."[4] This Lutheran superintendent had indeed given a picturesque description of imagined consequences of human folly, which humankind commits each day. Unfortunately, all this has nothing to do with comets. At the other extreme, we have David Hume, a Scottish philosopher. He too had his own ideas. Hume rightly shied away from a religious explanation. His imagination however surpassed Andreas Celichius'. He thought comets were like some sort of cosmic sperms and eggs, which produced planets![5] I wonder what Freud would have thought of it. If given no other choice, one would be hard put to choose between the two explanations.

However amusing it may all seem, in the past comets have always been thought of as the harbinger of misfortune. The Halley's Comet, which makes its appearance in the skies every 76 years, has been witness to many misfortunes, both of kings and of kingdoms. In the year 1066, this comet heralded the death of the Saxon king, Harold, at the Battle of Hastings. In 1466, a comet was to appear again before Constantinople fell to the Turks. In 1517, a comet was seen in the skies of Mexico. The Aztec Emperor, Montezuma, took it as a bad omen. When the time came in 1521, the entire Aztec Empire failed to stand up against only 400 Spaniards invaders. In the final analysis, it was not so much as the physical, but the psychological impact on the mind of Montezuma, which brought about his downfall. This shows how superstition can have a negative influence on the human mind. People who still believe in omens should stop for a moment and think what happened to have been a bad omen for the Aztecs was apparently a good omen for the Spaniards. Omens cannot be both good for one and bad for the other at the same time. It does not stand to reason.

Luckily for us, even in the past there were people to bring us back to our senses. The Roman Stoic philosopher, Lucius Annaeus Seneca (4BC-65AD), had long ago predicted, "Someday there will arise a man who will demonstrate in what regions of the heavens the comets take their way; why they journey so far apart from the other planets, what their size, their nature."[6]

Comets[7] are very small compared to other visible objects in the sky. The majority are only a few miles long (2 – 6 miles or 10 – 50 km). They are made of rocks, ices, and other debris. This forms the *nucleus* at the centre of the head a comet. It is the latter, which we see. Unlike the paths of the planets, the orbits of comets are highly elliptical. Most of their trajectories reach beyond Neptune, where it is extremely cold. As they close in towards the Sun they gather greater and greater speed. By the time they are a few AU's away from the Earth, they can be seen with a telescope in the reflected sunlight. As they approach the Sun, the material from their surface, such as ices, starts to evaporate in the Sun's warmth and in the process trapped dust and debris are released. Soon a visible tail develops. It is this tail, which forms the distinguishing image of the comets in our minds. All comets, however, do not necessarily develop a tail.

As the comet approaches it develops a *comma*, which consists of a cloud of material, which evaporates from the surface and surrounds the nucleus in molecular form. It can reach a diameter of 625,000 miles (1,000,000 km). The nucleus and the comma together form the *head* of the comet. Also, through this same process, the *tail* forms. As they come closer to the Sun, the tail grows and becomes visible to the naked eye. When they near the Sun, the tail in its full glory may reach millions of miles in length. When the comet recedes from the Sun, the tail gradually becomes smaller and smaller until it vanishes all together. The tail is the material of the comma blown away by the solar wind. This is why the tail of a comet always points away from the Sun as it is passing by. It is difficult to follow them all the way, as they lose their luminosity soon after passing perihelion.

The tail consists of dust, water vapor, gases and ions, as they evaporate the different components of the tail interact differently with the solar wind. Thus they contribute to its shape, direction and colour in different ways. The ions ejected from the comet move directly away from the Sun. The ions that come out of the comet are affected by the solar wind, both by its mechanical effects and at the same time they are also repelled by the ions in the solar wind. This causes the ions ejected from the comet to move directly away from the Sun. The molecules of gases, which mainly consisting of hydrogen, emitted from the comet envelopes the head and also forms a tail. These molecules start to fluoresce in the ultraviolet light of the Sun. This gives it an extra brightness and a bluish colour, but this can be only seen from space, because this light is absorbed by the Earth's atmosphere. Lastly, there is the dust which is comparatively heavier than the other components. Therefore they tend to lag behind the others. Thus dust tail usually appears curved. The degree of which depends upon the parallax effect.

Even though we associate comets with their tail, nevertheless all too often their tails do not develop to our full expectations. This may be due to the fact, due to repeated passages such "tail-less" comets have exhausted their store of ices and debris leaving only the central rocky core. Moreover not all comets are brilliant as we may think. Many comets can be hardly seen and present only as a hazy smudge in the sky.

Though comets are under the gravitational influence of the Sun their highly elliptical paths takes them beyond the orbit of the last of the planets. Some of them appear after a period of few years, while others take more than a lifetime. Thus, they are divided into short period comets and long period comets. The short period comets probably originate from the Kuiper belt or scattered disc objects, whereas the long period comets are believed to come from the hypothetical Oort cloud.

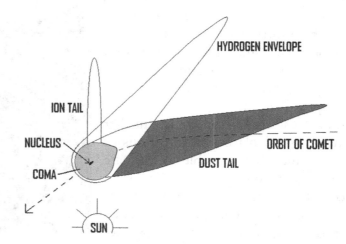

1.24 Anatomy of a comet.

THE HELIOPAUSE

The solar wind reaches out to about 95AU or 8,835,000,000 miles. Here front of this "wind" meets the interstellar winds. This region forms the *termination of shock*, which is an area of turbulence that encloses the solar system. It is known as the heliosheath. Far beyond this, the turbulence ceases to have any effect. This is known as the heliopause. It forms the outer boundary of the *heliosphere*, where interstellar space begins. Our space probes have not reached this region yet.

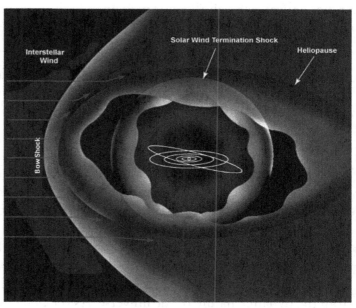

Credit: Courtesy NASA
Source: Cosmicopia (A service of the Astrophysics Science Division at NASA's GSFC. Website: http://helios.gsfc.nasa.gov/) » Sun » The Heliosphere » Diagram
Website: http://helios.gsfc.nasa.gov/heliosph.html

1.25 An artist's impression of the boundary of the Solar system.[Public Domain]

THE HYPOTHETICAL OORT CLOUD

The Oort cloud is thought to be made up of billions of icy objects considered to be situated beyond the heliosphere. Till now, we have no direct evidence of the Oort cloud's presence. It is conjectured, since these long period comets appear after centuries, they cannot possibly originate from the Kuiper belt, so we must assume they must come from much further away. By calculating their periods of appearance, it is possible to estimate their aphelion. Since many such comets appear from time to time, it was hypothesized, they originated from a region similar to the Kuiper belt, but from a place that lies at a much greater distance. Calculations place this region about 50,000 AU or possibly 100,000 AU away from the Sun. Though this distance is indeed very great, still such a distance lies within the boundary of the solar system. However, our probes are yet to reach there. It may be only then we will be able to confirm the existence of the hypothetical Oort cloud.

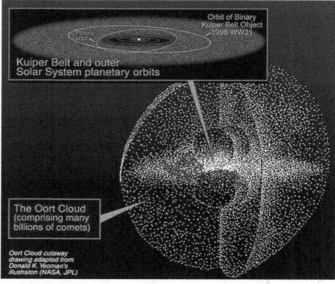

Credit: Courtesy NASA/JPL. (Adapted from Donald K. Yeoman's illustration)
Website: http://en.wikipedia.org/wiki/File:Kuiper_oort.jpg
1.26 Artist's impression of Oort cloud.[Public Domain]

THE LIMITS OF THE SOLAR SYSTEM

The Sun's gravitational field extends to about 2 light years or 126,480 AU on all sides. It may be considered the outer limit of the solar system. This boundary is known as the *Roche sphere*. Not much is known about this extreme outer region apart from what has been described.

CONCLUSION

Today we have certainly come to understand more about the universe than ever before. Nevertheless, we must not imagine as some do, after sending a few probes and looking through telescopes we should let things rest. Our quest does not end with the knowledge of how many stars there are in the galaxy or of how many planets there are in the solar system. Nor does it end with knowing the details of size, orbital velocity, orbital inclination, rotation, inclination of axis and other such information about the various planets. Though they go towards helping in our progress, such knowledge nevertheless should not deflect our attention from the dynamic nature of the universe. As the book unfolds, in the next part we will be able to appreciate that it is evolution, which makes the universe possible and thus everything in it is a part of this evolutionary process; it is an all-embracing process. Our quest has more to do with the understanding of the mechanism of this evolution. Those who think evolution is about life only, must reflect and think again. The evolution of life is after all only a part of the overall evolution of matter! Thus our mission should have to do more with the need for our understanding the answers to questions, such as how the universe came into being, how primitive matter evolved into galaxies, how stars formed the elements, why some planets only harbour life, while others do not and the many, many other questions related to the evolution of the universe. While we have come to know some of the answers to these questions; there is yet much more to learn, so that we might get some small glimpse of the meaning of it all.

Even though we have travelled to the Moon and back, it will be quite some time before we can travel to Mars and beyond. Journeying to the stars or travelling the galaxies is but a distant dream. Since it will not be possible to travel beyond our solar system for many, many years to come, we must study our neighboring planets well. Just like the serpent enjoined the man in the allegory. They will give us many of the answers we seek. Why did Venus not evolve like the Earth? Could Mars once have had conditions like primitive Earth? This is possible, because Mars had geological activity at one time in its earlier history. Today the Martian atmosphere consists predominantly of carbon dioxide. It is also possible the primordial atmosphere was converted to carbon dioxide by primitive organisms just as it happened on Earth. However, carbon dioxide is not the only ingredient that is needed for the next stage in evolution to pave the way for plants. They need nitrogen also. The nitrogen to be utilized by plants comes not from the air, but from the ground. It is soon used up by plants and has to be replenished. Nitrogen from the atmosphere has to be "fixed" to the soil. On Earth, this is done by certain bacteria; some are free living in the soil, while others live symbiotically in the roots of certain plants. They show their presence by the nodules they produce on the roots, as in the instance of leguminous plants, such as peas, beans and pulses. However, Mars' atmosphere does not appear to have nitrogen now. If there was ever any nitrogen in the Martian atmosphere in the past, then it must have leaked out long ago. The likelihood of Mars' low gravity holding on to the heavier carbon dioxide would be much more than holding on to the comparatively lighter nitrogen. The carbon dioxide is heavier than a nitrogen molecule by 16 times the weight of a hydrogen atom. So did Mars loose its potential for life because of its size? If this were so, it would indicate, smaller planets are not capable of harbouring life. What would have happened if Venus had been in Mars' orbit and Mars in Venus'? Could then Venus have evolved life? Or is it some thing to do with the Moon's presence, which has altered things for our Earth? Why is there a cut off point beyond the asteroid belt with the "gas giants" lying beyond? Were the inner planets a miniature version of the "gas'" planets at one time? Could it be the gases like hydrogen, helium, methane and ammonia were common to all planets in the past when they were formed? Is the methane and ammonia along with atmospheric conditions such as lightening and water responsible for seeding life? And it was only Earth that could harbour life, because it fulfilled many of the criteria as discussed before and the rest was moulded by life itself as we have seen in case of the changing composition of the Earth's atmosphere? Could it be the "gas giants" seeded the molecules required for rudimentary life, but they could not evolve further to find expression due the extreme cold and the fact that the hydrogen in the atmosphere could not leak out because of their massive size? These are some of the questions, which cross the mind. Such questions may lead to answers, but it is not yet time to ask them. We must have more facts. The quest must go on. It is only when we know more; we can ask the right questions. The answers will then automatically follow, if we take the right path. We have a very long way to go and our time is short for there are many distractions that lie on the way.

2 TERRA MUNDI

Of all the knowledge we have come to know about the Earth today, it was its shape that was first to be worked out. This was the quest, which started man on the road to where we are today and to what we would aspire tomorrow. If man had not made this endeavor, then it would certainly have been his original sin.

EARLY BELIEFS ABOUT THE EARTH' SHAPE

Since dawn of history man had considered the Earth to be flat. You cannot blame him. He believed it to be true, simply because the ground on which he stood looked more or less plane to him, except for the mountains and seas, which he thought had been placed there by some higher being. This last idea is reflected in the sacred song below.

> In his hands are all the corners of the Earth: and the strength of the hills is his also.
> The sea is his, and he made it: and his hands prepared the dry land.
>
> Psalm 95:1[8]

He also understood the Earth beneath his feet had to be stable and for anything to be stable, it had to have support. Man surmised that the Earth was held up in some way. Not knowing how, he resorted to his imagination. In the distant past snake worship had been prevalent amongst many cultures of the world. Amongst them were the Hindus. The word "Hindu" here is not used in the religious sense. In the distant past, the term was used for people who lived beyond (east) the River Indu (Indus). They thought the world was balanced on the head of a snake and to them, who could be more suitable than the divine snake Vasuki. The Hindus had never been satisfied with one idea; they came up with another idea. This time they believed the Earth was placed on the back of four elephants, each holding up one corner. After all the elephant were the strongest and largest creature know to them. But they realized the problem did not end there, since without any form of further support, the configuration would again become unstable. So they imagined these elephants in turn stood on the back of a gigantic turtle.

The Greeks borrowed the latter idea from the Hindus. The ancient Greeks were very particular. One version describes the turtle as floating on water. The concept of water is important here. It is said, once at a lecture a scientist while discussing this Greek concept, forgot about the water and queried what was below the turtle? This allowed an elderly woman, who was listening, to exclaim, "Its turtles all the way down"! It is of interest to note, when the Greeks brought in the concept water, they did not need to add further turtles or for that matter any other animal to complete the picture. It was implied, the water on which the turtle rested was infinite in depth and expanse. All this may appear to be ridiculous to us today. Nevertheless by bringing in the concept of infinite water, they gave their theory a sense of completeness. Something Newton would do for space many centuries later. So had the scientist mentioned the water, he would not have been open to criticism.

THE CUMULATING EVIDENCE

The history of finding out the shape of the Earth took a much smoother road than the knowledge of our place in the solar system. Here there was no great stumbling block of religion, because there was no mention in any scripture as to what the shape of the Earth should be. People had always assumed the Earth was flat, so they did not think of mentioning this obvious fact in their religious texts. The idea of the Earth being flat and the details of how it was supported were only described in the lore of the day. Though many other people, like the Hindus, who started off by believing the Earth was flat, later came to think otherwise.

After man learnt agriculture, he had surplus grain. He soon found a way to store it. This secured his source of food and it helped to tide over the lean months. This enabled man to expand his horizons. Elementary technology soon appeared and began to thrive. Weaving cloth and colouring them with natural dyestuff became popular amongst many cultures. Surface mining and working with metals also spread in diverse parts of the world. With surplus, the concept of trade was born. He had started off by bartering goods at local markets, but now he took to travelling long distances from his settlements to trade. These journeys took him further and further across the face of the Earth. The various observations the merchants brought back from their long journeys would provide people many information about their world; this would eventually make man reconsider his ideas about the Earth's shape. The merchants also took to sailing the seas for the purpose of promoting their goods and bring back other merchandise. Though these early sea voyages could not be classed in the same league with the later transoceanic voyages of the 16th and 17th centuries, nevertheless they were still able to give sailors an idea the Earth may not flat. Some of those experiences are summed here.

Drawing by Tarun Ghosh.
2.1 The picture on the top shows how the constellations would look, if the Earth was flat and below it, as they are actually seen.

Travellers from the north going south saw new constellations appear on the southern horizon as they progressed south. They appeared to come higher and higher as they penetrated southwards. This could not be so, unless the Earth was round. If the Earth was flat then one would be able to see the whole sky with all the stars in it.

When sailors on the seas spotted a ship on the horizon, it always appeared to rise out of the sea. This was because the mast appeared first and the hull later. The experience of people on the ship was the same when they came towards land. The tops of hills appeared first and the shoreline later. This they rightly concluded was due to the Earth's curvature. At the time it was considered to be an important evidence of the Earth's shape, but still it did not prove the Earth was a sphere, because it could be just some local effect or if the Earth happened to be a disc.

So it is no surprise when we hear the blind poet Homer (c.900BC– c.800BC) of Iliad and Odyssey fame, describing the Earth as convex disc. He also associated the appearance of the vault of the heaven with the Earth's curvature. He could not have been so blind after all! There is no record from where he got his ideas, but we know Homer felt strongly about the Earth's shape. As a bard he had travelled, so his ideas would accord with a man who had travelled much during his life time.

Again when we consider the Earth looked curved from every point, whether it was from a ship on the sea or from land, it could be assumed by extrapolation, the Earth was a sphere. There is no record, however, to show the civilizations of the past had thought of the matter in this way. Though Homer did not think the Earth to be a sphere, he at least based his observations on what he noted and did not stray into the realms of conjecture. This objectiveness of approach should make us respect him all the more.

Then there was the phenomenon of the eclipses. In the past it was the strongest evidence to be had for the Earth's spherical shape. In ancient times eclipses were a source of great concern and many thought it was the act of demons. The Hindus believed there were two brothers, Rahu and Ketu, who were giant demons. They wanted to swallow the Sun and the Moon. According to the story, they actually succeeded in doing so and the universe became plunged in darkness. The lesser gods, as ever when in trouble, ran to Lord Vishnu for succor. Lord Vishnu, recognizing the problem, promptly cut off their necks with his chakra, which was known as the Sudarsan Chakra. A chakra was a sharp rotating blade not unlike a small circular saw with a hole in its centre. It was sped up with the index finger of the right hand, inserted into the hole in the centre, thereby imparting to it a circular motion and then let go at the intended target. The chakra is depicted in the hands of many gods and goddesses in Hindu mythology. It is possible such a devise was in use in ancient times to wound an opponent, but it is difficult to believe it could cut off limbs or the head as described in Indian epics. Probably it had more psychological import than the potential for taking life. Such skill, if it had ever had existed, has not survived the ages. However, in the story, before the Sun and Moon could enter the gullets of the two brothers, the spinning chakra separated their bodies from their heads. As the Sun and Moon passed out of their severed throats and the universe came out of the darkness. Even to this day Rahu and Ketu or at least their heads keep looking for the Sun and Moon. Since the demons no longer posses their bodies, we no longer need to fear the world will be plunged in eternal darkness.

By the sixth century BC the true explanation of eclipses had been found. People in many parts of the world realized, the solar eclipse was nothing but the shadow of the Moon falling on Earth and the lunar eclipse was only the shadow of Earth projected on the Moon. During a total eclipse the shadow of the Earth was always circular, no matter whenever or from wherever it was viewed. Only a sphere could a cast such a round shadow always. The phenomenon of the eclipse was therefore a good indication of the Earth's shape.

The philosophers, sages and astronomers of some ancient civilizations, like the Chinese and Indians, had already concluded from eclipses that the Earth was a sphere. In Europe by the Middle Ages the people were also well aware the Earth was round. In fact, even as early as the 8th century AD, no person in Europe who considered himself to be an astronomer believed the world to be flat. This knowledge, as we already have seen, was not only confined to the natural philosophers of time, but was known to sailors, travellers and many laymen as well. The ideas about the shape of the Earth now came to be reflected in many ways in the life of the period. Emblems of the orb or orbs on sceptres of monarchs, which were used to represent the world, became common. Also many names in the Middle Ages referred to the shape of the world.

During the time of the crusades many Arabic words spread to Europe. Through this influence came the name Rosamund. The word "Rosamund" comes from two words – "rose" and "mundi". We all know rose is the name of a lovely flower and therefore the name Rose is given to girls, whose parents wanted to compare their daughter's beauty to this flower. Mundi is the Sanskrit means round or world. The Arabs in their heyday adopted the word mundi from Sanskrit and incorporated it into their language and brought it to the west through Spain. Thus Rosamundi or Rosamund, meaning "rose of the world", became to be quite a popular name during the middle ages.

It brings to mind the story of a young girl of that name with whom King Henry II of England became enamored. Henry II was the husband of Eleanor of Aquitaine and the father of Richard the Lion Heart. He was a powerful king who had lands in France and by marriage with Eleanor gained further lands there. If England was ever close to being an empire, it was then. Henry had a great appetite for affairs. There was the French king's daughter, who had been intended for his son Richard, but Henry had kept her for himself. It had been a scandal. So when it came to Rosamund he was more discreet. It is said, to avoid detection by his queen, he kept her hidden in a house surrounded by a maze. The story goes that Eleanor found out Henry's secret by attaching a ball of thread to his coat, which gave away the place of his clandestine rendezvous. This latter part of the story is probably not true, but Eleanor must have found out some other way, otherwise the story would not have come down to us today.

MEASURING THE EARTH

The Greeks also knew of these observations about the Earth from their own experiences. After all they had been sailing the Mediterranean long before the Siege of Troy. They must have also been aware of the experiences of others, like the Phoenicians and the Egyptians who preceded them on the seas. Amongst the Greek philosophers it was Pythagoras of Samos (c. 582-500 BC) who first suggested the Earth was round. This was as far back as

532BC. Whether it was a concept he himself had thought of and believed in or an idea he came to imbibe through contact with sailors and travellers is difficult to tell. It is on record, Pythagoras had travelled much and this no doubt had some influence on his thoughts. Later Aristotle (384-322 BC) and Hipparchus (fl. 146-127 BC), a mathematician and the greatest astronomer of Greek antiquity, also agreed with him. Though observational support was piling up, yet no one had conceived the momentous idea of measuring the Earth's circumference. It was left to Eratosthenes (276 - 194BC) to make the bold attempt. This made him the father of the science of geodesy. It is a branch of mathematics that deals with determining the shape and area of the Earth or at least large portions of it.

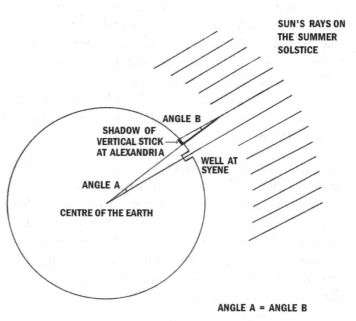

Drawn by Tarun Ghosh.

2.2 The geometric principle on which Eratosthenes based his measurement of the Earth. [Modified by author from Carl Sagan: *Cosmos*, Book Club Associates. London (1981). p. 15 . ISBN 0 354 045318.]

Eratosthenes came to know of a certain well in Syene, a place now known as Aswan, in Egypt. He had heard the Sun came directly overhead this well, at noon, on the summer solstice. So, one year, on the same day he used a vertical stick at Alexandria to cast a shadow. By using simple geometry, Eratosthenes worked out the angle of the arc that represented the Earth's surface between Syene and Alexandria. He found it to be 7.2°. He then had to find out the distance of this arc. Eratosthenes did it by determining the time it took a caravan to travel from Syene to Alexandria. The caravan took 50 days. He knew a caravan covered an average of 100 stadia a day. The length of the arc he wanted to measure was 5000 stadia. Accordingly he calculated the circumference of the Earth along the meridian over Alexandria as 360°/7.2° x 5000 stadia. This made the Earth's circumference 250,000 stadia.

The word stadia is derived from the measure of distance of the foot race in the Olympian Games held in honour of Zeus at Olympia a plane near Mount Olympus in Greece. A *stadia* was equal to 600 Greek (Olympian) feet, equivalent to 606 of our modern feet or 184.71 metres (185m rounded off). Later the word came to denote the race itself and then to the place where the race took place. The modern word stadium is derived from the Roman variation of the word stadia.

This makes the Earth's circumference equivalent to 46,250,000 meters, which is 46,250 kilometers or 28,906.25 miles. Today we know the actual distance of the Earth's polar circumference to be 40000 kilometers; for those who are familiar with miles it is 25,000 miles. Eratosthenes made an error of over 15 percent!

Many have criticized the work of Eratosthenes. Firstly, it is said he had assumed that the shape of the Earth was spherical. I personally cannot agree with this criticism, because the shape of the Earth's shadow on the Moon during a lunar eclipse traditionally has been put forward as an argument as proof of the Earth's shape being a sphere. It is difficult to believe a person of Eratosthenes' stature was not aware of this fact, before he planed his great undertaking. In fact, it was from this very basis that Eratosthenes set out to measure the Earth.

Secondly, critics point out Eratosthenes had only heard, but not verified, the Sun would come overhead this particular well in Syene and light up the bottom at noon on the summer solstice. The criticism is partly true. If Eratosthenes had verified this fact for himself, then he would have seen this was not correct. The position of Syene is 24° 6' N and therefore lies 36' north of the Tropic of Cancer, which we all know to be 23° 30' N. The Sun reaches, but never crosses to the north of this line. The critics therefore point out there is no question of the rays ever falling at the bottom of the well at any time of the year. This criticism is however open to question. The obliquity of the rays at Syene is 36' (0.6°), which is a little over ½° only. Thus, on the particular day in question, the rays could light up the bottom of the well, if the well happened to be wide enough and shallow enough.

In the past wells were shallow and would never reach a anywhere a dept that would cast a full shadow at the bottom at this latitude. Unfortunately the well no longer exists to verify my point. This point is, however, made clear in the diagram. Nevertheless it is equally true the rays did not fall exactly perpendicular and would not light up the whole bottom. This would therefore give rise to an error.

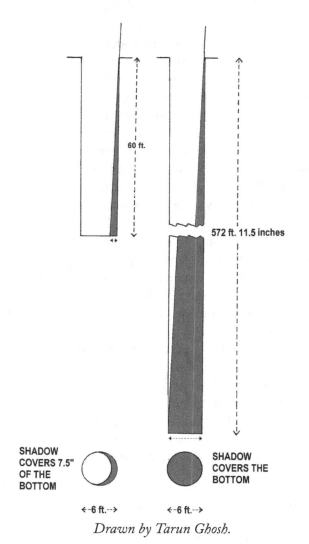

Drawn by Tarun Ghosh.

2.3 Lighting up the bottom of a well at Syene on the summer solstice.

A well 6 foot wide and 60 foot deep will cast a 7.5 inches shadow at the bottom. Only a well 572.93 foot deep will have a shadow that will fully cover the bottom!

His source of error can be ascertained from his mathematics. It will be remembered Eratosthenes' calculation was as follows: 360°/7.2° x 5000 stadia. Now in the first value of 360° there can be no mistake, because we all know the circumference of a sphere is 360°.

Regarding Eratosthenes' value of 7.2° there is an error. This value is actually 7.5° and not 7.2°, as Alexandria is 31° N of the Tropic of Cancer, which is 23.5°. Since according to Eratosthenes' calculations it was 7.2° we must calculate the angle of error as 0.3° and not 0.6°. Though this appears to be very small, it is magnified 50 times when we take into account the whole circumference. Thus 360°/7.2° x 0.3° = 15° and 15°/360° x 100, which is equal to an error of 4.17%, but actually 8.34% by our estimate.

Whichever value we take Eratosthenes' mistake as we have seen was much greater. So where did the error arise? This brings us to the third factor in Eratosthenes' calculations in which he took the distance between the two cities as 5000 stadia. Eratosthenes did not take into account that these two cities were not on the same meridian, Syene at 32° 55' E and Alexandria at 29° 55' E. However the two cities were the nearest possible in the north south direction according to his estimate; he knew his geography well. The caravan that he had employed to measure the distance could not possibly travel in a straight line from Syene to Alexandria for obvious reasons. We realize now where the rest of the error arose, but Eratosthenes could not have done better considering his circumstances. He was not rich, thus carrying out any measurement more accurately would have been beyond his means. In those days it would need a king's money to finance such a project and a king's authority to see it was executed properly. We should not therefore think less of Eratosthenes. He was well aware of the problems he faced and the assumptions he had to make, when carrying out these measurements. He was no fool. Thus his error of nearly 16%, though too much, is excusable. His scientific approach cannot be faulted even by his severest critics. Thus Eratosthenes' attempt to measure the Earth can only be described as an epoch making achievement.

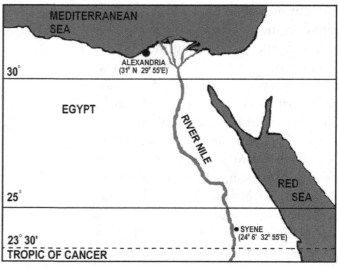

2.4 Map of Egypt showing the longitudes and latitudes of Syene and Alexandria, along with Tropic of Cancer.

Later Poseidonius (135 - 50BC) was to do a similar experiment using the same general principles as Eratosthenes. Except that he used the distance between Alexandria and Rhodes. Rhodes being an island he found the distance from the time it took for a ship to travel along the sea. Using this distance he computed the circumference of the Earth. Though his result was an improvement on Eratosthenes' estimate, Poseidonius' error of 11% was still too large. Nine centuries later Caliph Al - Ma'mun (813 – 833), the son of the famed Harun ar-Rashid (786 – 809) of the Abbasid Dynasty of Baghdad, did another experiment and found a figure, which had an error of only 3.6% too great; quite a feat for that time!

In spite of all their painstaking efforts, the critics could dismiss all these observations as being indirect evidence only. Since measurement based on geometry were dependent on assumptions based on other observations. As the actual act of measurement of the Earth's circumference is not a feasible option, the only way it could be proved the Earth was round in those days would be by going around it. Such a feat was not possible at the time of Eratosthenes or even for many centuries to come, because to do this one had to cross the oceans. Just like before 1957 people could only dream of going to space and look at the shape of the Earth from there. In Eratosthenes' time crossing the oceans was not a dream, but something beyond their dreams. The technology was not just there as we shall come to see a little later. The final proof would not come from any scientific intention like that of Eratosthenes, but from mans' motive of profit. Nonetheless gain has always been the impetus in solving many of our technological problems. In the end, very often, it has given man a broader understanding of science. Thus trade and economic considerations have often led man along the road to scientific progress.

LURE OF SPICES

Long before the Europeans, the Indians and the Chinese knew about spice, which were to be found in the *Spice Islands* or what later was to become known as the East Indies. The Hindu names still present in the region stands as testimony to the Indian influence over these islands more than fifteen centuries ago. Even centuries later, kings, who were Hindus or Buddhists controlled the spice trade, ruling the Shailendra Kingdom from their capital Borobodur in Java. Later they were eclipsed by the Srivijaya Kingdom from Sumatra who ruled from Palembang.

The hold on the spice trade had always been fought for over the ages. When the Arabs came to power they were at first restricted to the Arabian Sea and acted as middle men by buying from the Indians and selling to the Middle East and Egypt. This was in the eighth century. When the Chola kings of southern India became a great sea power, they spread

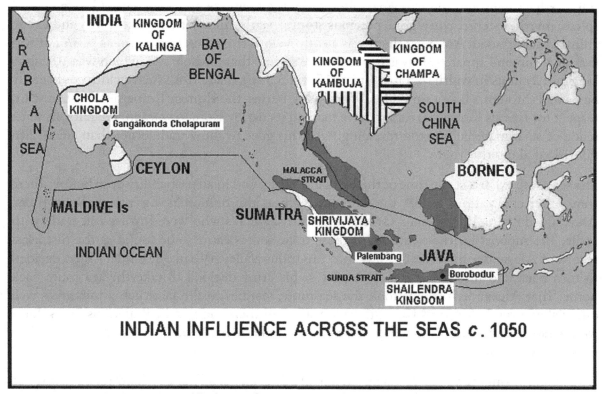

2.5 Indian influence across the seas. c. 1050

their influence up to the shores of the East Indies Archipelago. It is recorded that Rajraja I (985-1014) broke the rising influence of the Arabs in the Maldive and the Laccadive Islands. He also cleared the Arabs from the northern part of Sri Lanka. His son Rajendra I (1014-1044) crossed the Bay of Bengal to make a foray and break the Shailendra hegemony. At the same time he tackled the long standing problem of piracy in the Malaccan Straits. Rajendra's aim was to capture the spice trade by controlling this vital trade route, which touched Java, Sumatra and Malaya. Had he wished for an empire he could have easily obtained one! He was sensible, rather than to get into the intricacies of creating an empire, he was more interested in securing the trade.[9] When the power of the Chola's waned the Arabs stepped in again. Their influence extended from the Mediterranean to the Spice Islands. It was their power that converted the islanders to Islam.

The geography of these islands was vague in European minds of the period, which they only came to know through traders. Knowledge travelled slowly in those days. Thus the Europeans could be forgiven for calling them the Indies. They thought these islands to be a part of India. The Europeans were partly right. It was only sometime before, during the time of the Crusades, they had herd about spices and the Spice Islands, when the Indian influence had lain across these islands.

There was an already established sea route from the western coast of the Indian peninsula, through the Arabian Sea with a base at the island of Socotra and then along the Red Sea to Egypt. Once the goods reached the Mediterranean, traders from different countries in the region, who held sway at different phases of history, took over control. They bought the goods and sold them around the Mediterranean. At first it was the Phoenicians and Egyptians, then the Carthaginians and the Romans, followed by the Genoese and Venetians. Goods such as spices, myrrh, incense, ivory, gold, precious stones, pearls, fine cloth, tropical birds and exotic animals were traded. Amongst them was a cotton cloth from Bengal, known as *muslin*. It was used for women's apparels. It was reputed to be so fine that a whole sari, which was 10 cubits long and 2 cubits in width (4.5 m x 1 m or 15 ft. x 3.3 ft.), made from such a cloth could easily be passed through a lady's ring! Of all persons, Tiberius, the Roman Emperor, who had scant respect for morals himself, is said to have banned its use by Roman women. He felt garments made of such material was too revealing![10] All this goes to show trade other than spices also flourished along this route.

The sea route existed at the time of the Roman Empire and in all probability even before. There were other civilizations before the Romans, such as the Sumerian, Egyptians, Babylonian, Assyrians, Hittite, Phoenicians Greeks and Carthaginians who were involved in trade with India. The support for the existence of these trade routes can also be found in the historical record of Alexander the Great's foray into the Indus Valley. When Alexander's men decided to turn back and return to Greece. A part of his army decided to take the sea route back home. That Alexander's troops took the sea route, testifies to the fact such a route was well established even before 323 BC. Otherwise, after all their trials and tribulations, the solders would never have committed themselves to an unknown route. It is also on record they arrived home safely.

At this point one may well ask why merchants did not favor the overland route to the sea route. The reason probably does not lie in the risks involved. As far as the merchants were concerned, the chances of loss of goods and life were equal in both the instances. There were pirates on seas and robbers on land. Hardships over long routes were probably greater over longer distances, when travelling over high mountain passes and deserts. The existence of the Silk Route over the centuries shows that men were willing to face these risks. But overland, merchants had to pay taxes at every border. These borders were not the borders of countries that we see on the map today. Many of those borders were that of tribes and petty kings, whose control could change hands frequently in those uncertain times. So merchants travelling overland did not know what to expect each time they came to cross any border. Once man began to master the seas, the ships were able to carry greater amounts of merchandise in a shorter time. Thus the trade through the Silk Route gradually dwindled till it vanished into history, but before all this came about man had to master the oceans.

MASTERING THE WATERS

No one knows which nation was first to make sailing ships, many think it may have been the Egyptians. Ships were sailing the waters of the Mediterranean some 5000 years ago. We must however remember there were other great river valley civilizations on the Tigris and Euphrates, Indus valley and along the great rivers of China. Though it could have been any amongst them, but unfortunately records are lacking to verify the facts.

THE PRINCIPLE OF SAILING

Sailing requires wind power. The principle behind this was simple. Anything that floats freely on water will be pushed along by the wind. The direction, which the floating body takes and its rate of progress, will depend upon the resultant forces of the power of the wind that drives the body and the resistance imparted to it by the water.

WATER CURRENTS

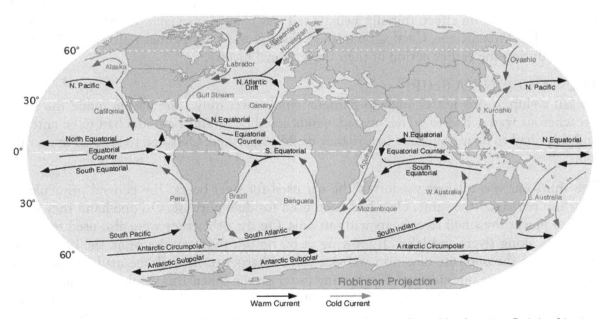

Credit: Courtesy United States Federal Government for releasing it to the public domain. Original image made by Dr. Michael Pidwirny.

2.6 A map showing ocean currents of the world. [Public Domain]

Water currents do matter also. They are significant when, say, sailing along or against named ocean currents like the Gulf Stream. So sailing between Europe and America may cause a difference of a few days in the journey. The American sailors were the first to notice this effect. It was easier to sail from America to Europe, if they sailed with the current of the Gulf Stream. Sailing from Europe however took longer, but it was quicker if this current was avoided. We must remember the Gulf Stream had not been discovered at that time. When the American sailors mentioned their experience to the British sailors, they were ignored at first. After all it was the British who ruled the seas. Thus they were above taking advice from others regarding navigational matters. Only later the British came to realize the American sailors were right. Like the Gulf Stream there are other currents in the oceans and seas of the world. In contrast local currents could be erratic. The latter are usually caused by weather and local conditions, such as underwater reefs, which by themselves may affect the progress of a ship adversely. These were all important considerations in navigation during the days of the sail. We will however ignore the effects of water currents here in order to make our discussion simpler.

WATER RESISTANCE

The shape of the part of the floating object lying under the water's surface determines the resistance to the progress of the vessel. Through experience of using boats and small crafts, the design of the hull had been more or less standardized quite early in the ships' development. They were made long and narrow at both ends to make them stream-lined.

THE POWER SOURCE

In past there was no coal or oil to provide power, there was only wind power and muscle power. Muscle power could not be sustained for long. In those days ships depended on the wind to supply most of the power for longer voyages. They kept muscle capacity for short bursts of power needed to tide over emergencies like a lull in the wind or during a navel engagement. Sails were designed square so as to catch the maximum wind to drive the vessel forward to maximum effect. Initially a single sail was hoisted on a single mast. The mast was straight. The Phoenicians, Carthaginians, Greeks and the early Romans all used ships with one sail during the earlier period of their navigational history.

MANEUVERING A SHIP

To sail a ship, it was not enough to harness wind power only. One had to guide the ship. Treacherous waters had to be negotiated, winds could be erratic, harbours had to be entered and ships had to be maneuvered before they could be berthed. Oars had to be used to do this job.

The ship's oar was a larger version of the oar used for river boats. The general principle of their use remained the same. Oars had been used for duel purpose. On one hand they were used to power the small river boats without sails. On the other they were also used to guide the boats along rivers and lakes. By rowing faster on one side, say the right, the boat could be turned to the opposite direction and vice versa. Sharper turns could be made if the rower stopped rowing on one side and kept rowing faster on the other. This would have more effect if the idle oar was held firm in such a way that it would offer the maximum resistance in water. This latter technique of keeping the oar steady to maneuver the boat was the origin of the oar being used as a rudder for ships. It became the standard steering device to maneuver ancient ships. In its earliest form it was a single oar on one side of the stern. It was known as a *stern oar rudder*. Later there was another oar added to the other side of the stern. This gave better control as ships became longer and longer. Some ships during Greek and Roman times had an extra pair of oars installed as rudders, one on either side of the bow as well. Now they could be maneuvered equally well in either direction, as the ships were pointed at both ends. Another advantage has been suggested. If during a storm the ship pitched to the extent that one pair was heaved out of the water, then the pair at the other end would be able to keep the ship on course.

Credit: Courtesy Maler der Grabkammer des Menna (Artist)/ Reproductions compiled by The Yorck Project
Website: http://en.wikipedia.org/wiki/File:Maler_der_Grabkammer_des_Menna_013.jpg#filehistory
2.7 An ancient Egyptian riverboat with a stern oar rudder. (c. 1422 BC) [Public Domain]

Soon the blades of the oars used as ship's rudder were enlarged to obtain better effect. In ships from ancient till medieval times, such an oar was attached to the stern, usually on the right hand side. The right side of a ship thus came to be called the *steerside*. The word *starboard* is derived from it. The other being the *port side*, the side presented to the quay, when the ship was berthed. In this way the oar which was being used as a rudder would not foul during any maneuver within the harbour.

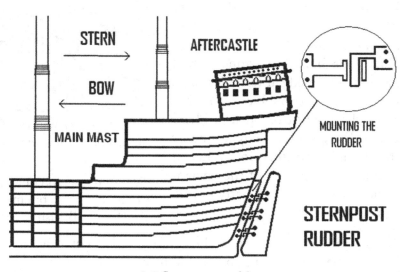

2.8 Sternpost rudder.

Eventually by the 14th century AD, the stern oar rudder was replaced by a rudder placed at the centre of the stern. It was suitably attached to the sternpost and incorporated into the steering mechanism of the ship. It came to be known as the sternpost rudder. That the rudder has its origins in the oar is also confirmed by its etymology. The word rudder comes from the old English word *rother*, which means *rower* or oar.

The mechanism controlling the rudder was initially a handle known as the *tiller*, which was probably derived from the colloquial English word for the handle of a spade. Later the tiller was replaced by a wheel, known as the *helm*, which turned the rudder. It is in this shape, it is more often seen today. The word helm comes from the Anglo-Saxon word *hillf*, a handle, which harks back to the word tiller.[11]

THE GALLEYS

The ancient *galleys* that were being used for trade and war by the Phoenicians and Greeks, about 3500 years ago, had long hulls. They had a ratio of 6:1 of length to the maximum width at their middle. Galleys were narrow at both ends with a shallow draft and a mast having a square sail. They were designed to slice through the water at speeds hitherto unknown. By modern standards it may not have been much, but in those days meant going at about 6 or 7 knots. Turning was not easy for them. Nonetheless as we have seen they could be made to move forward or backward with almost equal ease.

A tier of oars was known as a *bank*. A galley with one bank of oars on either side was known as *unireme*. These were in vogue at the time of Homer. At that period the galleys were quite small and were armed with a battering ram at the bow, known as an *embolon*. This word is derived from the Greek word *embolos* meaning "peg" or "stopper". Embolons have long gone out of fashion. The medical terms *embolus*, means a clot or bubble of air that stops the flow of blood in a blood vessel. Though many may not know the etymology of this word, today it stands as a reminder of the ancient device used in naval battles long ago.

Credit: Courtesy User: "PHGCOM" (CC-BY-SA-3.0). Website: http://en.wikipedia.org/wiki/File:AssyrianWarship.jpg

2.9 A photograph of an Assyrian warship 700-692 BC from Nineveh, a bireme galley with embolon, from the British Museum. [CC-BY-SA]

The more oars a galley had the more speed it could achieve and this was useful in naval battles. The seafaring nations came to understand, they who ruled the seas ruled the trade. Trade meant money and money power. This was a simple equation, which all understood. Soon a scramble began to attain higher and higher speeds. Thus, as the knowledge of ship building grew, more banks were added with time. By the 8th century BC galleys with two banks came into use. They were known as *biremes*. These galleys were about 100 ft. (30m) long and about 15 ft. (4.5 m) wide. The *triremes* with three banks appeared on the seas by the 6th century BC. They were about 133 ft (40 m) long and 20 ft. (6 m) wide and still had a draft of about 40 inches (1 m). They were manned by an average of 180 rowers, which makes it about 30 rowers per bank. The *quadreme* with its four banks were in use by the 4th century BC and the *quinqueremes* with five banks were soon to follow. This goes to show the fierce competition between the seafaring nations of the Mediterranean countries. Ptolemy IV of Egypt attempted to build a ship like a huge catamaran, which would have 40 banks! It was too grandiose a project for the day and not surprisingly it never materialized. It must be appreciated to work all these oars in unison it needed great precision and skill. All this goes to show man's craze for speed is not something new.

NAVIGATION

These galleys or even the larger Roman ones were still too small to be considered fit for transoceanic voyages. They were more fitted to keep in sight of land. At no place would such ships venture out for more than a day or so from land. However in Europe there was an exception; the long ships which were used by the Vikings of the north. These were open ships with a single square sail and used oarsmen, somewhat like the galleys used in the Mediterranean. They sailed to Iceland, Greenland and even to far off Vineland (Newfoundland). This last was a remarkable feat by the Norse seamen in the 9th century, when Leif Ericsson discovered America. This was possible, because the world's weather had become warmer and Iceland and Greenland could be used as bases, where permanent colonies and agricultural settlements had become possible. This was many years before Leif Ericsson ventured across the unknown ocean. Their ships had almost certainly used the Shetland Islands, Faroe Islands, Iceland and Greenland in their progress westwards.

Credit: Courtesy Wikipedia User: Wikid77
Website: http://commons.wikimedia.org/wiki/File:Viking_ship_stylized_gray_sky_narrow.gif

2.10 A Viking ship. [CC-BY-SA-2.5]

The reason why their long journeys were possible was because the design of the long ships was somewhat different from the ships in southern Europe. The hulls of their long ships were made of overlapping planks, which were held together by iron nails. This latter was the reason they came to be known as clinker ships. The word *clinker* comes from an obsolete word "clink", which meant to "clinch". It implies securing anything sideways by nails, screws or rivets. The Vikings used this technique, as they did not know about the carpenter's saw. They used axes to hew out long planks. The overlapping planks were fixed with iron nails driven into their sides. The gaps between the planks were filled by *caulking*, which is the method used to stop the gaps between the planks with *oakum* and molten pitch. Oakum was the loose fibers got by picking old rope, usually a work done by slaves or prisoners. Whereas around the Mediterranean the hulls were made of planks fitted edge to edge. This was possible due to their use of the saw. Even though without the use of the saw made the Vikings technologically less advanced, nevertheless it gave them the advantage, because their long boats leaked a lot less than their southern European counterparts.

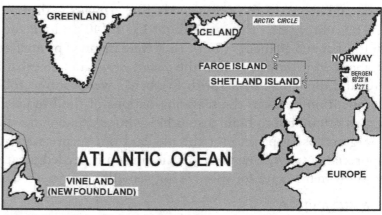

2.11 Map of the North Atlantic showing the stages the Vikings took in their progress west.

To do what the Vikings did it needed a wider concept - the concept of navigation. It is one thing to be able to sail a ship on sea along the coast, in sight of land. It was quite a different ball game all together when it comes to navigating a ship across the open ocean. Ideally it means steering a ship from any given point to another on the surface of any water body. The word "point" implies the intersection of a specified longitude and latitude. In the past such accuracy was not possible. In those days all it meant was going from one port to another without loosing one's way. It was a science that would evolve slowly over the centuries, eventually to become a state of the art technology. Only then point to point navigation would become truly possible.

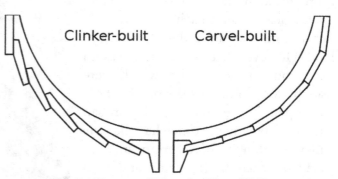

Credit: Courtesy Wikipedia User: Willhig. Website: http://en.wikipedia.org/wiki/File:Clinker-carvel.svg

2.12 The difference in assembly between a clinker ship and a caravel (southern European ship). [Public Domain]

The concept of navigation was very easy. Since any one point on the Earth's surface could be projected on a corresponding point on the celestial sphere, then the map of the world could be projected on the vault of the heavens also. Thus the heavens and the Earth could now be united! Instead of looking at the featureless sea, one could look up at the stars at night or the sun during the day and tell exactly where the ship was situated on the sea. By introducing a grid, which we call longitude and latitude the sailor could now tell the direction he had to take.

Knowing the longitude in this way would be very simple if the Earth did not rotate on its axis, but the Earth does rotate and we cannot stop it from doing so! Thus the reference points kept moving overhead. However a solution was at hand. The Earth rotated once around its axis in 24 hours. Since a circle has 360°, the celestial sphere rotates 15° in 1 hour or 1° every 4 minutes. The longitude is thus linked to the celestial clock and thereby to any mechanical clock on the planet. If a connection could be made the solution would become simple. Such a connection was difficult to arrive at during this early juncture, simply because the technology for making an accurate clock, which was also seaworthy at the same time, was not achievable at the time! Other means, such as dead reckoning, had to be used as we shall come to see. It was much later that a carpenter's son, who himself started off as a carpenter, was able to achieve what other great minds of the time, including Isaac Newton failed to do. But that is another story altogether, which we will recount elsewhere.

The other problem was determining the latitude. The only star which appeared to remain truly constant in the ever moving night sky was the Pole Star. This was because the Earth's northern axis passed very close to it. It could always be relied on by the sailors. By measuring the elevation of the Pole Star from the horizon one could know the latitude of the ship. At this early stage it was not necessary to know the latitude in degrees. It was enough to know, if the ship was deviating from its course with respect to the latitude. It was only later the technology for determining the latitude's degree with accuracy became necessary. Much of the history of navigation revolves around the technical development of such devices for determining longitude and latitude

THE DEVELOPMENT OF THE TECHNOLOGY FOR POINT TO POINT NAVIGATION

All these were vexing issues for uneducated mariners of the past. In ancient times the mariners guided their ships by keeping sight of land. By observing landmarks along the way they were able to reach their destination. Sailing beyond the sight of land, the seas presented a vast bewildering featureless expanse of water to them. Very few ventured across the waters for fear of loosing themselves. However this was not always possible to avoid. Such situations were forced of sailors for various reasons. At times due to bad weather they were compelled to travel in fog and thus could be lost. At other times, while chasing enemies or fleeing from them, they ventured further and further till they lost sight of land.

Taking shorter routes meant shorter time and shorter time meant more profits. In time people became bolder and took the shorter routes through open seas, taking their cue from those who sailed the Arabian Sea. Here the winds were more or less constant. They blew in one direction for six months and in the opposite direction for the other six months. Though the monsoons are equated with the rainy seasons today, it originally came from an Arabic word *mausim* meaning "season". It later came to indicated seasonal winds. What was it that enabled them to take this greater risk of sailing into the open seas? They realized by observing the position of the sun by day and the stars at night they could keep their bearings in open seas. Even though at first it may have been for short distances only, nevertheless it was the beginning of celestial navigation. It is interesting to note that the Vikings also used a form of celestial navigation, using a technology suited to their weather conditions. Viking way of navigation was something unique at the time.

People from the tropics, who have never been far north, do not always appreciate the conditions of light in the extreme northern latitudes. At these latitudes the mid-winter nights are long and dark and long night usually meant winter to them. In late autumn and early spring the Sun hardly ever rises above the horizon during the day and that only for short periods. Travelling in the dark seas with icebergs floating was fraught with danger. Thus they were excluded from navigating by the stars.

Thus the Vikings availed of the long summer days. These journeys would take them across the oceans to Iceland, Greenland and even to far off Newfoundland. At 10° south of the Arctic Circle the Sun remained in the skies for nearly 14 hours in the day. During summer the daylight hours, extended far into the night. This made the night almost non-existent at the height of the summer season. They therefore could not see the stars during those summer nights and had to depend on the Sun for navigation.

There was an added problem, which was specific to these northern latitudes. More often than not, even during summer, when the cold air of early morning or late evening met the sea, fog covered the waters. Sometimes fog banks would enfold the ship altogether bringing down visibility to zero, but the skies overhead usually remained free on a cloudless day. Under such conditions the Sun could not be seen on the horizon, thus the ships position could not be gauged until the fog lifted later in the day. By that time however the sun was no longer on the horizon. Thus even a rough idea of the latitude was difficult to make out.

The Vikings had not developed any instruments of navigation, but they had a very good idea of their latitude from estimates of the elevation of the Sun on the horizon. Also from experiences handed down through the generations they knew about the prevailing winds and currents. Birds flying and floating vegetation would alert them that they were nearing land. They had another advantage. The Norwegian coast for about 150 miles (240 km) north and south of Bergen was aligned almost on 5° east longitude. Then by going due west and keeping their latitude they could reach Shetland Islands. By going another three days or so north and then west again, they could reach Faroe Islands. If they continued further west they would reach Greenland and thence sail on to America.

It was alright once you knew about these islands already, but for the discoverers of these lands, it was no easy matter. An idea of how those intrepid sailors achieved this feat may be had from one of their legends. The story goes that Floki Vilgerdarson, an early explorer, took three ravens on his voyage of exploration. After he travelled a while he let one free. Floki noted it flew homewards. The next one he released after a few days flew around the mast. The bird then perched on it and could not be coaxed to fly anywhere else. This meant they were too far away from land. Again later when the last one was set free it flew away; this time in a new direction. Now all that Floki had to do was to follow the bird's flight and in this way he reached Iceland. Thus Floki Vilgerdarson came to be renowned as Floki the Raven.

In time things changed for the Vikings. One of them stumbled on a stone, which held a mystery. It may have been a mere child or it may have been some mighty chief. Who knows how they came upon it? In time, it would come to be held in awe, because it possessed a mystic quality for them - something to be revered.

The stone was a clear natural crystal. Today it is know to us by the name of *optical calcite* or *Icelandic feldspar*. Chemically, it is calcium carbonate, the stuff we take to keep our bones strong. At first it was like a toy to them. Looking through they could see two images. Playing about with it, they soon were able to deduce that it possessed a strange property. A property, which could be put to use in finding the Sun's direction, even in dense fog, provided the skies overhead were clear.

It was found if a spot of tar was placed on one face of the crystal it produced not one, but two shadows on the opposite face. Holding it aloft and rotating the crystal side to side, one spot remained stationary, while the other moved around it. Unbeknown to them, what the Vikings were seeing the phenomenon of *polarization*. This was responsible for the two shadows. The explanation for this phenomenon lies in the fact that sunlight is non-polarized. This means the orientation of light waves are not in one plane, but lies in many different planes in the direction of travel. When this light enters the Earth's atmosphere some of the rays scatter causing some of them to become polarized. This scattered light coming from the sky overhead becomes polarized in two directions as a result of refraction. One set enters the top surface of the crystal and pass straight on through to emerge from the other side. Thus when the crystal is rotated from side to side the shadow produced by the spot does not move. Whereas, the rest of the unpolarized light, which enters becomes polarized in a different direction due to refraction after entering the crystal and is deflected from its course to produce another beam, which runs parallel to the former as it emerges from the opposite face. Thus a second shadow is produced by the spot of tar. On rotating the crystal horizontally this beam, the second shadow rotates around the unmoving first shadow. The former beam is known as the *ordinary ray*, whereas the deflected one is known as the *extra ordinary ray*. The explanation I have given is a simplified explanation.

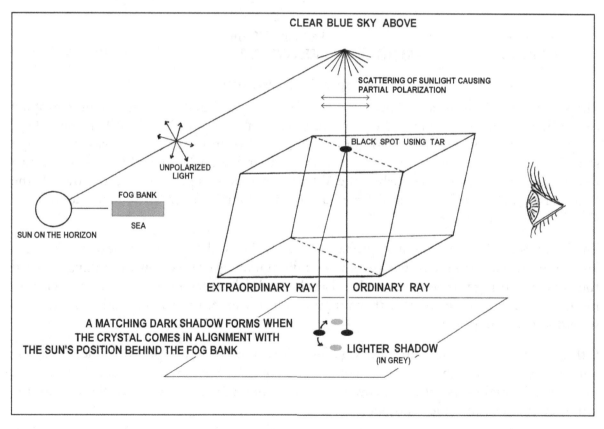

Credit: Leif K Karlsen. Modified by author from his Website: http://www.oneearthpress.com/pdf/nav_notes.pdf

2.13 Principle of the Sunstone.

(Modified and redrawn from Leif K Karlsen's Viking Navigation article on "Viking Navigation using the Sunstone, Polarized light and the Horizontal Board in Navigational Notes". Issue 93, p. 8, fig. 9. One Earth Press.)

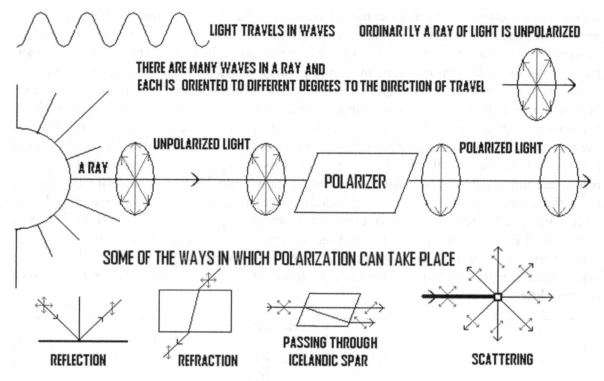

2.14 Explaining polarization.

The Vikings named it the sunstone. It is not to be confused with the semi-precious gem stone of that name, which is composed of sodium calcium aluminum silicate. When the Viking sunstone, as it should be known, so as to distinguish it from the gemstone, was rotated, the shadow thus produced by the extraordinary ray varied in the quality of its brightness. The two images were of equal darkness when the Viking sunstone was pointed directly towards the Sun. In this way the true direction of the Sun could be found even in dense fog, as long as the sky above remained clear.

The Vikings rarely spoke about it, probably because they did not want its secret to become known to others. Until even a few years ago we did not know exactly how the Vikings crossed the oceans by using the Sun, but the late Leif K. Karlsen's researches have made this clear in his article.[12] Thus not unnaturally they revered their sunstone. The Vikings did what the bees were doing all along.

Other nations also use the Sun for navigation during the day, but not in the same way as the Vikings. In their case the sun would come well above the horizon. They waited for the sun to come at its highest point in the sky, which was at noon. By measuring the elevation of the Sun they could deduce their latitude.

At night they used the Pole Star. All the ship's captain had to do, during navigation in its most elementary form, was to measure the vertical height of the Pole Star from the horizon when setting out. This in fact gave him an estimate of the latitude of his home port. He then set sail when the tides and winds were right. If for example the port of destination was to the south, he would travel south keeping the Pole Star aligned to the ship's path. This ensured he maintained his longitude. All through the voyage, from time to time, he would continue to measure the height of the Pole Star from the horizon. By this he would make sure he was travelling in the right direction. If the ship had for some reason drifted north, the height of the Pole Star would increase. If his ship had gone too far south the height of the Pole would diminish. The captain then had to correct his ship's direction accordingly. When the ship reached the expected latitude of the port of destination, the captain would know this from the height of the Pole star. Of course the captain would have to know this from previous experience. Then he would turn the ship either east or west according to the where his port of destination happened to be. All he had to do after turning his ship in the right direction was to see that the height of the Pole Star from the horizon remained the same during the rest of the journey. In this way he would eventually arrived at his destination. This technique was an approximation, which held true for short distances and over short time only. This was because of the Earth's shape, no matter from where one looked at the Pole Star, he or she would be always looking north. Thus one could not distinguish from this which longitude he or she was on. To keep their bearings the mariners had to take help of the other stars as well and also allow for the time of the night and the seasons. So they had to be well versed in the pattern of the stars around the year. Thus in those times navigation was an art, rather than a science.

THE ASTROLABE

At first such estimates of the height of the Pole Star may have been made by eye only. Later they used their thumb or fingers with their arm outstretched. There was no standard. A devise was used to solve this problem. The simplest contrivance was a *latitude hook*, used by the Polynesians. It was a piece of wood or stick with a string attached to it. This was an improvement on the finger or thumb. A slightly more sophisticated version of this device was invented by the Arabs. It was known as a *kamal*. Here the stick was replaced by a card and the string had knots at specified lengths.[13]

To use the kamal, the captain had to hold out his hand with the bottom of the card aligned to the horizon. Then by adjusting the length of the string, the top was matched to the object of attention, such as the Pole Star. A rough estimate of its elevation from the horizon could be got by counting the knots on the string. This could only give a relative estimate of whether the ship was keeping to the latitude or not, but could not measure the degree in any way. If during the voyage, the Pole Star happened to be higher or lower, it would make the captain aware that he was off course. The kamal was later followed by the *cross-staff*, which was similar in principle to the kamal except that it was rigid frame. The cross-staff would be eventually replaced by the *astrolabe*.

Credit: Image courtesy of Peter Ifland[1]. Used with permission from his talk on "History of the Sextant" given at the amphitheatre of the Physics Museum under the auspices of the Pro-Rector for Culture and the Committee for the Science Museum of the University of Coimbra, Portugal, the 3rd October, 2000.
Website: http://www.mat.uc.pt/~helios/Mestre/Novemb00/H61iflan.htm

2.15 The kamal. [By written permission]

---

1 Peter Ifland, author of "Taking the Stars" and "Celestial Navigation from Argonauts to Astronauts".

The principle of the astrolabe was derived from the fact, the stars in the heavens were fixed with respect to each other and their pattern overhead always looked the same from a particular place at any given time in the year. The stars as a whole have movement, they appear to rise in the evening and set at dawn. This was because the Earth revolves around its axis. The view also appeared to shift a little to the north and then south as the year progressed, only to come back again to the same position after one year. This happened because of the tilt of the Earth's axis as it goes around the sun and is the same reason for which the Sun is seen to move northward and southward along the ecliptic. [Please see Fig. 3.1] Thus, if a device could be made to match the pattern of the stars overhead for any time of the year, it could give the month, date and time from a specified place. Conversely it could tell the latitude also.

It is thought that around 150 BC Hipparchus, the Greek astronomers, invented this clever devise or it may have been developed even earlier by Apollonius of Perga (c. 240 BC). The name astrolabe is derived from two Greek words. "Astro" comes from star and "labe" means to "take", meaning to take measurements. Initially it was an instrument used for astronomical work, like locating stars at particular times in the year or even for teaching astronomy.

Later it came to be used by mariners for navigation. The Antikythera mechanism found in an ancient shipwreck in the Mediterranean Sea just south of Greece in 1900AD may have been such a device. Many believed it to be much more sophisticated version of the astrolabe. It is thought to have been used around 80 BC. It appears to have a gear mechanism, which makes the use of gear wheels that made its creation far in advance of its time.

After the Greeks, Europe forgot the use of the astrolabe for some centuries. In its original form it was not useful to sailors, who were sailing greater distances now. They found the instrument useless for places other than for which a particular astrolabe was designed. It was with the coming of Islam there flowered a vibrant civilization, which spread from Arabia to the west around the south Mediterranean to Spain. In the east it stretched to Sind in the Indian subcontinent and later beyond. [Please see Fig. 3.4] They were interested in the various arts and sciences of other cultures that had preceded them. The Arabs came to know about the astrolabe used by the Greeks and they developed this instrument further.

In its more sophisticated form, the instrument was a flat disc made of brass with a ring at the top edge, so that it could be hung when measurements were taken. This brass disc, known as the *mater* or mother disc was about 6 inches (15 cm) in diameter and ¼ inch (6 mm) thick. Its circular edge was raised to form an elevated margin, which created a hollow that held a thin circular plate, known as a *tympan* with circular markings. Over it was placed the *rete*, which looked liked some ornate filigree ring. Once the rete was fitted over the tympan it became level with the raised edge of the mater. The alidade was mounted over this. It could turn around the centre and read the elevation of any object from the markings on the edge of the mater in degrees. A small metal peg at the centre was used to hold all the parts in place.

On this tympan was carved, offset, many eccentric circles and arcs. They represented the latitudes projected on a plane. The offset centre, around which these circles and arcs were constructed, represented the Pole Star. The central point of the tympan itself represented the point overhead the observer.

Credit: Photo by Andrew Dunn / Whipple Museum of the History of Science in Cambridge. Website: http://commons.wikimedia.org/wiki/File:Astrolabe-Persian-18C.jpg

2.16 An 18th century Persian astrolabe. (CC-BY-SA-2.0)

Within the circular peripheral ring of the rete there was another smaller, but eccentric ring. The two were held together by what looked like fretwork, between whose gaps there was enough space left to allow for the markings on the tympan to be seen easily. These two rings together formed the rete. On it were flame shaped pointers, which represented the position of the prominent stars and the Sun. The rete was made in such a manner it could be fitted on to the mater and be rotated as required. By turning the rete, the flame shaped markers could be aligned to the marking on the latitudes on the tympan. But before this the elevation of the star in question or the Sun had to be obtained. This could be obtained by a rotatable arm, the *alidade*, with sights mounted at the centre. It could be turned round within the circle of degrees marked on the edge of the mater.

One of their other interests in developing the astrolabe was that it would give them the direction of Mecca. This was essential was for them, as they had to face the direction of their Holy City when they prayed. At the same time it was also important for them to be able to tell the time of sunrise and sunset, so they knew when to begin their prayers. All this formed an important aspect of their religion.

To take a reading of time form the astrolabe, one had first to take the elevation of a known star or the Sun. This was done by suspending it on a string and allowing it to hang vertical. The elevation then determined by rotating the alidade and aligning its sights to the celestial object concerned. The elevation was read off as described before. Now by turning the rete the appropriate flame shaped marker, which represented the celestial body concerned, was aligned to the latitude marked on the tympan. From this the time and conversely the position of the observer could be obtained.

The other modification the Arabs made to the astronomer's astrolabe was to make it simple, in order to be useful to the mariner. It could determine the latitude within the limits of their usual voyages. Like the *quadrant* before it, it had to be pointed to the Pole Star. At the same time it had to be held vertical, so that the *zenith distance*, which implied the angle in degrees from the point of observation to the point overhead where his feet were planted, could be measured. Subtracting 90 degrees from the value would give the latitude. It could be used for the Sun as well.

The principle on which this instrument was based allowed any one instrument to be used from one place only. However, the up-market astrolabes had a set of tympans, each for particular town. Thus by interchanging them they could be used from different locations. In this way these high priced astrolabes could be used for different places to where the travellers and merchants usually went.

The astrolabe was like a medieval computer. It could tell time, find a direction, confirm one's position and even measure the elevation of celestial bodies or buildings.

Just by using of the astrolabe and the sextant, which came after it, it was not possible to know the longitude. To pin point the east west position of a ship the mariner had to determine the longitude as well. This was still the missing element, which was essential for navigation. To solve this problem the medieval sailors used a method known as *dead reckoning*. It was the only easy way to measure the longitude, before the invention of the marine chronometer. Famous captains like Christopher Columbus (c.1451 – 1506), Giovanni Caboto or John Cabot (1450-1498), Vasco da Gama (1460 – 1524) and Ferdinand Magellan (1480-1521) used this method when they undertook their great sea voyages across the oceans.

DEAD RECKONING
Dead reckoning was a roundabout way to find the longitude. In using this method the captain of a ship had to start from a fix on a map. This usually meant the place where the voyage started. The captain then had to obtain another fix at a given interval, which he obtained from various data, such as the speed and the direction of the vessel. The information of the speed of the vessel was recovered from a *knotted line* paid out, which gave the distance travelled by the vessel against time as shown by an hour glass. While the direction in which his ship was travelling had to be obtained from a *magnetic compass*. Thus by obtaining fixes at given intervals of time, the captain would guide his ship to his destination using a map. In doing this he was not only keeping a check on the north-south position (latitude), but also the east-

west position (longitude) as well. To keep track all the information required had to be fed onto a *traverse board* first. It was a simple device, which even an illiterate sailor could use.

THE TRAVERSE BOARD

A traverse board consisted of two parts. The top, which was circular, on it was drawn a compass rose. The bottom was a rectangle marked with four rows of equally placed squares. Each part was used for a different purpose.

The bottom part with rows recorded the rate of the ship's progress in knots per hour. Each square represented one knot, which was the unit of measure of the speed of a ship at sea. It was equal to one sea (nautical) mile per hour or 1 1/6 statute miles/hour or 1.9 km/hr.

The top part was complimentary to the bottom part and was used to trace the direction of the ship's journey with respect to the cardinal points of the compass. Thus in this way fixes could be obtained, both for distance and direction with respect to time.

The information to be placed on the traverse board was applicable for one watch only. A *ship's watch* consisted of a four-hour spell of duty on a ship. Thus there were six such four hour watches every 24 hours. During each watch the information was transferred and the recording on the traverse board every half an hour. When the next watch started recording on the traverse board was started anew again.

Though there was no dividing line on the lower part of the traverse board, it was understood that the rows of squares were divided into equal parts vertically in the middle. Each row of squares, in each half, with their corresponding holes represented one half hour of a watch. The first, half hour, of a watch started from the top left hand square and then down one line every half hour till the first two hours of the watch was completed. The tally for the rest of the watch now started from the right half and completed one row down, in a similar manner as before. In this way the first two hours of the watch was represented on the left half from the centre. The last two hours of the watch was represented on the right of the centre. During each watch the progress of the ship in knots was reported to an officer, who marked the

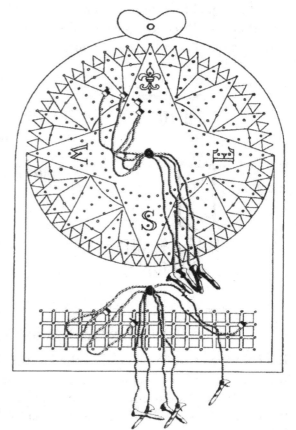

Credit: Courtesy Wikipedia user SV Resolution. Website: http://en.wikipedia.org/wiki/File:Traverse_Board.jpg

2.17 A traverse board. (CC-BY-SA-3.0)

ships progress by counting off the squares horizontally from the left and inserting a peg to represent the appropriate speed. Thus if the ship was travelling at three knots the peg was placed on the corresponding hole for third square from the left for the half hour concerned.

Note there was a hole before the first square, a peg in this hole represented the ship had not progressed. This would happen during a lull.

THE KNOTTED LINE

The data on the ship's progress was obtained by a long line with knots at fixed intervals of 1/120 geographical miles. This line was attached to a log of wood, also known as a *chip log*, which trailed behind the ship and helped to determine the number of knots given out. The thumb rule, as far as the uneducated sailor was concerned, was the number of knots paid out at half-a-minute intervals indicated the rate of progress of the ship in nautical miles per hour. Therefore if two knots were paid out in half a minute the rate of progress was 2 knots per hour or if five knots were paid out the ship's speed was 5 knots per hour and so on.

Simple arithmetic shows, if two knots on the line was paid out in any 30 seconds then the rate converted to one minute would be 2 knots on the line * (30 seconds * 2) = 4 knots on the line per minute. Now if this was converted into the number of knots paid out in one hour it would be equal to 4 knots * 60 minutes = 240 knots / hour. We already know in one nautical mile there are 120 knots on a line for 1 nautical mile. Therefore the rate of progress of the ship would be 240 knots / 120 = 2 nautical miles per hour.

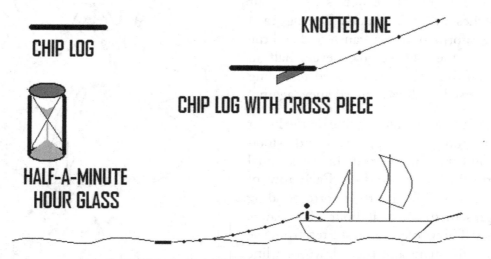

2.18 Chip log with knotted line and hour glass.

The chip log at the other end of the knotted line helped to keep it afloat. Later, to keep the line taut, a wooden crosspiece was attached to the float. This would provide the resistance and ensure a better measurement of the ships progress, than a line that happened to become slack every time the ship slowed. Any of the ship's hand could be placed in charge of this system. All he had to do was see how many knots were paid out by using a half-minute sand glass. If the line ran out, then he would gather up the line and repeat the process again. After that all he had to do was report his findings every half hour to an officer, who would record it on the traverse board.

THE WIND ROSE

The top of the traverse board was represented by a *rosa ventorum* or a *wind rose* as it is sometimes known. The concept of a wind rose goes much further back than the compass on

Source: http://commons.wikimedia.org/wiki/File:Europe_Mediterranean_Catalan_Atlas.jpeg
2.19 A Portolan chart of the Mediterranean from the Catalan Atlas of 1375. (Author unknown) [Public Domain]

which it is displayed today; as far back as the days when Andronicus of Cyrrhus (c. 100 BC) built a *horologium* in Athens for telling time. It is also known as the Temple of Winds. Part of the building still stands today. The building had an octagonal shape with eight figures at each corner representing the winds that blow from each corner. Starting from the north and going anticlockwise they are now known respectively as Tramonta, Greco, Leventer, Sirocco, Ostro, Africo (or Libeccio), Ponente, and Maestro. It will be noted the names are Italian and not Greek. This is because by the time the maps were being developed in Europe during the medieval period, they were being done by people from Italy, like the Genoese and the Venetians. These maps came to be known as *portolan charts*. They were fairly accurate maps of the Mediterranean on which lines were drawn to represent eight directions from which the eight winds blew. These lines crisscrossed the chart to form a pattern with a point at the centre from which a pattern radiated out. This to the imagination looked like a rose. Thus the name wind rose. Though the lines represented the direction of the winds, it should not be construed that they were used to predict the winds. They only helped to point to the directions of the compass with respect to the ships position on the chart. They were the precursors of the top portion of the traverse board, which allowed the sailors to plot the direction of the progress of their ship. Thus the alternative name of *compass rose*. Today the rose is no longer used even on maps. It only survives on the compass as an art form as an embellishment in the form of fleur de les, which marks north, on a compass. It is derived from the original 'T', which meant Tramonta; indicating the direction from which the north wind blows.

The 8 point compass rose on the traverse board was further divided into north northeast, northeast, east northeast and so on, thus turning it into a 16 point wind rose. Before this the Romans had used a 30° rose which meant dividing each quadrant into three parts. This made more sense, because the circle made up 360°. However, it fell into disuse after their empire fell. The reason for this was division by three proved too complicated for the uneducated sailors of the Middle Ages. Their ideas of division went as far as dividing by two only. So they divided each quadrant by two and again by two and so on. Thus unintentionally they bettered the accuracy of the Romans, because their smallest segment came to be 22.5°.

The more points you had on a compass rose the more accurate you could be, but accuracy of navigation depended on other factors as well. There was no purpose in making a 128 point rose if you could only measure the direction, say, to a one point out of 16 points.

At the centre of the wind rose on the traverse board there was a hole. Along the main lines of the directions of the compass points there were eight holes equally placed as they radiated out from the centre. The difference between a pair of holes along any one radiation represented half hour of a watch, the time of which started from the centre. Readings taken from a compass to determine the direction of the ship every half hour were again transferred here and represented by a peg, which was inserted appropriately. This would show the direction in which ship was travelling.

All the pegs were secured to the traverse board by a string, so they would not be misplaced or lost when the ship was tossing on rough seas. When the watch ended, the whole information was transferred to a slate. The captain could then collect all this data and note them in his log book. This would enable him to plot his progress from time to time on the map.

The traverse boards proved to be a fairly foolproof devise. It was so simple it could be used with very little training. It would work in all but the roughest of weathers. Its use, which lasted till the early 20th century, when steam had already replaced the sail, stands as a testimony to its success.

THE MAGNETIC COMPASS

Many nations lay claim to inventing the magnetic compass. Today it is attributed to the Chinese. The Chinese claim rests on a legend. Their emperor, Huang-ti, in the year 2634 BC fought against some of his rebellious subjects. To confuse the emperor's troops the rebels created a smoke screen. The emperor sent in chariots into the fray. On each of the chariots was placed a human figure made from a magnet pointing to the south. Thus they were able to find their way in the smoke and fall on the rebels and defeated them.

One is intrigued as to how these compasses helped. I always wondered, since I had read the story, how the charioteers could see these ingenious direction finders as they drove through all the smoke with their eyes burning and watering. Nevertheless these ingenious compasses must have worked somehow, because their emperor won the day. The Chinese are a very clever people.

However, the earliest recorded use of a compass by the Chinese comes from a work entitled P'ing-chou-k'o-t'an during the end of the 11th century. Though its use as long back as 2634 BC may be doubted, nevertheless its use must have certainly predated the P'ing-chou-k'o-t'an by decades, if not centuries. In those times there were no prizes for announcing an invention, so inventions usually made earlier were recorded much later by writers.

The loadstone, which is a certain magnetic oxide of iron known as magnetite (Fe_3O_4), occurs in many parts of the world. It is to be found from New Zealand to the Americas. In Europe it is found in Sweden, Norway, Germany and Italy. It is to be found in South Africa. In Asia magnetite is found in China as well as in India. Australia has its own deposits of this material. Many cultures in these lands had maritime histories dating back to many millennia. In those days the knowledge of a magnetic compass would have been a guarded secret amongst sailors. As such it may not have been mentioned. It is only when the knowledge became common that people started talking and writing about it. Therefore today we cannot tell with certainty when and where natural magnets were first used as compasses for navigation.

Once the magnetic property of magnetite was known, it is not far fetched to imagine someone realized that a small piece could be floated on a thin slice of wood in a bowl of water and used as a direction finder. Later it was used to magnetize a small piece of iron in the shape of a needle. This could have happened in any country that had a seafaring tradition. It could also have happened independently anywhere from China to the Mediterranean countries. In fact the Arabs had perfected the compass with a 32 point rose, which they used during the Middle Ages. It had an accuracy of 11.25°. It is thus quite possible the magnetic compass may have had its origins elsewhere other than China.

The compass was used both in the Arabian Sea and in the Mediterranean Sea by the Arabs. But it seems they did not use the traverse board at least on the Indian coast. Here their requirements were met by a good compass and accurate maps. This is an observation from a record that goes as far back as July 1498, when João de Barros, a Portuguese historian, described a map of all the coast of India, which was shown to Vasco da Gama by a Moor from Gujarat. It showed bearings laid down after the Moorish style "with meridians and parallels very small, without other bearings of the compass; because, as the squares of these meridians and parallels were very small, the coast was laid down by these two bearings of N. and S., and E. and W., with

Website: http://commons.wikimedia.org/wiki/File:Jorge_Aguiar_Wind_rose.jpg

2.20 Replica of a wind rose from a chart of Jorge de Aguiar, 1492. Original is in the Beinecke Library, Yale University, USA. [Public Domain]

great certainty, without the multiplication of bearings of points of the compass usual in our maps, which serves as a root to others."[14]

The earliest known reference to the magnet, as a nautical direction finder in Europe, was by a monk called Alexander Neckham. In the year 1187, he describes how he had seen it seven years earlier. Nearly a hundred years later in 1270 Alfonso X of Spain (1252-1284), known as the Learned, decreed that all ships should carry this needle.

A contemporary compass rose has two concentric rings. The outer ring represents the cardinal directions, while the inner ring gives the magnetic directions. This is because the Earth's magnetic pole is not truly aligned with its geographic pole. It was Pedro Reinel, a Portuguese cartographer who lived in the 16th century, was the first to draw the standard compass rose.

THE STAR TABLES

There was another method of navigation available to ancient sailors - the *star tables*. These were tables of stars as seen overhead at different times from various places on the Earth. Since the celestial sphere was almost concentric with the Earth, by consulting the tables the mariner was able to get his bearings at sea from a "fix" obtained from the position of any known star at a particular time. Ptolemy of Alexandria had made one such table in the 2nd century AD, but it did not turn out to be popular amongst sailors. This was because most of them could not read or write, so only a very few ever used them. Later more detailed and more extensive star charts were to become an important part of the navigator's repertory from 18th century onwards. Eventually the star charts became obsolete with the advent of the global positioning system.

DEVELOPMENTS IN SHIP DESIGN

The developments in navigation went on side by side with the development of ship. The latter became essential as the merchant ships had separate requirements from ships of war. They had to fill the maximum space available with goods for the maximum profit in a very fiercely competitive world of commerce. At the same time speed was also required. The more voyages they could make the more money there was in it.

After the fall of the Roman Empire in the middle of the fifth century there was a lull in the trade. It was after this brief interval Byzantium (Constantinople), became the centre of trade in the Mediterranean world. After Constantinople fell to the Turks, the Venetians and the Genoese fought for supremacy for the Mediterranean trade.

In spite of all the changes to ships, which were being introduced by merchants, the war galleys descended from the ancient design still held their place and continued to be used in sea battles. The great sea Battle of Lepanto in 1571AD, the combined forces of Venice and Spain under the command of Don John of Austria broke the power of the Turks in the Mediterranean. Each side had had 300 ships, out of which the Europeans had 208 galleys and the Turks had 250. It was considered to be one of the greatest sea battles, where galleys would engage in large numbers. At the end, not counting other religions and the dead, 15000 Christians, mostly oarsmen were liberated from the Turkish ships. This gives us an idea of how much space was taken up by oarsmen. When the ship's cannon appeared on the scene the oar had to give way for their placement. By then the techniques of sailing and sea battles

Website: http://en.wikipedia.org/wiki/File:Battle_of_Lepanto_1571.jpg

2.21 Battle of Lepanto 1571 by Yogesh Brahmbhatt at National Maritime Museum, Greenwich, London. [PD-ART.]

had changed a great deal. Thus by the latter part of the 16th century the use of oars had all but disappeared on sailing ships. Though there were some instances of galleys being used in later naval battles, their numbers were never that significant nor did their presence have any effective impact on the course of the engagement. Some countries like Sweden still retained galleys in their navy even during Napoleonic times. Their last galley was decommissioned in 1835, which was after the introduction of steam power

Between 1350 and 1450 the design of ships had undergone radical changes that would ultimately allow them to undertake transoceanic voyages. In spite of the speeds galleys could achieve, they had many disadvantages. Human power was not of much use for really long voyages that were being attempted at this period. Wind power had to be utilized to the full. The navigators of the time realized this need. Thus sails had to be added. This would more than compensate for the loss of oars. In order to get the maximum benefit more masts

Credit: Courtesy Wikipedia User: Denelson83 (Artist).
Website: http://commons.wikimedia.org/wiki/File:Flag_of_Saint-Pierre_and_Miquelon.svg

2.22 The emblem of Saint-Pierre and Miquelon depicting both 'forecastle and after castle on a carrack. [Public Domain]

had to be introduced. At first there were two and then came three masts, the fore-mast, the main-mast and the mizzen-mast. Later another would be added. This and other innovations, such as breaking up the square sail into many parts and adding extra-wide sails also helped. Without oars the hull of the earlier galleys became rounder. This gave them more stability in rougher seas. This in turn allowed the stern to be widened; allowing the *aftercastle* to be placed in its upper part. It served as quarters for the captain and his officers and made longer journeys comfortable. [Please see Fig. 2.23 also] Later *forecastles* came up on the bows also to accommodate sailors. The large oar which had acted as a rudder had also been replaced by a stern post rudder over a century earlier. The rudder now being at the centre, the ship now became easier to maneuver.

In spite of all this all was not well with the sailing ship. The maneuverability of the ship with square sails left a lot to be desired. A square sail could never be made to sail against the wind and ships would stall if the wind blew from contrary angles. Something had to be done.

THE LATEEN SAIL

The sailors of Europe imported the idea of a new sail from the east. It was the *lateen sail*, which was copied from the Arabs, who in turn had either got it from India or from far off China. Though with this new sail the ship could not sail directly into the wind, it still allowed the ship to advance forward. This is how it could be done. The lateen sail could be moved from side to side and so, unlike the square sail, it had the option of catching the wind from side. As the wind struck the sail at an angle, its force would tend to propel the ship sideways. At the same time the hull met with water resistance. Now with the help of the rudder, as long as the wind was not coming directly from the front, but only up to 45° on the side, the ship would be "squeezed" forward between the two forces. Thus a ship with a lateen sail could sail "against the wind", in a zigzag fashion, with winds up to 45° on either side coming opposite to the direction of movement.

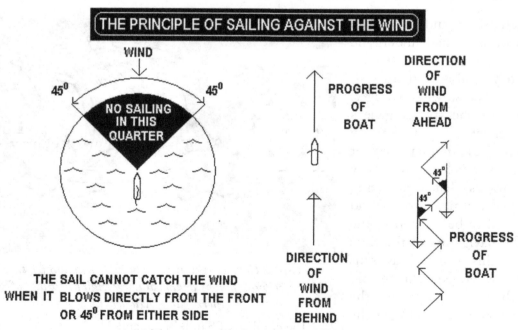

2.23 The principle of sailing against the wind.

All these devolvements evolved around the *caravel*. The caravel was originally a small fishing craft, used by Portuguese fishermen. Under Prince Henry the Navigator (1394-1460), it was developed by the Portuguese for their early explorations of the West African coast. In its perfected form it had three masts. The foremast and mainmast had square sails, while the mizzenmast had a lateen sail. This type of ship thus could sail into the wind. It had two decks and quarters suited for long voyages.

A typical caravel of the late 15th century was a lateen rigged vessel had a displacement of around 50 to 60 tons and was about 75 feet long with three masts the largest being forward. Columbus' Santa Maria was a carrack, but her companions the Niña and the Pinta, were caravels. Vasco Da Gama and other Portuguese explorers used the caravel to reach India. John Cabot was trying to find a northwest passage to India and rediscovered Newfoundland on a caravel named Mathew. Magellan had one when started on his voyage of circumnavigation. The caravel was a smaller and a lighter ship compared to the *carrack*. The carrack would come to supersede it and in its turn, the carrack would be replaced by the Spanish galleon.

The trade which the Arabs had snatched from the Indians was soon to be wrested from them by the Portuguese in the east. In the mean while the Mediterranean was abuzz with ships carrying trade. Venice controlled this trade. Marco Polo (1256-1323) the Venetian traveller had been as far as distant Cathay (China) and what Marco Polo could do overland could also be done over sea routes. The European traders knew about these routes through the information that filtered from their counter parts in the orient through trade with Egypt and Middle Eastern countries. Under the direction of Prince Henry the Navigator (1394-1460), Portuguese captains started exploring the archipelagos of east Atlantic in the hope of reaching the east. They made progress along the West African coast line. However, at first their progress was halted by violent seas at Cape Bojador (26° 07' N, 14° 29' W). Many ships during earlier European expeditions had been lost here, giving the region a bad name.

2.24 "Resolution in a gale", by William van de Velde the Younger, c. 1678. [PD-ART.] Note the triangular lateen sail. Website: http://commons.wikimedia.org/wiki/File:Van_de_Velde,_Resolution_in_a_Gale.jpg

Credit: Courtesy Wikipedia User: PHGCOM
http://en.wikipedia.org/wiki/File:PotugueseCaravel.jpg

2.25 A model Caravel displayed at Musee de la Marine, Paris. [CC-BY-SA-3.0]

Unknown to them at the time this was due to underwater reefs and the peculiarities of the sea floor causing the seas to be unruly in that region. After their first failure, they learnt to go around the area the next time, thus becoming the first Europeans to cross this region successfully. Next, the Portuguese captain, Bartolomeu Dias (1451-1500) successfully rounded the Cape of Good Hope during his expedition of 1487-88. This paved the way for Vasco da Gama to reach India ten years later.

Credit: Courtesy Wikipedia User: Arne List.
Website:: http://commons.wikimedia.org/wiki/File:Faroe_stamp_226_Discovery_of_America_-_Kristoffur_Kolumbus.jpg

2.26 Picture of a carrack on a postage stamp from Faroe Islands. [Public Domain]

In the mean time, the Spanish, the other nation with seafaring ambitions, soon came into the fray. The Pope Alexander VI on 4th May, 1493 to prevent conflict between the two great Catholic nations drew a line on the map to separate the lands each could colonize. This line ran north to south 100 leagues west of Azores and Cape Verde Islands. All lands west of this line went to the Spanish and the Portuguese got the east.

However the Portuguese remained dissatisfied with this arrangement. In the explanation, which follows the reason for their being unhappy will become clear. After Vasco da Gama's discovery the Portuguese had organized another expedition to India. This time da Gama

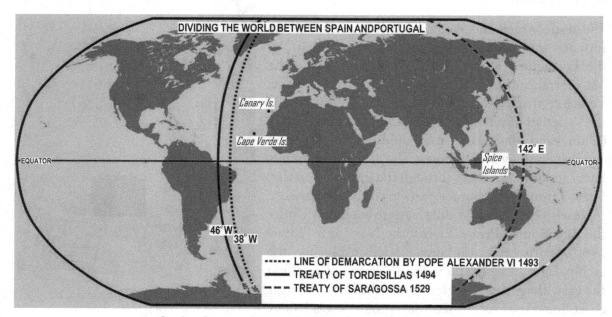

Credit: Courtesy Wikipedia User: Lencer (CC-BY-SA)
Website: http://en.wikipedia.org/wiki/File:Spain_and_Portugal.png

2.27 Dividing the world between the Portuguese and the Spanish. [CC-BY-SA-3.0]

personally drew the course this expedition would take. To avoid the calms of the Gulf of Guinea, he advised the new expedition to skirt this region by going further west. In trying to do so the Portuguese armada travelled much further than they had intended. In doing so they reached a new land, which they claimed for themselves. Today we know it as Brazil. According to Pope Alexander's line of demarcation Brazil would go the Spaniards. This is what made Portugal unhappy with Pope Alexander's line of demarcation. By the Treaty of Toresilla with Spain the following year, Portugal finally obtained a new line of demarcation, sanctioned by Pope Julius II in 1506. This new line was 270 leagues west of the previous line of demarcation. This brought Brazil under the Portuguese. The Spanish got the lands west of the new line and the Portuguese the east as their share.

There was an already existing trade rout between India and Madagascar Thus, in 1498, Vasco da Gama was able to make it to the orient with the help of an Indian pilot from Madagascar. In due time the Portuguese became established at Goa, Daman and Due, which they held long before the British set foot in India and were to hold for 10 years longer, after the British left in 1947. The Spanish took possession of large tracts of the Americas after Magellan had shown the way.

THE FINAL PROOF

The doubters who still persisted in the belief the Earth was flat, for them the final proof came with Magellan's voyage.

Ferdinand de Magalhães (c.1480–1520) or Magellan, the son of Pedro de Magalhães was born in Sabrosa, in Villa Real district of the Traz-os-Montes province of Portugal. His family belonged to the lesser nobility. As a boy he was one of the pages to Queen Leonor, wife of King John II the Perfect.

Source: http://en.wikipedia.org/wiki/File:Hernando_de_Magallanes_del_museo_Madrid.jpg

2.28 Ferdinand Magellan. Original at Museo naval de Madrid. (Artist anonymous) [Public Domain]

When John II died in 1495, Manuel I the Fortunate became king of Portugal. The Portuguese hoped to take over the control of the spice trade from the Arabs, so they wanted a strong presence in India. Thus King Manuel decided to send Fransisco de Almeida, as first Portuguese Viceroy of India. Magellan volunteered to go east with him. This was the start of Magellan's carrier. They sailed on 25th of March 1504. On reaching India they took control of the harbour at Cannanore on 16th March, 1506. During the fight Magellan was wounded. This would become the hallmark of Magellan's life. He would be wounded many more times, before finally being killed in battle at a comparative early age of about 40 years. It may have been his daring or his passion for being at the thick of things, which made him prone to injury or may have been just bad luck. It is interesting to note many others, like Alexander, Hannibal and Julius Caesar, who had often been in the forefront of battle, yet they did not receive any major wounds. Whereas, others like Nelson and Magellan were repeatedly wounded and eventually died in battle.

Magellan later helped to build a fortress at Cannanore. He took part in the naval Battle of Diu on 3rd February, 1509. The Portuguese fleet defeated the combined strength of the Egyptian, the Ottoman fleets and the fleets of two local Indian kings, the Zamorin of Calicut and the Sultan of Gujarat. It was a crucial engagement, which gave the Portuguese the control of the Arabian Sea and opened up the route to the Spice Islands. Magellan was again wounded in this naval engagement. By August that year he seemed to have recovered and he joined in a voyage to look for the source of spices in the East Indies archipelago. On their way they narrowly escaped a treacherous plot by the Malay people in Malacca. The following year he was raised to the rank of captain. In the rainy season of 1511, under the command of Admiral Afonso de Albuquerque (1453-1515), who was now the second Portuguese Viceroy of India, Magellan was there at the taking of Malacca. Under instructions from Albuquerque, Antonio d'Abreu went in search of spices and Magellan went with him. They left Malacca in December 1511 with a squadron of ships. They traced their path through the various islands passing Java and winding their way between Java and Madura, Celebes, skirting Gunong Api volcano, touching at Bura. They finally reached Amboina and Banda. As some of these places were east of the 125^0 longitude, this journey would later turn out to be significant for Magellan being the first person to cross all the meridians,. At Banda they found what they were seeking - cinnamon and nutmeg. Their mission accomplished they returned back.

In 1512 Magellan returned to Portugal. In July he was given the rank of fidalgo escudeiro. This implied he was a man of noble birth, who had been given the rank of esquire. The following year he set out with the Portuguese expedition to Morocco against Azamor. The city fell at the end of August 1513. Here, a little later, Magellan was again wounded in a raid and became lame for life. At this time Magellan became involved in trading with the Moors. This earned his king's displeasure. Though the charges leveled against him had been dropped, Magellan was out of favour. Magellan now realized there was no future for him in his own country. He renounced the citizenship of his country of birth and turned his future hopes to Spain.

He arrived at Seville on 20th October 1517. He hoped to meet King Charles I of Spain at Valladolid. Charles would later become the Holy Roman Emperor as Charles V. Magellan met some high officials and important people. Amongst them was an influential Portuguese, Diogo Barbosa, who had become a naturalized Spanish citizen. Barbosa had a daughter, whom Magellan married. It was through his help Magellan obtained this first audience with Charles I.

By now Magellan firmly believed, what many people had already been thinking, the South America continent was not joined to the Great Southern Continent (Antarctica). He therefore proposed to reach India by crossing the southern tip of America. Since the Portuguese held the eastern routes to the lucrative spice trade, the Spanish lent an ear to Magellan's proposal. Charles and his minister Juan Rodriguez de Fonseca took interest in the scheme and gave permission to proceed with his preparations. Ruy Falerio, an astronomer and a Portuguese exile aided Magellan in planning his project. Financially he had help from Christopher de Haro. On 22nd March 1518 Charles V signed an agreement with Magellan that he would be joint captain-general with Ruy Falerio and they would receive one-twentieth of the net profit from the expedition. Moreover they and their descendants would be invested with governorships of these new lands with the title of Adelantados.

They left Seville on 10th of August, 1519 with five ships. Passing down the river Guadalquivir where they halted at Sanlúcar de Barrameda and remained there for five weeks. Ruy Falerio however decided not sail. Having had his horoscope cast by astrologers, he was told the voyage he was about to undertake would be fatal. So Magellan was now left in sole charge of the expedition. Before embarking on his voyage, Magellan wrote a letter to Charles. In this letter he affirmed his intension of settling the demarcation-line the Pope had drawn and define the regions of the Spice Islands that would go to Spain and those which would go to Portugal. He would also see to it that Moluccas came to Spain. He drew up his will and set sail. Thereafter he set sail on 20th September, 1519 and reached Teneriffe on 26th September, 1519. From there he again set off on 3rd October, 1519 to reach the coast of South America at the Cape St. Augustine, near Pernambuco on 29th November 1519. He then followed the coast southwards seeking a passage to the west. At the mouth of the River Plate he thought he had found what he was seeking. He searched this area from 11th January 1520 to 6th February, 1520 with the hope of accomplishing his dream. He realized that this was not a passage, but only the mouth of a huge river, unlike any seen in Europe. Disappointed he turned back and went south once more. Now it was autumn in the southern hemisphere and winter would soon be at hand. So when he and his men found a natural harbour in Patagonia, they settled in for the winter. Today it is known as Puerto San Julián or Port St. Julian (31st March 1520). On 1st April a mutiny broke out there, however it was quelled by the next day. Here the explorers came in contact with the local people whom, because of the large size of their feet, they named Patagonians.

With spring arriving in the southern hemisphere the expedition left on 24th August 1520 sailing further south. At last on 21st October 1520, on the feast day of St. Ursula, they found what they were looking for - a passage that would take them beyond the American continent. They called this the Cape of the Eleven Thousand Virgins in her honour.

The story goes St. Ursula, who was a daughter of a British chieftain of Roman Britain, went to join her future husband in Brittany, France. She was accompanied by 11,000 virgins! On the eve of her marriage she wished to go on a pilgrimage around Europe. On their way they were eventually murdered by the Huns. What the Huns did may not be in doubt, but the rest of the story certainly is not true. In fact the story does not have the backing of the Church today.

Little did Magellan and his men suspect the trials and tribulations, which lay ahead of them! Through this treacherous strait they made a very difficult passage. The scenery was stark, both sides were lined with high cliffs and bleak snow peaked mountains. At places the passage was very narrow and very tortuous. On the south lay a cold and hostile land with fires burning. Magellan named it the "land of fire" or Tierra del Fuego. It took them thirty-eight days to travel the 360 miles of this forbidding straight. This amounted to a progress of less than ten miles a day. One can only imagine the difficulties the expedition was going through. During the passage the San Antonio deserted them. Subsequently the rest of the captains met on the 21st of November to discuss whether to proceed or not, because of the difficulties that they had to face. On the 28th of November the little armada passed the Cabo Deseado (Cape Forward) at the western limit of the straits, which is now known as the Straits of Magellan. In front of him lay the calm waters of what they thought was a great sea, which he had named the Pacific.

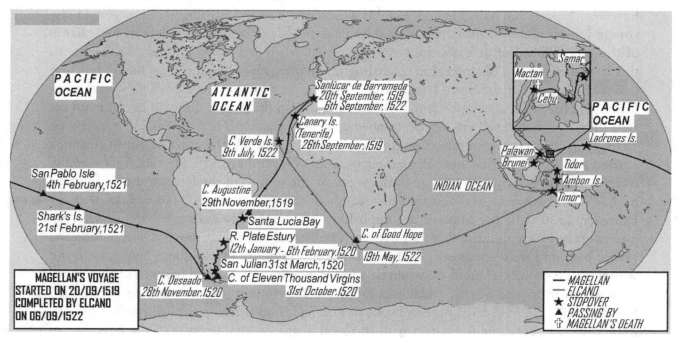

Credit: Courtesy Wikipedia User: Sémhur/Derivative work by User Uxbona
Website: http://en.wikipedia.org/wiki/File:Magellan_Elcano_Circumnavigation-en.svg

2.29 Map showing Magellan's voyage and proof of his circumnavigation of the world. (The original labeling modified for clarity.) [CC-BYSA-3.0]

Magellan had no inkling of the vastness of this ocean that he was looking upon. Little did he imagine an even more horrendous experience lay in wait for him and his men! It was during his passage across the Pacific, Magellan and his crew was tried to the very limits of human endurance. Hunger beset them. When their supply of food had run out, they ate stale biscuits. When that was finished, in their hunger they were forced to eat rats. After hunting down all the rats down, they were reduced to eat boiled hides and some of the crew even ate the sawdust on the ship! As their journey progressed whatever water was left had turned putrid. They were beset by disease and scurvy was rife. Not surprisingly many of his crew died. After the ninety eight days it took to cross the Pacific, they finally sighted Ladrones on 6th March 1521; so called by Magellan, because of the thievish habits of the natives there. But Guam was probably the first place where he eventually landed on 9th March. They replenished their stores and repaired their ships and started again. They sighted Samar Island on 16th March, 1521, which is a part of the Philippines. On 7th April 1521 they arrived at Cebu, an island to the south west of Samar. The King of Cebu Island professed Christianity in order to utilize Magellan and his men to conquer his enemy, the King of Mactan. Magellan met his death in a fight here on 27th April, 1521. Some of his men fell into the hands of the King of Cebu and were murdered.

One of the three surviving ships was burned and the other two set off for Moluccas. On their way they stayed in Borneo from 9th July to 27th September from whence they reached Moluccas on 6th November. At Tidor they collected a cargo of cloves. The Trinidad had become unworthy for further sailing, because of leaks and was left behind including her crew. The Vittoria sailed on under the command of Juan Sebastian del Cano on 21st December, 1521. While crossing the Indian Ocean, in order to double the Cape of Good Hope, the Vittoria met with adverse weather. Now the crew again suffered from starvation and scurvy. Eventually they rounded the cape and reached the Cape Verde Islands on 9th July 1522. Here thirteen of the crew members were detained by the Portuguese for a while. After this final ordeal they set out on the last lap of their voyage on the 15th July, 1522. They reached Seville on 9th September, 1522. Only eighteen men survived this epic odyssey. They had brought back 26 tons of cloves, which was more than enough to compensate the costs of their expedition.

Though it is true that on this voyage Magellan did not circumnavigate the globe, because he had died fighting on the Island of Mactan before he could complete his journey. However, it will be remembered Magellan had been to the Spice Islands before in 1511. He had therefore crossed east of the 125^0 meridian to reach Moluccas at the time. Magellan was the first person to have crossed all the meridians of the globe. Thus rightfully he circumnavigated the globe, even though he had done it in two phases. Although it had not been his intention, Magellan had proved the Earth was round. This was not however the Earth's true shape. The quest for the Earth's true shape was to be taken up later by others on a scientific footing and thus the quest for the Earth's shape did not end with this epic journey.

THE ACTUAL SHAPE OF THE EARTH

Magellan's voyage established the Earth as being a sphere. The work of Isaac Newton and Christiaan Huygens showed if a sphere like the Earth spun around its axis then the equator would bulge at the cost of the poles, which would be flattened somewhat. Such a shape is known as an *oblate spheroid*.

Today we know the true shape of the Earth is neither a sphere nor even an oblate spheroid, but something that is peculiar to itself. We call it a *geoid*. Being a geoid means the shape that the average surface the Earth has over sea and land does not match an oblate spheroid shape, but distorted somewhat to accommodate the many factors, which influences this shape. It is the result of the way in which the Earth is made. The inner core, outer core, the mantle and the crust all have different consistencies. Superimposed on this there are deforming forces at work due to gravitational influence combined with the effects of the Earth's rotation around its axis. The crust and the superficial part of the mantle together form the lithosphere. They make up the oceanic and the continental plates, on them lies the oceans and the mountains respectively. Again all these have a somewhat deforming effect on the surface. Thus the combined result is a geoid.

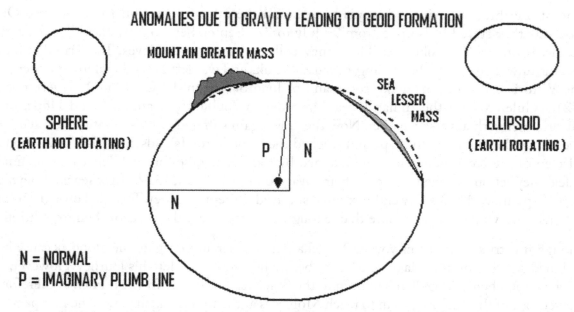

2.30 A sphere, an oblate spheroid and a geoid. (Redrawn by author from Encyclopaedia Britannica Vol. 10 p. 132 Fig. 4a 1962)

It was initially thought the crust was in hydrostatic equilibrium, but later it was realized it could not be so. If it were, then the water levels at different points on the Earth's surface would match. This is not the case. It was Leonardo da Vinci, the great genius, who was the first to recognize this fact. He suggested the Earth's crust was in isostatic equilibrium rather than hydrostatic equilibrium. However, the detailed practical idea was developed and established by the British in India during the mid-nineteenth century through survey work.

3 KNOWLEDGE FROM THE HEAVENS

The earliest attempts to observe the skies and deduce the paths of planets would eventually culminate in one of the greatest triumphs of human intellect, which would eventually lead to the understanding of the place of our Earth in the cosmic order and one day, also, pave the way to a broader knowledge of our world.

INTRODUCTION

Today we often proudly announce, "Man has carved out his destiny on the planet." or "We hold our destiny in our hands." To many, in the limited context of our role on this planet, this appears to be true. On reflection, the sensible will feel, such thoughts are only a manifestation of our ego. With hindsight, when we look back across time, we can trace some of the many events that brought us to where we are today. Until recently the significance of such events had been ill understood, but nonetheless we will find that they have been crucial to the position we hold on the planet today. The reality is we have hitchhiked on the shoulders of such chance factors to attain our present status.

If a giant meteor had not led to the extinction of the dinosaurs and paved the way for mammals. We would be still foraging on forest floor. In time, due to evolutionary pressure, our ancestors took to trees. They became the primates. The forelegs of their quadruped ancestors turned into forelimbs, with hands instead of feet. The anatomy of the upper limb also changed. The movement of their shoulder, elbow and hand now gained an exceptional range of movement to cope with life in the forest canopy, something that no tiger or deer could think of. In time, the climate of the planet changed. Large portions of the forests gave way to vast expanses of grassland. When our ancestors came down from the trees to utilize this new niche, their forelimbs were no longer suitable for walking on the ground. They had to take on an erect posture. With the hands now freed, they could be put to greater use. The thumb now faced the other four fingers. The opposing thumb helped man to do many things and as we shall see later, this allowed his brain to develop. This in turn facilitated acquiring of knowledge. However to acquire knowledge, one has to reflect and reflection needs time. This was not possible when man had to run around for food. It was at this point, providence in the form of the lowly grass came to man's help. It must not be construed it was for man that the grass did it, it was rather for its own interest. The selfish gene works for its own interests and not for others. When some grasses turned to wheat, they left the propagation to man. Man only took advantage of this fact. In between such chance occurrences, we can think of many other innumerable individual examples, which are too many to enumerate here.

When the last Ice Age receded some twelve thousand years ago, the harsh climate eased and life flourished. The grasses thrived across the planes and the valleys. Amongst them were *goat grass* and *wild wheat*. They both had 14 chromosomes each. Sometime about ten thousand years ago, by chance and through a quirk in the genetic process, they combined to form the *Emmer*, which now came to possess 28 chromosomes. It had a more substantial seed and became the forerunner of the modern wheat. Again, the genes of this particular wheat would join hands with human destiny. This time another variety of goat wheat bred with the Emmer. The newly bred product had 42 chromosomes! Its seed was even more generous, but just like the mule derived from a he ass and a mare, it would not have been fertile. Once more, unbeknown to man, the unseen hand of providence came to man's aid. Amongst the 42 chromosomes, there was one, which contained a gene that held the key to whether this new variety would be fertile or not. This gene, which was the *right* gene, underwent the *right* change at the *right* time, thus ensuring the new breed would be fertile. This wheat could be grown, husked, ground into flour and kneaded into dough, made into bread over the fire and above all well digested. Man was now poised to take advantage of this fortuitous change. [15]

By now, the hunter gatherer in many regions like the Middle East had become nomads. They followed their herds wherever they went. These heards grazed on the grasses; once they had finished the grass in one place, they had to move on. Therefore, man had no other choice, but to follow his flocks from place to place. Until then man had picked wild wheat as a supplement to his diet. Occasionally when he had stopped in one place long enough, he may have tried his hand at growing wheat like the emmer, but probably not on a regular basis. Man at this period was very much dependent on his animals, not only for food in the form of milk and meat, but for other things as well, such as the cloths on his back and shoes for his feet and their skins for his tents. They were used for transport as well. His culture revolved around his animals.

The coming of agriculture saw the gradual fading of nomadic way of life, except for few pockets where they still exist even to this day. Now with agriculture, finally, man was able to settle down. Soon great civilizations grew up along the fertile river valleys. Now it became important for man to know the right time to grow his crops. He came to realize that he could tell this by observing the movements of the Sun. It helped to know the exact time of the solstices. Man chose the winter solstice, because it was from this day, the days gradually became longer. He could then calculate the time for sowing by counting off the days from here. At first, he used natural land marks, later he created stone circles, in order to know the date of the winter solstice. The telling of the winter solstice became interwoven with religious symbolism of the regeneration of life and also attempts at reconciling the earlier reliance on lunar time with the new dependence on solar time. As a result great edifices, like the Stonehenge came up. We seldom appreciate the ethos involved, when we look upon them today. Soon man would be able to master enough mathematics to dispense with such cumbersome structures. He started to study the heavenly bodies and make calendars. Therefore, man's interest in the sky grew more and more. He could now predict the movements of the various celestial bodies with fair accuracy and even forecast eclipses.

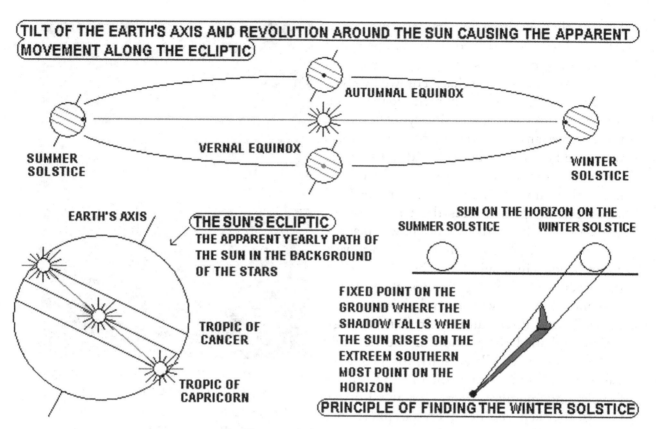

3.1 The principle of the stone circle.

Soon man came to have definite ideas about how the Earth should look like and what its position should be in the order of the universe. He assumed the Sun, Moon, the planets and stars to be smaller than the Earth, just because they looked small. He also considered the Earth as stationary, since it did not appear to move. From all this he deduced the heavenly bodies went around the unmoving Earth. He thus concluded the Earth must be at the centre of the universe.

At the same time, man continued with his imaginary ideas about the heavens. He thought the sky was like a roof over the Earth with all the celestial bodies placed on it, which had to be held up, so that it may not fall down on the Earth. Such thoughts were common to most cultures and ingrained in the psyche of their times. They continued down the centuries to the ancient times when man started keeping records.

The Greeks though, two pillars held up the sky and that it was Atlas' responsibility to attend to them. Homer referring to Atlas in the Odyssey says, "…one who knows the depth of the whole sea, and keeps the tall pillars which holds the earth asunder". According to the Greek idea, these two pillars stood beyond the Straights of Gibraltar, where the world ended for them. Beyond was Ocean, a great body of water, which surrounded the Earth. Here Atlas has the connotation of some sea deity who had the duty of tending to the pillars, which held the Earth and heaven apart. The reader must have spotted that this is in variance with the usual depiction of Atlas holding up the heavens or sometimes even the Earth. Then mythology is not history and stories changed from period to period. It was through Hesiod and not from

Homer that we have the modern picture of Atlas.[16] They also believed every day, preceded by Aurora the goddess of dawn, the sun god drove across the sky on a chariot drawn by seven horses. By the time when mythology passed into literature, no one seriously believed in these stories anymore. We might expect things would be better now, but this did not happen. The natural philosophers of the day had other ideas. They thought the heavens were made of seven crystal spheres, thus the term "seventh heaven". Each of the then known planets, Mercury, Venus, Mars, Jupiter and Saturn were all embedded in a sphere of their own, while the other two remaining spheres were for the Sun and Moon. They believed the spheres were turned by the Three Fates of Greek mythology, which explained the motion of the heavenly bodies. Later with Christianity, the people of Europe borrowed much of their ideas from the Greeks and modified them to suit Christian beliefs. They kept the idea of crystal spheres, but replaced the Three Fates with angels fluttering their wings to keep the celestial spheres moving. They awoke from such pleasant dreams when they heard the noise created by Tycho Brahe, smashing the idea of crystal spheres.

Image source: Website: http://en.wikipedia.org/wiki/File:Paradiso_Canto_31.jpg

3.2 The angels fluttering their wings to keep the celestial spheres moving. {From a picture, "Rosa Celeste" in Alighieri Dante, a book by Henry Francis Cary, (ed.) (1892)} / Artist Gustave Doré (1832–1883) [Public Domain]

SETTING THE SCENE

In spite of all these fanciful ideas, by the time we come to recorded history, man had already achieved a lot. Having secured his supply of food, man had mastered many things, such as language, the art of writing, mathematics, poetry, literature and music. All these lead to subtle changes in his mind. When he looked up at the skies, like our man in the Allegory, he glimpsed the workings of the heavenly bodies. Man now started to draw conclusions through his power of reasoning, ushering in the science of astronomy. This made him confident enough to take his first step towards interpreting his world, but this time it was through reason.

However, confidence was not enough. Correlation between facts was important in the progress of science. This could come only with accumulation of knowledge over time. At that early stage, many of the facts that he needed for building up the picture of his universe had not yet come within his grasp. Therefore, for the ancients the various individual entities of knowledge remained isolated. They were thus precluded from knowing the true picture. The whole picture would unfold only when many more facts could be pieced together. It was from such shaky attempts man began slowly, but surely, acquiring the individual pieces of the great puzzle of nature.

The ancient philosophers realized the mystery of the heavens was in some way related to their lives on Earth. Different people viewed this mystery in different ways. Those who were practical realized it related to an accurate calendar. This was necessary for any civilization that lived by tilling the ground. In contrast, many mystics came to believe that stars influenced destinies of men. Their ideas gave birth to astrology. With civilization, trade began to spread and the hardheaded sailors realized the importance the accurate knowledge of the stars. Lastly, there were the purists who pursued knowledge for knowledge's sake.

In the pursuit of knowledge, at times man would go astray, but over time man's collective effort would eventually lead him back to the right path. Gradually a picture of the universe would emerge over the ages. In this way, concepts about the Earth's place in the order of the universe came to change amongst some civilizations. Some of these ideas arose independently, but others as we have seen came through influence of trade. In those days, trade not only stood for exchange of goods, but it was also a vehicle for transfer of knowledge. Under the Ptolemy's of Egypt, ships that came to Alexandria were searched for manuscripts. If any manuscript was found, then the ship was detained, the manuscript copied before allowing the ship to leave with the original. In this way, the library at Alexandria expanded its collection. In it's hey day, before its destruction, the library is said to have contained as many as half a million scrolls! Compare that with how many books there are in your local library or the library of many towns and cities today. This gives us an idea of the value some ancient civilizations put to knowledge.

EARLIER GLIMPSES OF THE HELIOCENTRIC IDEA

The ancients had the opportunity of looking up at the sky every day, so they were more intimately aware of many things about the skies than we are today. Even during the time of fanciful thinking, some felt that not all was quite right. They realized there was more to the skies than just stories that went to enrich their lives.

To astute observers of the time and many others who came after them found there were a number of observed phenomena in the skies, which cast doubts about their earlier beliefs. They concluded that there could be alternative explanations to their earlier ideas. In India, there are records of mention in various Vedic texts that the Earth was a sphere, which spun on its axis and that it was the Sun and not the Earth, which was stationary. Later somewhere between the ninth and eight century before the coming of Christ, about the time of Homer, it is on record that the sage Yajnavalkya

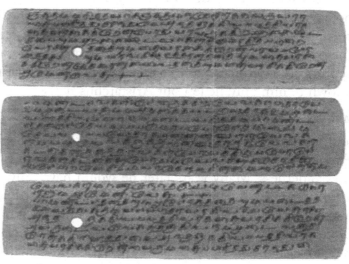

Credit: Picture by Vadakkan / Library of Congress.
Website: http://en.wikipedia.org/wiki/File:Christian_prayers_in_tamil_on_palm_leaves.jpg

3.3 Puthi or palm leaf manuscript. [Public Domain]

(Jag-ya-balkyo) wrote a book on astronomy, the *Shatapatha Brahmana*. By "book", one should not imagine that it was printed or even written on paper. In India in those distant days, books were written not on paper, but on dried leaves of certain palms. They were known as a *puthi*. Here he mentions that the Sun was much larger than the Earth, the Earth was a sphere and that it went round the Sun. However, his words were couched in the metaphysical terms of those times, so that today it is sometimes difficult to make out what was actually implied. It is also recorded he described an accurate solar calendar.

In what follows I will only trace the Greek experience mostly, because we can trace a continuous link between our present knowledge to their efforts, through the Arabs. Though much of their records are lost, still much abides. Moreover, their records are accurate and reliable. In many cases we are able to cross check them through the writings of others.

A few Pythagoreans thought that the planets went around the Sun. Their ideas arose not from any scientific work, but rather from concepts that were of a mystic nature. The Greek astronomer and mathematician, Aristarchus (c.310 – c.230BC) born on the island of Samos, was the first Greek to proposed that the Sun and not the Earth was at the centre of the universe. Unfortunately, his work has been lost to posterity. Luckily we know about it indirectly from a reference made by Archimedes (287-212BC) in one of his works called *Archimedis Syracusani Arenarius* or simply *Arenarius*, otherwise known as *The Sand Reckoner*. Here by using Aristarchus' idea, he attempts to find the number of sand grains that could fit into the universe! I wonder how many of us would have such a grand idea. Those who are curious to know, his answer was 10^{63} grains. The size of his universe, however, was limited by the ideas of his time.

Many have pointed out that his teacher Philolaus (c. 470- c.385BC) thought to have born in of Kroton, Italy may have influenced Aristarchus. However, he excelled his guru by being the first known person who attempted to obtain the distance of the Sun and Earth as a ratio of the distance of the Earth to the Moon. According to his estimate, the Sun was the larger body. He deduced that the Earth was smaller than the Sun. Therefore, Aristarchus thought it reasonable to assume that the Earth being smaller, it revolved around the Sun. He was one of the first in Greece to put the known planets of his day in their proper order. After eighteen centuries, Copernicus was to acknowledge him in his earlier work. However, later for reasons known to him only, he left out Aristarchus' name.

Aristarchus' work looked like it would take western astronomy out of the stage of speculation and mysticism to bring it on a scientific footing. Sadly, his influence would not last long. Amongst the Greek astronomers of old, Aristarchus was like a lone voice crying out in the wilderness. Many years later he would have only one follower, Seleucus of Seleucia (c. 190BC - ?) of Babylon, who was a Chaldean astronomer. He is thought to have taught around 150 BC. We know about Seleucus only from the writings of Plutarch (c.46-120), the Greek biographer. It was he, who recorded that Seleucus was the first to have argued for the heliocentric theory through use of reason. Unfortunately, for science his proof has been lost.

The forces at play then were much the same for Aristarchus, as what Copernicus, Kepler and Galileo had to face nearly two millennia later. Human attitudes do not change easily over time. This is what one of Aristarchus' detractors had to say. Cleanthes, the head of Stoics and a contemporary of Aristarchus, opined that it was the duty of all Greeks to accuse him on the charge of impiety for proposing that the Earth went around the Sun. This sentiment was echoed across the centuries, when Martin Luther said of Copernicus, "There is talk of a new astrologer (read astronomer) who wants to prove that the Earth moves and goes around instead of the sky, the sun, the moon, just as if somebody were moving in a carriage or a ship might hold that he was sitting still and at rest while the Earth and trees were moving. But that is what things are nowadays: when a man wishes to be clever he must invent something special, and the way he does it must needs be the best! The fool wants to turn the whole art of astronomy upside-down. However, as the Holy Scriptures tells us, so did Joshua bid the sun to stand still and not the earth."

This opinion symbolizes the controversy between, what is looked upon by many, as a religious dogma on one hand and scientific rationality on the other.

It is amusing to note that in criticizing Copernicus, Luther trips up. Any sensible person can see his criticism happens to agree with the very thing that Copernicus is trying to say. Luther in giving the example of what the observer sees from a moving carriage does not realize that he is concurring with him. An observer on the moving Earth would see the Sun moving, just like as observer in a carriage sees the ground and the trees moving. In fact, today children are given this very example, which explains why we see the Sun moving instead of the Earth, except, now the "carriage" is replaced by a "car" or "railway carriage". Luther was no fool, but he must have been beside himself when he came to know what Copernicus was trying to say. In spite of everything Luther and the others were saying, the "fool" did turn the science of astronomy upside down.

THE RISE OF GEOCENTRISM
THE HINDUS
The heliocentric theory never had any real hold in the west any time before Copernicus or even for another fifty years, after his book was published. In the east, in India the theory was put forward from time to time by various sages as we have seen earlier. The sporadic appearances of such ideas in the history of Indian astronomy were like flairs lighting up the darkness, only to go out again to usher in the blackness. These works were never accepted in a true scientific spirit; consequently, like flairs they did not have any lasting effect towards any concerted progress in the field of astronomical science. Like the Mayans who had missed out on the wheel, the Indians never really got to grips with science, because they were never comfortable with its practical aspect. Detailed observations, keeping records, devising experiments in order to obtain proof and thereby working towards a goal, was not widely practiced. Nor is it the forte of the people of the subcontinent even now. Thus the ancient Hindus took one-step and once they got what they wanted they looked no further. This was the result of two factors. One was the caste system, where artisans did things by rote. There was not much change from father to son, as society did not demand more from them than what tradition expected. With the caste system, one could not transcend his status in society in the present life, though according to their beliefs he or she may do so in another life. Second, their patrons were more engrossed in the arts and the metaphysical to give any serious thought to science. They therefore relegated such matters to the Brahmins. Unfortunately, caste does not dictate intellect. The truth was thus lost amongst the myths in the minds of people. It

is true that in India there were no serious criticisms about the heliocentric theory; but that may be explained by the fact that in the Hindu ethos the elders, especially sages, were outside the purview of any from of criticism. Their word was taken as gospel. However, there were a far greater number of other sages, who believed in the geocentric theory and for the similar reasons their ideas were also accepted. This resulted in a dualistic approach, which is not congenial to a scientific atmosphere.

THE ARABS

The Muslims, whose golden age around ninth century AD to the thirteenth century AD, held sway at a time when Europe had just emerged from the Dark Ages. Their empire stretched from Spain, around the southern Mediterranean and extending east as far as Sindh, now in Pakistan. They nursed the knowledge of Greece for the future of mankind, which they had acquired through Egypt and Byzantium. To this, they also added the wisdom from India through translations of Aryabhatta (476-550) and others texts. In this way, they came to know of both the geocentric and the heliocentric ideas. This knowledge they would pass on to Europeans through translations from Arabic. However, at best, their approach to the heliocentric theory can only be considered as lukewarm. Amongst their earlier astronomers, who came closest to the supporting the heliocentric hypothesis, al Biruni in his *Indica* 1030 AD was neutral. Abu Said Sinjari, a contemporary of his thought it might be possible that the Earth went around the Sun. While Quib al-Din Shirazi (1236 - 1311) remarked in the same vein, saying that the Earth moving around the Sun could be true. Abu Ubayd al-Juzjani (died c. 1070) was the first to point out discrepancies in the Ptolemy's system. This is would raise questions, which in time would contribute to the heliocentric theory. Ibn al-Shatir (1304-1375) proposed a system, which was only approximately geocentric. In the west, after Aristarchus, nothing is herd of the heliocentric theory except silence until the arrival of Copernicus.

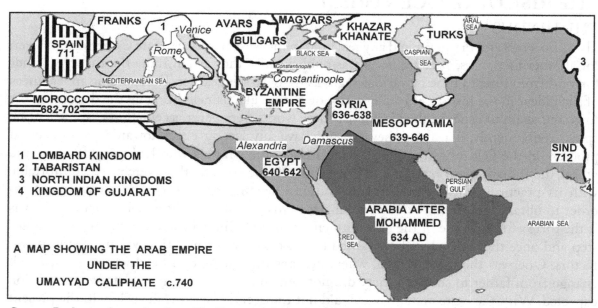

Source: Redrawn by author from The Penguin Atlas of Medieval History by Colin McEvedy. Publishers – Penguin Books. Reprint 1986. ISBN 0 14 051152

3.4 Arab Empire 737 AD.

THE GREEKS - EUDOXUS, ARISTOTLE, CLAUDIUS PTOLEMY AND THE GEOCENTRIC THEORY

Eudoxus and Aristotle were both students of Plato. Their teacher, Plato (427–347 BC), believed that the Earth was at the centre of the universe and that the Moon, Sun, Venus, Mercury Mars, Jupiter, Saturn and stars were each placed on a sphere in the order mentioned. They were moved around by the Three Fates and attended by the Sirens of Greek mythology. All these religious connotations proved too much for his pupil Eudoxus of Cnidus (400-347BC).

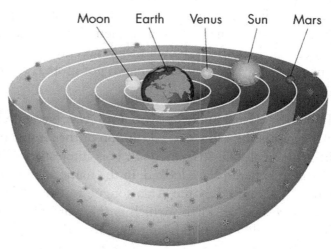

3.5 Website: http://www.redorbit.com/education/reference_library/universe/geocentric_model/204/index.html

Below the image on the above website is given the cryptic credit -

"Credit: Cutaway view of the geocentric model of the Solar System according to Eudoxus."

Eudoxus developed a more practical model based on mathematical principles, but did not digress too far from his teacher's model. He, however, went on to mention the movements all heavenly bodies could be explained by circular motion. This was in a way a step in the right direction. Nevertheless, the assumption that the heavenly bodies moved in circles was not correct. Even Copernicus after eighteen centuries would make the same mistake. It would take another after him to correct this error. Aristotle (384 –322 BC), on his part went on to add further metaphysical touches to Eudoxus' practical approach. He proposed that these celestial bodies mentioned were attached to more than one sphere each, thus making up fifty-six spheres in all! He also came to think that the Moon being the closest to the Earth was contaminated. That is why, he thought, the Moon, unlike the other celestial bodies had phases and dark blemishes on its surface. We can only conclude from all this that he certainly had been influenced by his teacher to have such thoughts.

By the time Cladius Ptolemacus (c. 150 AD), otherwise known as Ptolemy of Alexandria, came into the picture in 2nd century AD, the heliocentric theory had faded from peoples' minds and geocentric theory held sway. Ptolemy elaborated on the geocentric hypothesis espoused by many Greek philosophers, astronomers and mathematicians before him. He conceived of the Earth as immovable, situated at the centre of the then known universe with the Sun, Moon, planets and stars all moving around it. With the rise of Christianity, the Church brought many nations into her fold. They had to perform both religious as well as secular functions. One such was the need to give a calendar to the people. To do this they needed to predict the equinox, eclipses, and the position of the planets and know the solar and lunar cycles. They selected the Ptolemaic geocentric hypothesis, because it was the most complete theory available about the heavens at the period. This established Ptolemy's geocentric theory.

Even though the Ptolemaic theory was *not* simple with its unwieldy deferents, epicycles and equants along the added problem of complex calculations required to predict the paths of the various celestial bodies, it was still accepted. Simply, because it gave the best results at the time. The sheer magnitude of Ptolemy's work was imposing, but with this theory came the geocentric hypothesis, which meant accepting the Earth was at the centre of the universe. In buying this hypothesis, the Church got more than it had bargained.

THE ALMAGEST

Ptolemy's contribution to astronomy was his encyclopaedic work called the *Almagest*. The Almagest is the Latinized corruption of the Arabic name *al-kitabu-l-mijisti*, which translated literally means the "*Great Book*". It is interesting to note that though it is not mentioned in any place where I have looked; in the name *al-kitabu-l-mijisti* we can distinguish two familiar words, each gifted by the Arab tongue to two different languages. The first is "kitabu" that is closely related to "kitab", which is the word for "book" in Urdu. It is from here it has been incorporated in the Hindi language. The other word is "mijisti", which curiously sounds like the word "majesty" or "majestic" in English. There were other books written by Greek astronomers, but they were dwarfed in contrast to Ptolemy's monumental work. So they came to be known as the "Lesser Books" by the Arabs to distinguish them from Ptolemy's "*Great Book*".

The Almagest was translated from the original, which was written in Greek called $μαθηματικὴ$ $σύνταξις$ or *Mathematike Syntaxis* or *Mathematical Treatise*, later to be known as *Hè Megalè Syntaxis* or *The Great Treatise*.

The date of this treatise has been recently established from which it appears to have been completed sometime between 147 and 148 AD. Ptolemy had labored for over twenty years before this book was produced. Sadly, the original has been lost to us, but the Arab translations still survives. It was again later translated to Spanish, Latin and even back to Greek.

In his book Ptolemy gave order of the celestial bodies as the moon being closest to Earth, next in order of sequence came Mercury, Venus, Sun, Mars, Jupiter, Saturn and lastly the stars. This was somewhat different from Plato's idea. Ptolemy stated that the Earth was a sphere and lay immovable at the centre of the universe. The stars were placed on a celestial sphere, which moved around the Earth. He rightly guessed that it lay at so great a distance from the Earth that the Earth was to be treated as a point. Ptolemy's Almagest became a great success.

EXPLAINING THE PATH OF THE PLANETS THROUGH THE GEOCENTRIC MODEL

The geocentric theory had its own interpretation about the apparent looping movements of planets across the sky. Ptolemy assumed that each planet was moved by five spheres, but sometimes he had to use more to suit his theory. Like in the same way as theoretical physicists do today with regard to their theories on the early universe, where additional dimensions are taken into consideration to fit their theories. In both instances, whether it be additional spheres or additional dimensions, all depended on the mathematical requirements on which the theories are based. So these five spheres were all necessary to explain how the looping movements of the planets took place. Here we will only deal with the two spheres that are appropriate to our discussion.

> ## Contents of the almagest [4]
>
> 1. Book I contains an outline of Aristotelian cosmology, a set of chord tables, and an introduction to spherical trigonometry.
> 2. Book II covers problems associated with the daily motion attributed to the heavens, namely risings and settings of celestial objects, and the length of daylight.
> 3. Book III covers the motion of the Sun.
> 4. Books IV and V cover the motion of the Moon, lunar parallax, and the sizes and distances of the Sun and Moon relative to the Earth.
> 5. Book VI covers solar and lunar eclipses.
> 6. Books VII and VIII cover the motions of the fixed stars, including precession of the equinoxes. They also contain a star catalogue. The brightest stars were marked of the first magnitude ($m = 1$), while the faintest were of sixth magnitude ($m = 6$), the limit of human visual perception (without the aid of a telescope). Each grade of magnitude was considered to be twice the brightness of the following grade (a logarithmic scale). This system is believed to have originated with Hipparchus. The stellar positions too are of Hipparchan origin, despite Ptolemy's claim to the contrary.
> 7. Book IX addresses general issues associated with creating models for the five naked eye planets, as well as the motion of Mercury.
> 8. Book X covers the motions of Venus and Mars.
> 9. Book XI covers the motions of Jupiter and Saturn.
> 10. Book XII covers stations and retrogradations, which occur when planets appear to pause, then briefly reverse their motion against the background of the zodiac. Ptolemy understood these terms to apply to Mercury and Venus as well as the outer planets.
> 11. Book XIII covers motion in latitude (the deviation of planets from the ecliptic, the apparent path of the Sun through the stars).
>
> *Website: http://en.wikipedia.org/wiki/Almagest*

Box 3.1 The thirteen sections of the Almagest from a copy printed in 1515, which has 152 pages.

First, there was the *deferent*. This was a sphere, which was originally centered on the Earth. But for a long time, even before Plato, it was centered away from the Earth making it an eccentric sphere. In this way, calculations of the predictions of the position of the various celestial bodies were nearer to the actual observed values. Next was the sphere that allowed for the *epicycles* of the planet. This sphere was centered so that the surface of this sphere always remained just within the sphere of deferent. The planet moved within the sphere of the epicycle. It had the same sort of relation to the epicycle as the epicycle had with the deferent. It was with the epicycle the supporters of the geocentric theory could explain the apparent *retrograde motion* of the planets as seen from the Earth. That is to say slowing, stopping, turning back, again slowing, then stopping and finally going on.

Ptolemy's main contribution to the geocentric model was to fix the eccentricity of the deferent sphere. He stipulated that the centre of this sphere lay on a point, which was exactly midway between the Earth and another point called the *equant*. Ptolemy claimed that if one stood at the equant then the motion of a planet would appear to be uniform, something that would otherwise not be apparent from any other point. Whatever was the truth of the matter, the astronomers in accepting the position of equant, as Ptolemy had proposed, saw it gave them better results. After all this is what theories are about. Even though his calculations were rather intricate, the Ptolemaic view of the physical world was accepted by the Church, as it gave many right answers to important practical problems in astronomy.

Ptolemy's Almagest was so successful that many of the older texts became redundant and soon fell into disuse. The work was so complete that much of our knowledge of Greek astronomy and Greek astronomers has been handed down to us through the details given in the Almagest.

Through his book, we have come to know how Ptolemy had inherited the idea of the geocentrism from earlier Greeks astronomers. Apollonius of Perga (c.262 - c.190BC), had used the concept of eccentric deferents and epicycles. We also learn Hipparchus (c.180BC) was aware that Babylonian astronomy gave accurate predictions. He worked out the mathematical model for the movement of the Moon and the Sun

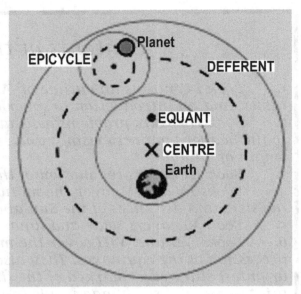

Credit: Courtesy Fastfission.
Website: http://en.wikipedia.org/wiki/File:Ptolemaic_elements.svg

3.6 The epicycle deferents and equant, which made up the basic elements of movement of planets in Ptolemaic astronomy. [Public Domain]

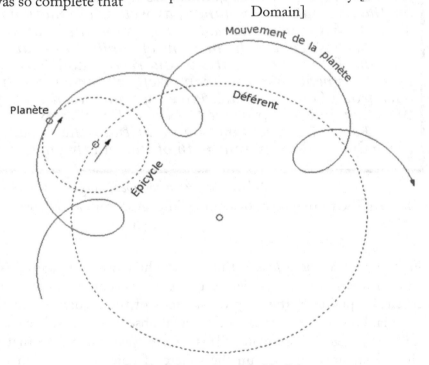

Credit: Dhenry / modified by Julo.
Website: http://commons.wikimedia.org/wiki/File:Epicycle_et_deferent.png
3.7 How the epicycle works. [CC-BY-SA-1.0]

to a high accuracy, but he failed to obtain similar levels of accuracy in the case of the planets. In adding his own concept of the equant Ptolemy succeeded where Hipparchus had failed. The *Almagest* was essentially a book on mathematical astronomy with conceptual models to explain the observed motions of planets. It may thus be compared to a work on theoretical physics. Ptolemy wrote another book the *Planetary Hypothesis*. It was complementary to the *Almagest* and explained how his geometrical models could be transformed to what he thought to be the reality of the cosmos.

PTOLEMY'S INFLUENCE

The importance of the *Almagest* before Copernicus cannot be overemphasized, because of its details and accuracy. It is because of this the book had been translated at various periods. At one time, it was translated by the order of Caliph Al-Ma'mun (813 - 833), who we have met in an earlier chapter. A Latin translation was again made from Spanish by instructions of Frederick II (1194 – 1250) the Holy Roman Emperor. He was an enlightened person. His life was of enough interest for us to pause for a moment and know a little more about him without having considered our time as wasted. Frederick II knew six languages and was a man of education. He was conversant with philosophy, mathematics, natural history, medicine and architecture. He kept a collection of animals and wrote a book on falconry in which he recorded with remarkable accuracy the habits of birds. He also kept a harem, which was secluded and guarded by eunuchs in the eastern fashion. Many scholars were attracted to his court. He was no religious bigot as one can imagine from his credentials and all creeds were welcome to his court. Some years after his death no less a person than Alighieri Dante (1265-1321), the greatest of the Italian poets, mentioned that it was in Frederick's court that Italian poetry was born. However, his fame rests as a lawgiver. It is said, he gave Sicily "the fullest and most adequate body of legislation ever promulgated by any western ruler since Charlemagne". In 1248 AD towards the end of his life he was involved in a protracted seize. While he was away hunting, the besiegers in an unexpected foray came out and attacked his camp. The result was an unmitigated disaster for Frederick's soldiers. His army routed, the imperial insignia taken, his treasury plundered and his harem captured along with a few of his trusted ministers. Some say, the imperial crown of the Holy Roman Emperor was then placed on the brow of a kyphotic dwarf by the victors, who took him back to town on the shoulders of a cheering crowd.[17]

Another Latin translation of the Almagest was by Gerard of Cremona who had come upon the book in Toledo in Spain. There were many other translations, including one at the behest of Pope Nicholas V (1328AD-1330AD) which again indicated the importance of the Almagest in the lives of the people of that time. In the 15th century at the instance of Cardinal Basilios Bessarion, Regiomontanus (1436-1476), whose real name was Johannes Müller, completed a shorter version of the Almagest, which was begun by Peurbach.

It is of interest to know that apart from their scientific pursuits, the astronomers of the period used star tables for astrological work as well. It was a side business for them. They cast birth charts of people who were interested in their destiny. This usually meant the rich and powerful. Their patrons included kings, princes, nobles, the gentry and later the rising, but powerful, merchant class. The contrast between the lofty ideals of astronomy and the pursuit of the occult art of astrology, which many astronomers of the day did not believe in, could not be reconciled by some. Kepler was aware of this contradiction. His approach was however a philosophical one. Although Kepler himself had no particular interest in casting horoscopes for the wealthy, he believed that it supplied the means to provide a living for needy astronomers. One of the results of such work was *the Rudolphine Tablets*. They were essentially tables that gave the positions of stars, planets, Sun and the Moon over a period of time as seen above a certain area. The star tables had other uses. They had been the means of finding the position of the ships at sea and had been in use in earlier centuries by Greek astronomers as well. These tables however proved to be too complicated to be used by unlettered sailors of the time, as we have observed before.

All was going very well with the geocentric theory until Copernicus came and upset the apple cart. He proposed the heliocentric theory in 1543.

UNDERSTANDING THE CONTROVERSY IN ITS PROPER PERSPECTIVE

When Copernicus proposed the heliocentric theory he realized, like Darwin three centuries after him, it would create a storm. Nevertheless for about fifty years after publication of his book, even though every one came to know about the heliocentric theory, things lay more or less quite. This was because he presented his theory not as a challenge to the geocentric theory, but as an alternative to Ptolemy's method. It was only when things came out into the open with Galileo and others proclaiming it loudly that the Church reacted. It was a panic response, just like an insecure person overreacts when he or she feels threatened. Now the Church was pitted against the intellectuals of the period and they had to respond aggressively since reason was not on their side. Bruno was burnt at the stake and Galileo was put under house arrest.

The reason why the Church behaved in this brutal fashion was that their power lay in the total belief the people had in the Scriptures and the Scriptures implied that the Earth lay unmoving at the centre of everything. If the Church accepted the heliocentric theory, it would be contrary to what the Bible said. Once a belief was questioned and proved wrong, then people would start questioning other beliefs in the scriptures as well. This could not be allowed to happen. The very foundation on which the Church's influence rested would be undermined forever. Those in the Church who held power knew that too much was at stake for them to give in so easily. This was the politics behind the controversy.

The average person in the streets or in the fields, however, was not concerned about the politics and arguments that were going on amongst the intellectuals of the day. To them it was pointless to know whether the Earth was going around the Sun or if the Sun was going around the Earth. As long as people had a calendar, they were satisfied. They need this to conduct their daily lives. The farmer needed to know when to sow seeds. The servant had to know the day he would get his wages. The artisan had to know when to deliver his product. The merchants had to know the dates and tell the stipulated times to complete their transactions. Those who borrowed had to know the dates when their loans would fall due. In their turn, the financiers had to calculate the interest thereon along with their dates of repayments. The governments had to know the dates when taxes had to be collected. The people had to know the saint days; and separate the days of prayer from the days of work. At that time, it was the Church's responsibility of giving the people a calendar and seeing to it that it was correct. Thus for secular work a solar calendar was used.

The Church was also responsible for seeing that the various festivals were predicted correctly. An important part of the process was to predict the time for Easter, which was dependant on the lunar and not the solar calendar. Many of the other festivals and saint days depended upon the lunar calendar too. Thus, a luni-solar calendar was used and is still followed to this day. They got the necessary answers by using Ptolemaic method. When Copernicus introduced his hypothesis, even though it had made the calculations much simpler to perform by taking the planetary orbits as circles, but his predictions of planetary positions calculated in this way were still not wholly correct. The reasons for this will become clear later.

THE CAUSES OF GROWING RESENTMENT TOWARDS THE CHURCH

At this time in the 16th century, Europe became involved in a great controversy, which arose out of corruption in the Church. This state of affairs went back a long way.

Celibacy amongst the sex-starved churchmen was the exception rather than the rule. It extended from the Popes right down to the friars. The general population came to be well aware of their clandestine sex lives, because the objects of such activities came from their own families.

Then there was the practice of sale of *indulgences*. These were documents sanctioned by the Pope that guaranteed "the remission before God of the temporal (as opposed to spiritual) punishment still due after the spiritual aspect of the guilt had been forgiven". The concept behind this was that the spiritual aspect of the sin committed by a person could be absolved by the infinite merit of what was good in Christ and from the surplus of good, which the saints had done during their lives. The temporal matter however remained. The restitution for this had to be made in a material way. The person had to go through a process of discipline, which meant penitence under the direction of a confessor. Only then, the sinner could be absolved for his or her sin. Otherwise, he or she had to face purgatory after death.

Such an idea in itself was sound, because even if the person truly repented his or her sin from the moral point of view, still their sins could not be forgiven from the material point. The question of the material consequences of their actions thus needed consideration. Unfortunately, the Popes deviated from this principle when in lieu of temporal punishment the Church took to the practice of taking money by the sale of indulgence by papal agents or *quaestores*. This allowed the rich to get away with murder, in the literal sense, by paying money, whereas the poor had to suffer under similar circumstances.

The money thus collected by the Church was used initially to make hospitals, universities and buildings of worship, but later this money found its way to the pockets of greedy churchmen. The sale of indulgence was stopped after the Council of Trent, but it was too late by then the system had already become corrupt. The things had come to such a pass that the average conduct of the laypeople was better than those who were supposed to set an example to them.

Again, there were the Church holidays. During these "holy days", the milder forms of amusement had gradually deteriorated into revelry, carousing and wanton behaviour. The Church had been blind to all such riotous behaviour. The sensible saw that this was no better than what the pagans had been doing before.

Many distinguished churchmen going back to St. Bernard in the early 12th century and the nine great world-councils that were convened from the 13th to the 16th century reiterated the need to rectify matters. Finally, in the Council of Trent in 1545, even though the practice of indulgence was abolished, but everyone agreed to the helplessness of dealing with the other problems.

Things had been brewing for many years. There was little religious in Popes. By meddling in politics openly, they behaved just like politicians. The Popes competed with various kings, princes and dukes of the period to project their grandeur and wealth. All this did not go unnoticed by the intellectuals of the day. Men like John Wycliffe (1320–1384) and Jan Hus (1369–1415) had questioned the Church's policies and its ways. From time to time people rebelled. The Church branded them as heretics. Instead of hearing what people like them had to say, many were subjected to the inquisition.

All came to a head when on the last day of October in the year 1517 Martin Luther (1483–1546) had nailed his 95 thesis on the power of indulgences on the door of the church Wittenburg Castle Church. Whole of Europe was soon divided right down the middle, from kings and their courts; bishops and friars; in places of learning; and even in families between husband and wife, parents and children. In all this confusion, due to the religious divide, the science that lay behind the controversy between the heliocentric and the geocentric theories was lost in the feelings that ran high amongst the people. Thus apart from the interests of the true men of science, the merits of the geocentric theory verses heliocentric theory would ultimately be reduced to a question of Catholics verses Protestants in the minds of people.

THE SCIENTIFIC DETAILS OF THE CONTROVERSY
To understand the technical aspects of this controversy we must go back to the days before the telescope. It is only when we look at details of the geocentric theory through the ideas of the time and by comparing them with our point of view today; we will know what contributed to the controversy. With this, we may come to understand why they rejected the heliocentric interpretation.

The victors, as always, are derogatory toward the vanquished. When Copernicus showed that the Earth went around Sun, the refrain was taken up by many. By doing so, they wanted to show they were rational and supporters of science. Even today, there are those like them who look upon it as one more victory over religion. The truth is that many of us today do not understand this controversy at all! The modern supporters, such as these, are not even aware of the historical perspective of the times in which these ideas were fought out; sometimes so viciously, as to raise our sentiments to a pitch that echo even to this day. To appreciate the conflict between these two ideas, we must know more about the points on which each theory rests, before we start taking sides or form any opinion about the matter. Then only we will come to know why and how it all came about. It should not appear that with the advantage of hindsight we condemn the geocentric theory just because it is the fashion to do so today.

SEEING IS BELIEVING
A person looking up at the skies from the northern latitudes sees the stars above the equator move quicker across the heavens than the stars that lie further north. The movement of the stars would appear to slow down more and more until his eyes settled on a particular star,

which appears not to move at all. This was the Pole Star, which lies almost directly above the North Pole. The reason is the Earth is a sphere and as it turns on its axis the points projected on the celestial sphere appears to turn quicker at the equator.

We know this happens because of apparent motion. Apparent motion is an illusion, so there is no way of knowing whether the Earth is spinning on its axis or the celestial sphere that is moving? Today we all know for a fact the Earth spins around its axis. If we did not have this proof of the Earth's rotation and took the observation at its face value only, we might equally argue the celestial sphere is moving, not the Earth. Even in those early days, people were aware of relative motion. The evidence was thus inconclusive either way. In the past even if one thought the Earth was turning on its axis, one could not prove it in any way. Thus the concept that the Earth was moving would be an assumption only. Most thought the Earth was too big and therefore it could not be moved. Each camp thought their own theory to be correct and each believed in their idea deeply.

Again, the majority continued to believe, the celestial bodies were smaller than the Earth. They knew their geometry well and were well aware of the parallax phenomenon, but they had no proper way of measuring distances without telescopes. Therefore, they could not calculate distances of the Moon or planets, let alone that of the stars or deduce that the Earth revolves around the Sun. This would only become possible with powerful telescopes in the 19th century.

THE INERTIA FACTOR

The misunderstanding that the Earth could not move arose out of a mistake in the idea of inertia amongst the ancients. In the past, they reasoned that if a heavy object could not be moved it was because of inertia. At the time, they made no distinction between resistance due to inertia and resistance due to friction. So the term inertia was not used in the same sense as we do today. To the ancients for all practical purposes the heavier the body the more inertia it would have and the less chance it could be moved. A big rock could only be moved with great effort, but a hill or mountain could not be moved at all. Thus to the ancients' way of thought, since the Earth was very much more massive than a mountain, it was good reason for the Earth not to move. To the supporters of the geocentric hypothesis, because of this false idea about inertia the Earth could not move, the idea of motion, whether relative or any other, was out of question. Thus any argument, which involved the Earth's motion whether on its axis or around the Sun failed to catch on! It was only later with Galileo and Newton people came to understand the true nature of inertia. So in the mean time things had to rest.

PREDICTABILITY

Copernicus' theory approximated to reality better, but it had its flaws. Even though it was simple, as all scientific theories should be, it still could not predict the exact positions of the planets as borne out by actual observations. Predictability is the hallmark of any scientific theory. The reasons for this would become clear with Kepler's work. So in a way Copernicus' efforts did not prove to be much superior to Ptolemy's method, which had been giving almost equally good results except for the more complex calculations.

SOME OTHER VERY SUBTLE CLUES ABOUT THE EARTH'S POSITION

There were some observations, which would have taken the ancients forward in the right direction. Some went ignored by them, while in others they were constrained due to lack of appropriate instruments, such as telescopes, clocks and accurate measuring devices for angles.

EARTH CENTERED SKY VERSUS A SUN CENTERED SKY

If one looks closely at any constellation that arose on the horizon at a particular time on a particular night and if he or she continued to do so for a few consecutive nights, they would notice that it would rise on the horizon a little later on each subsequent night. Today we know the exact time of this delay to be 3 minutes and 56 seconds. This is easy to detect, because we have accurate clocks today. This observation therefore creates a discrepancy in the period of the Earth's revolution. This should not happen. Each time the Earth rotates taking one full turn on its axis, the star seen rising on the horizon should do so at the same time on each consecutive night. The actual explanation for this phenomenon as to why this does not happen is because the centre of the observed celestial sphere is *not centered on the Earth* as the geocentric theory proposes, but is *centered on the Sun*. If it were possible to view the celestial sphere with the Earth in place of the Sun, it would prove the geocentric theory wrong. This experiment is however not possible!

Today we can do this with the help of space probes with cameras that would be locked in a polar orbit with respect to the Sun or by computer models. In those days, there were no such devices. So, in the past one might argue, the rate of movement of the celestial sphere did not do a complete full circle each day, but a little less. However, they were not so simple minded as all that. They compensated for the eccentricity of observation by the use of an eccentric deferent and the introduction of the equant. This for the supporters of the geocentric hypothesis would account for the experience that I have described.

THE PARALLAX PHENOMENON

Then there is the problem of retrograde motion of the planets that we have mentioned earlier. We now know this to be a due to parallax, which is the result of different rates of motion of the planets as they go around the Sun as viewed from the Earth. Even though this was the correct explanation, the phenomenon of the apparent retrograde motion of the planets could also be interpreted as evidence for epicycles by those who supported the geocentric theory. Moreover by the use of epicycles the planetary positions could be predicted with equal accuracy in those times. This was another point against adopting the accepted heliocentric theory.

Parallax implies the difference in the

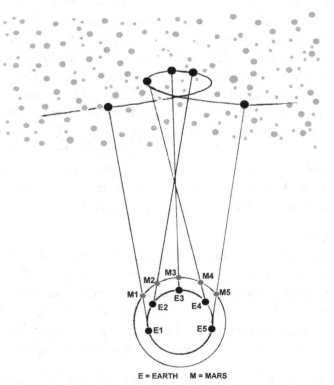

3.8 Retrograde motion of planets – the parallax effect.

apparent position of an object when seen against a background or foreground, which is caused by a change in position of the point from where it is viewed. A reference point in the background or foreground is required to detect this effect. Also the angle subtended at any point, on the object in question, seen by imaginary straight lines from any two points of observations must be appreciable to detect this parallax effect.

The easiest way to demonstrate parallax is to look at any object in the room and quickly blink the eyes alternately. In this case, each eye acts as two separate points of observation. You will see the object selected for the demonstration changes position against the background each time you do so. It brings to the mind stereoscopic vision. Yes, this is same the principle on which our stereoscopic vision is based.

The phenomenon of parallax is applicable in another situation in astronomy. It is used to measure distances of celestial bodies and to judge their motion.

If you want to know the distance of a body lying at a very great distance, like say a star, then it must be subject to detectable parallax from the Earth. Only the Moon is close enough to be measured by the parallax method using the unaided eye. As long back as 150 BC, Hipparchus determined the lunar parallax as 58 minutes arc, which made the ratio of its distance from the Earth about 59 times the Earth's radius. This comes to 236,000 miles or 377,600 kilometers, which was nearly correct. Even though there may have been many errors in his observations, they probably cancelled out. Nevertheless, it was a remarkable achievement, because the modern value is a little over about 60 Earth radii. Any object that is further away, such as the stars, the angle of parallax becomes too small to be determined from the Earth with the naked eye. The only conclusion that drawn from naked eye observations from the Earth is that the distances of the stars are much further from the Moon, the planets and the Sun.

The distances of planets are such that they can be measured by increasing the base of the triangle whose two ends form observational points, such as two different points on the Earth, which are hundreds or even thousands of miles apart. Distant stars or galaxies, however, are so great that they can only be measured from two diametrically opposite points on the Earth's orbit around the Sun.

Hipparchus' work showed that the ancients were well aware of the parallax factor. However, without powerful telescopes they were unable to put this knowledge to wider use. It was only in the eighteenth century with the aid of powerful telescopes that William Fredrick Herschel (1738-1822) would make people aware of the possibility of worlds beyond our galaxy. He

Drawing by Tarun Ghosh
3.9 Explaining the parallax phenomenon.

started work to show through the phenomenon of parallax that the Andromeda Galaxy was a separate galaxy. His work however remained incomplete. The German astronomer, Friedrich Wilhelm Bessel (1784-1846), in 1838 was the first to measure distance of a star. He measured the distance of the star 61 Cygni as 11 light years from us. So when the ancients failed to notice any parallax through naked eye observation of the stars, they assumed that the Earth must to be stationary.

EFFECT OF THE INVERSE LAW

Lastly, there was the small problem of varying brightness of the planets when seen from the Earth. This again was due to the fact the distances of the planets from the Earth varied as they moved around the sun. According to the *inverse square law*, the amount of light would fall by the square of the distance from the source. If a planet receded from the Earth, it would appear less bright. On the other hand, when it came closer it would look brighter. This phenomenon can be observed with the naked eye sometimes as in the instance of Mars and Jupiter. I do not mention Venus as its changing brightness with its phases could be confusing to the naked eye observer.

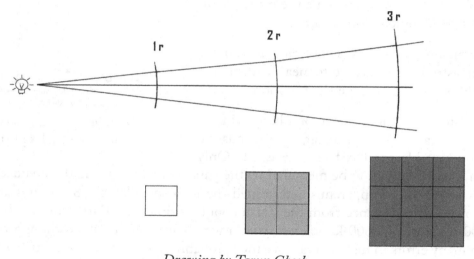

Drawing by Tarun Ghosh
3.10 The inverse square law.

We should remember in the pre-telescopic era people did not know that the planets shone due to reflected light from the Sun. At the time, the whole point about the varying brightness seemed to have been ignored by the supporters of both theories. However, as with the other points discussed, this phenomenon could be explained by the epicycles as well, but in the latter event, it would not tally with observational data. The actual brightness of the planets when seen from the Earth during their orbital revolution would vary much more than if was a result of epicyclical movements. Since in the latter, the distance from the Earth would vary much less, therefore the variation in brightness would also be much less than what was actually observed. However, at this stage as the inverse square law, developed by Newton, on which this observation rests was not known. Therefore, this point was overlooked.

These then were the pros and cons of the arguments for and against the geocentric theory and the heliocentric theory before Copernicus stepped into the picture.

A CARDINAL SUMS UP THE CASE
Cardinal Robert Bellarmine (1542-1621) declared that the exposition on the heliocentric theory made "excellent good sense" on mathematical grounds, because of its simplicity. He thought that after all it was, but a hypothesis only and went on to say, "If there were a real proof that the sun is at the centre of the universe, that the Earth is in the third sphere, and that the sun does not go round the Earth but the Earth round the sun, then we would have to proceed with great circumspection in explaining passages of Scripture which appear to teach the contrary, and we would have to say that we did not understand them than declare an opinion false which had been true. But I do not think there is any such proof since none has been shown to me."[18]

This shows that the Church did appreciate the progress that had been made, but for them to commit them selves to any new theory was not possible for reasons discussed before. In the background of all this there was politics. It was this, which tended to mask the issue in the minds of people.

This sums up the state of affairs in Galileo's time.

4 THE TRIUMPH OF THE HELIOCENTRIC THEORY

"Of all discoveries and opinions, none may have exerted a greater effect on the human spirit than the doctrine of Copernicus. The world had scarcely become known as round and complete in itself when it was asked to waive the tremendous privilege of being the center of the universe. Never, perhaps, was a greater demand made on mankind - for by this admission so many things vanished in mist and smoke! What became of our Eden, our world of innocence, piety and poetry; the testimony of the senses; the conviction of a poetic - religious faith? No wonder his contemporaries did not wish to let all this go and offered every possible resistance to a doctrine which in its converts authorized and demanded a freedom of view and greatness of thought so far unknown, indeed not even dreamed of."

Johann Wolfgang von Goethe
(1749AD to 1832AD)

THE ARCHITECTS OF THE HELIOCENTRIC THEORY

There were three people, who each in his own way contributed to the knowledge of the Solar system. This put us on the road to modern science. Copernicus was able to establish the Earth and the other planets revolved around the Sun. It brought us out from the fallacy of thinking the Earth as the centre of the universe. Nevertheless, his idea of circular orbits did not explain the discrepancies relating to the predicted and the observed positions of the planets. It was Johannes Kepler, who correctly showed the planets went around the Sun in elliptical paths. However, he could never have succeeded without help of the huge data accumulated by Tycho Brahe through meticulous observations carried out over many years.

4.1 A map of the places where Copernicus, Tycho Brahe and Kepler lived and worked.

Today the average man in the security of a relatively stable society, living in centrally heated or in air conditioned houses and having excellent means of communication, can never know the experiences of these pioneers of science. They had to struggle under adverse conditions of weather, Spartan living and working conditions, prey to diseases and caught up in socio-politico-religious upheavals of their times; including war. Also, the prejudices of their times lay heavy on their backs. In those days, it was not a case of someone making a discovery and announcing it to the world, as we would do today. Instead of applauses, one might be burnt at the stake for one's efforts. I have therefore detailed their stories, so that we may have some understanding of the struggle they had to face and the circumstances they had to go through to achieve what they did. It was people like them, who gave their lives to science, so that we may reap the benefits we enjoy today. This is their story.

NICOLAS COPERNICUS

Mikolaj Kopernik or Nicolas Copernicus (1473-1543) was born on 14th February 1473 at Torun in Poland. It is probable that his ancestors came from a Silesian village named Kopernic or Koperniki near Nysa. The term "kopernik" in Polish implies one who works with copper. Even though his ancestors, possibly, may have been coppersmiths, his father now belonged to the merchant class. Trading in copper, he had become a man of some substance and standing in his community. He became involved in struggle for independence by the Pomeranian cities, against the Teutonic Order.

The latter were a German Catholic military cum holy order, created during the third Crusade. Their aim was to care for the injured crusaders and aid pilgrims who went to the Holy Land. They ranked as knights, who fought in the Holy War and so came to be known as Teutonic Knights. In time, they turned materialistic and managed to incorporate lands, including Pomerania into their monastic domain.

Copernicus' mother was well connected also. Her brother, Lucas Waczenrode the Younger, would later become the Prince-Bishop of Ermeland (Warmia). This implied his see or rather his domain was almost an independent state under the Polish Crown with its own army, currency, treasury and Diet. The word "Diet" is derived from Latin word "dieta" or assembly or the day's work: In Germany it meant an assembly of dignitaries or delegates to debate or to decide on important questions.

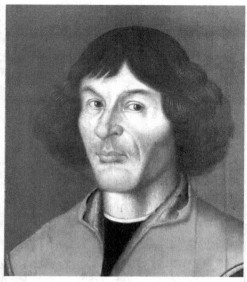

Website: http://en.wikipedia.org/wiki/File:Nikolaus_Kopernikus.jpg

4.2 Nicolaus Copernicus portrait from Town Hall in Thorn/Toruń, Poland – 1580. (Artist unknown.)[Public Domain]

At eighteen, Nicolas went to the Krakow Academy, which is now the Jagiellonian University at Kraków. There he came under the influence of Wojciech Brudzewaski, who was one of his teachers. Like most people of the time Brudzewaski, believed in the geocentric hypothesis, but it was through him Copernicus developed his taste for astronomy.

By the time, Copernicus returned home after three years at the Krakowac Academy, his maternal uncle was already established as Prince Bishop. He offered Copernicus' the position as canon of Frauenburg (Frombrok). Copernicus however had other ideas. He politely declined and took his uncle's consent to study law at Bologna in Italy. Here amongst other subjects he studied Greek. He thus came to know about the Greek philosophers and their teachings. It was also here he met Domenico Maria Novarra da Ferrara, a well-known astronomer. da Ferrara noted his interest in astronomy and encouraged him to observe the heavens. Copernicus took his advice sincerely. It is recorded, he observed the occultation of the star, Aldabran, on 9th March 1497 for the first time. He was only twenty-four. The same year he was elected Canon of Frauenburg by proxy.

In 1500, at the age of twenty-seven, he was invited to Rome and lecture on mathematics. This indicates his rising prominence as a mathematician. The following year he came back and took up his post as canon at Frauenburg Cathedral, but he soon left for Italy again. This time he went to join the University of Padua to study medicine. By the age of thirty-one, he had also become a Doctor of Canon Law. His renown as a mathematician, his knowledge in theology, astronomy and medicine and his connections through his uncle allowed him to move in the higher echelons of society.

RETURN TO POLAND

He finally returned home in 1503 and worked as his uncle's advisor till the year 1512, when his uncle died. He was asked to attend the fifth Lateran Council in the year 1514 and to give his opinion on calendar reform. There he declined to give his views, as he felt that the position of the Sun and Moon in the solar system had not been adequately established, so it was not possible to give an opinion on the matter. Sensibly, he did not go on to elaborate his ideas about heliocentrism. In fact, at the time these ideas had been consolidating in his mind. In 1516, he went to Olsztyn Castle, where he would stay until 1521 as economic administrator of Warmia, Olsztyn (Allenstein) and Pieniezno (Mehlsack). During the Polish Teutonic War of 1519-1521, he was in-charge of the defenses of Olsztyn and Warmia. When the conflict ended, he was one amongst those who were involved in the peace negotiations. Copernicus also advised Sigismund I (1506-1548) King of Poland on monetary reforms. He also took part in the East Prussian Diet on coin reform. Here he was one of those who solved the delicate question as to who had the authority to mint coins. At the time, many people who did not have the authority, but possessed some sort of power, were minting debased coins. This led to people to hoard the genuine coins and pay their taxes with those debased coins. As would be expected in such a situation, chaos ruled in the monetary system of the day. Copernicus sensibly advised that the debased coins should be withdrawn and instead genuine coins be issued. His advice was not heeded due to lack of bullion. In 1526, Copernicus wrote *Monetae Cudendae Ratio*, where he stated that bad coins push out the good ones from the market, but he had been thinking about the problem for much longer. He had already written a draft on the subject, De aestimatione moneta (On the Value of Coin). He anticipated Gresham's Law by nearly 40 years, when Sir Thomas Gresham (1519-1579), a financier of the Tudor period was to point this out in a letter to Elizabeth in 1558. However, they were not the only ones, Nicole Oresme (c. 1323-1382), one of the most original thinkers of his time had noted the issue. Also, as far back as 5th century BC, Aristophanes (448 BC -388 BC) mentions the problem it in his play, *The Frogs*.

THE REVOLUTIONIBUS ORBUTUM COELESTIUM

By 1504 Copernicus' thoughts about the solar system were slowly taking a shape in his mind. Some have attributed his ideas to his earlier knowledge of the Greek philosophers and also to Arab works, which he may have accessed subsequently. This may well be true, because as we have mentioned earlier Copernicus was already aware of the heliocentric hypothesis. He may also have come to know about the precession of the equinoxes through references about Hipparchus' pioneering work nearly seventeen centuries earlier, which he was to mention in his work. Again his book had pictures that had remarkable similarities to drawings by the Persian polymath and astronomer Nasir al-Din al-Tusi (1201–1274).[19] Tusi had been the first to have questioned the authenticity of Ptolemy's equant and offered an alternative to it. So, it is quite probable that Copernicus had been influenced by all this earlier knowledge, but his conviction certainly came from his own labours.

During all this time he had continued with his astronomical observations, which like Tycho Brahe he pursued far into the night, but unlike Tycho he seems to have done his work entirely on his own. All this he did after carrying his duties, as demanded by his official position during the day. He gradually came to a point, when he became dissatisfied with Ptolemy's way of calculations and his system; especially with the concept of equant.

In the course of his experience he found that if he took the path of the planets, including the

Earth, was taken to be going around the Sun in circular orbits, his calculations would agree more closely with their observed positions. He thus came to concluded it would be a mistake to assume the Earth was at the centre of the solar system. Moreover, while doing calculations, by taking the Sun to be at the centre, he found he could dispense with the cumbersome concepts of deferents, epicycles and equants, which were an integral part of the Ptolemaic system. Copernicus wrote *De Revolutionibus Orbium Coelestium* (On the Revolutions of Celestial Orbs). No one knows when he started on his book, it may have been by 1517, but it is certain he had finished by 1530.

In his book, Copernicus proposed the Earth and the planets moved around in circular paths with the Sun at the centre. He placed the planets in the right order, with their periods in ascending order as one moved away from the Sun. Here he observes that the Earth rotated on its axis and this caused the apparent motions of the Sun and stars. He also noted the Earth's axis precessed. This meant its axis wobbles in a retrograde fashion like a top, which is losing its speed. Today we know this to be true. It is a slow process, which takes about 26,000 years to complete each cycle.

Copernicus however retained some of the features of Greek cosmology. He still believed that the stars were embedded in a celestial sphere formed the outer limits of the universe. In spite of such misconceptions Copernicus' heliocentric theory explained many of the observed phenomena, like retrograde motion of planets, occurrence of day and night and the yearly motion of the Sun and the stars through the heavens. The truth was now out of the bag! Who would bell the cat?

Though Copernicus had written a summary of his ideas in the *Commenta Riolus* (Little Commentary) as early as 1514 and had shown this manuscript to some of his friends, he did not however publish it. At the time, he was still working on the more detailed version. Later, in 1530, Johann Albrecht Wimanstadt presented the shorter version of Copernicus' hypothesis, the Little Commentary, to Pope Clement VII. This discourse appeared to have been appreciated by the Pope. Nevertheless, Copernicus continued to delay the publication of his main work, the *De Revolutionibus Orbium Coelestium* (On the Revolutions of Celestial Orbs). Being a churchman, he was well aware of the possible consequences it could have on both him and his work. His hesitation and how it was subsequently published, reminds us of Darwin's predicament some three centuries later. In the mean time, people came to know about his work and many became interested.

It was not until 1539 when one such person, Georg Joachim Rhaticus (1514–1576), a 25-year-old German mathematician, arrived at his door that things would change. Rhaticus had initially intended to stay for only a couple of weeks; instead, he ended by becoming his pupil and remained as Copernicus' houseguest for a couple of years.

It was Rhaticus, who would bell the cat for him. After reading Copernicus' work, Rhaticus wrote a summary, *Narritio Prima* (Primary Narration). He then sent it to Johannes Schöner (1477-1547), a German polymath, who had been his teacher. When Copernicus found it had not received any adverse criticism, he agreed to publish the complete work. Rhaticus took Copernicus' manuscript and attempted to publish it at Nuremberg, but met with vehement opposition from Martin Luther. He then went to Leipzig. There he met Andreas Osiander

(1498-1552), a German Lutheran theologian, who agreed to oversee the publication of the book. It was Andreas Osiander, who would take the edge out of any criticism. Otherwise, we may not have known about Copernicus' work at all.

When Andreas Osiander read Copernicus' work, he realized the implication of what he was about to do and what would follow if the book was published. He had the wisdom to add a few words in the beginning. In essence, it implied what was written in the book about the Sun being at the centre of the universe need not be taken as the truth. Things were only *assumed* to be so, in order to simplify calculations. He went on to reiterate, in no way did the author suggest the Earth was not at the centre of the universe. He left it unsigned to make the readers think Copernicus had written it. This was a clever stroke indeed! However, in spite of this comment, it was plain to those who read the book that Copernicus was suggesting the heliocentric system as the actual model. Now with Osiander's amendment the book was finally published. It is said Copernicus received a copy of his book on his deathbed. We would all like to believe the story, but it should probably be best left to the realms of myth.

Copernicus' work however was not perfect, as we have mentioned before. If one looked for a planet, using Copernicus' methods of calculation, it could still not be located in the exact position in the sky, as predicted. This was because according to his theory the orbits of the planets were assumed as circular, which was not the case. It would be left to Kepler's genius to work out their true paths. This however did not diminish the merit of Copernicus' work in any way, because it was he, who for the first time in the history of mankind took the first step to show through accurate observations and records over years, the Sun and not the Earth was at the centre. In doing so, he abolished the idea of the Earth being at the centre of man's universe forever.

TYCHO BRAHE

The next player in this act was Tyge Ottesen Brahe (1546 – 1601). His name Tyge had been Latinized to Tycho, by which he is familiar to us this day. He was born on 14th December, 1546 in Skane, now in present Sweden, but was then a part of Denmark. He was the eldest son of Otto Brahe and Beatte Bille. Both families were rich and prominent members of Denmark's nobility. When Tycho was about one year old, while his mother was giving birth to another son, his paternal uncle, Jörgen Brahe was said to have kidnapped him. This was a fitting beginning to the colourful life of one of the most vivacious of characters in the history astronomy, if not of all science. The incident, however, did not appear to have disturbed the parents. Jörgen Brahe adopted him, as he had no children of his own and brought him up to make him his heir. At that, time for a child of Tycho's standing the only future for him lay in becoming a member of the king's council. It was thus imperative that he should be versed in law. So

Website: http://en.wikipedia.org/wiki/File:Tycho_Brahe.JPG

4.3 Tycho Brahe by Eduard Ender. [Public Domain]

with this in mind his uncle arranged for him to learn Latin at home from the early age of six. At the age of twelve, he was sent to the University of Copenhagen to learn philosophy and the art of rhetoric. In the year 1562, he went to study Law at Leipzig, accompanied by his instructor.

This was not to be, the fates had other plans for Tycho! At the age of fourteen, he purchased a book on astronomy. A new love was born, which would last for the rest of his life. The book can be viewed at the Kronborg Library today. With his interest in astronomy now growing, two years later, he took up the challenge of studying the passage of Mars. Later that year he observed the conjunction of Jupiter and Saturn. Being a meticulous observer he realized that there is an error in prediction of timing in Mars' path and this left him dissatisfied. Here we have the first glimpse of his passion for precision. He was probably the first astronomer to realize the need for accurate instruments in an era when astronomical observations were made with the unaided eye. We must remember that the invention of the telescope was not yet due for another forty-five years. He was only eighteen, when he built his first astronomical instrument, which was a *radie* or a pair of wooden calipers, which he arranged to be accurately calibrated. His astronomical pursuits were interrupted, when in the following year his foster father recalled him. Before Tycho could reach home, Jörgen Brahe died of pneumonia. This unfortunate illness was a result of his catching cold. He had jumped into one of Copenhagen's canals in order to save the life of Frederick II of Denmark, who was about to drown.

TYCHO'S NOSE
Next October Tycho enrolled at the Rostock University. After a few days of his stay there, he predicted the occurrence of a lunar eclipse and prophesied the death of Sultan Soliman also. His astronomical prediction was right, but his astrological prediction proved wrong. The Sultan had actually died seven weeks before! Less than two months later he attended a party, where he became involved in a duel with another Danish nobleman named Manderup Parsburg. The cause of which is not clear. The encounter apparently took place in the dark and it appears that Tycho had the worst of the exchange. Parsburg's rapier had badly injured Tycho's nose. Eventually three weeks later he would lose it. Luckily, he does not seem to have suffered greatly from any ill effects of his loss. In this, he was fortunate. The injury could have easily turned out to be fatal. Not that he would have been at risk from exsanguination, but because the injury had involved what is surgically known as the "danger area" of the face. It is so named, because the veins there do not posses any valves and connects this area directly to the brain. Any infection in this region can travel to the brain with serious consequences to life, especially in the days before the era of antibiotics. We should all be grateful to providence that nothing serious happened to Tycho. Three months later, he was well enough to observe a solar eclipse, which he recorded.

Characteristically, Tycho made himself a golden nose to cover the defect. There has been some controversy as to what metal was used, because when his body was exhumed, the margins of the area covered by his artificial nose had a green line. This indicates that the prosthesis was made of either copper or some sort of copper alloy. Some have suggested this was so because it was cheaper. Knowing Tycho, my feeling on the matter is that he was not the sort of person to use some thing because it was cheap. It was not in his character. He was rich, so a gold prosthesis would not be outside his means. In fact, it would probably be his first choice. The reason that he did not use a gold one may have been due to the fact that pure gold would have been too heavy for routine use. To hold the prosthesis in position, some sort of glue would be needed. Of course, a string could have been used to hold it also. We do not know if he took recourse to any such device. Moreover, Tycho was a stickler for perfection. It may have been that he was using some sort of alloy of gold, silver and copper in trying to get a colour match with the rest of his face. Copper gives a reddish hue to the colour of gold, if silver is added it will make it paler. Remember, we must not take the standard mixtures used by jewelers to make gold ornaments, but rather a mixture that obtained a reasonable colour match. Thus if copper had been used for this purpose, a greenish mark could be left behind with prolonged use.

RISING FAME

On 11th November, 1572 Tycho observed the new star in the constellation of Cassiopeia and published it in the following year. It made him well known in astronomical circles of his day. Others had also seen this phenomenon in the sky. It seems that Wolfgang Scholer of Wittenberg had recorded this new star a few days earlier. That Tycho had not observed it earlier may be attributed to weather conditions, because it is known he observed the skies regularly. Neither of them knew the significance of what they had seen. This "new star" was a supernova, which is an existing star in its violent death throes. Be that as it may with this his fame gradually spread. Later in 1574, he was invited to give a course of lectures on astronomy at the University of Copenhagen.

THE NEED FOR ACCURATE OBSERVATIONS

Once more, he travelled to Germany, but this time to meet other astronomers on equal terms. As early as 1563, when Tycho Brahe was 17, he had read many books on astronomy. Tycho had already grasped the need for accurate observations of the skies at an early age. He wrote, "I've studied all available charts of the planets and stars and none of them match the others. There are just as many measurements and methods as there are astronomers and all of them disagree. What's needed is a long term project with the aim of mapping the heavens conducted from a single location over a period of several years". Now, after meeting the various astronomers, his earlier conviction about the need for accurate observations assumed paramount importance in his mind.

Website: http://en.wikipedia.org/wiki/File:Uraniborgskiss_90.jpg

4.4 TychoBrahe's observatory at Uraniborg, Hven, built c. 1576-1580 from Blaeu's Atlas Major. (An original woodcut or engraving in 1598). [Public Domain]

On his return Tycho must have convinced the king about his ideas, because he accepted the Island of Hven (Ven) in the Sont, near Copenhagen to hold as a fief. Not only this; but the king, out of gratefulness for his life being saved by Jörgen Brahe and conviction in Tycho's genius, laid at his disposal a fortune equal to a king's ransom, so Tycho might build a state of the art observatory. To have an idea of this enormous amount at his disposal, some think it was equivalent to ten per cent of the gross national product of Denmark at that time! Tycho built a toy castle to harbor his observatory. The building itself was only 50 feet by 50 feet (15 meters x 15 meters) and its gardens were tastefully laid out. This miniature castle was named Uraniborg, in honour of the ancient Greek elder god, Uranus, who was the sky god. Uraniborg became the finest observatory of the day. Many scholars and the famous came to visit, some even from distant lands.

True to his form Tycho feasted and entertained during the day. It is said that he even kept a dwarf as his jester, who would play pranks on the guests. Whatever Tycho did during the day did not matter; he made it a point to carry out meticulous observations of the heavens on every suitable night that the weather would allow. After some time, he realized his huge instruments needed more stability for accurate observations. To this end, he built a new observatory, partly placed underground, so the colossal apparatuses might be supported more suitably. He named his new observatory Stjerneborg, meaning Castle of the Stars. Sadly, both Uraniborg and Stjerneborg have been lost to posterity. They were destroyed a few years after Tycho's death. During this period, Tycho designed many instruments and had them built to his specifications. He used several clocks to keep accurate time. In this way, Tycho was able to obtain the positions of celestial bodies to an accuracy of one minute of arc, which is equal to $1/60^{th}$ of a degree. This was truly a remarkable feat before the era of the telescope. He also saw to the repair and maintenance of his instruments on a regular basis. Unlike most at the time, Tycho had realized that this was an important chore in order to obtain accurate observations.

Website: http://en.wikipedia.org/wiki/File:Uraniborg_main_building.jpg
4.5 Uraniborg main building. (Copper etching in Blaeu's Atlas Major 1663). [Public Domain]

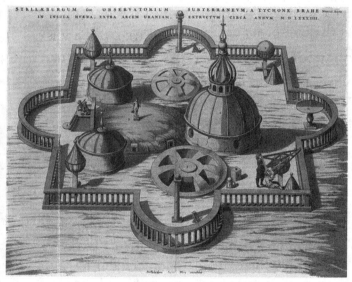

Website: http://en.wikipedia.org/wiki/File:Tycho_Brahe's_Stjerneborg.jpg
4.6 TychoBrahe's observatory at Uraniborg, Hven, built c. 1576-1580 from Blaeu's Atlas Major. (An original woodcut or engraving in 1598). [Public Domain]

4.7 A sextant used by Tycho Brahe

4.8 A bipartite used by Tycho Brahe

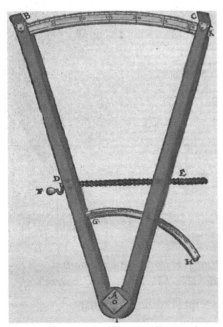

4.9 An arc used by Tycho Brahe

4.10 A quadrant used by Tycho Brahe]

All images from Astronomia instaurata mechanica, Wandsbek 1598. [Public Domain]
Website: http://fr.wikipedia.org/wiki/Fichier:Tycho_instrument_sextant_mounting_19.jpg
Source of 4.9, 4.10 and 4.11: Wikimedia commons
http://commons.wikimedia.org/wiki/File:Tycho_instrument_bipartite_arc_15.jpg
http://commons.wikimedia.org/wiki/File:Tycho_instrument_augsburg_quadrant_20.jpg
http://commons.wikimedia.org/wiki/File:Tycho_instrument_arc_18.jpg

In his pursuit of excellence, Tycho came to understand that atmospheric refraction lead to errors in observations. It was necessary to correct these to obtain precise results. This made him the first astronomer to account for this type of error. Again, before Tycho's time it was customary for astronomers to record only the salient positions of planets along their orbits. Tycho was not happy about this situation. He recognized the need to trace their paths more meticulously across the heavens. However, as it was not possible for even an energetic person like Tycho to carry out such a Herculean task on his own, he employed many assistants. There is no doubt that he supervised them, as his observational records have proved to be very accurate indeed. They were ten times more accurate than that of his contemporaries, sometimes even more. Such degree of accuracy revealed a number of abnormalities in orbital positions, which Tycho began to correct.

TYCHONIAN SYSTEM AND HIS OTHER CONTRIBUTIONS

Through the accuracy of his observations, he hoped to build up a picture of the universe that would confirm with his new idea. Here in keeping with the earlier concept that the Earth could not move, he proposed the planets went around the Sun and the Sun with the Moon moved around the Earth. This came to be known as the *Tychonian system*. In doing so, he missed the fact that the Earth was just another planet and the Moon its satellite. Nonetheless, this did not diminish Tycho's merit as the greatest observer before the invention of the telescope. Tycho's records would be fundamental to Kepler's work on planetary orbits and in turn, Kepler's discoveries would enable Newton to find

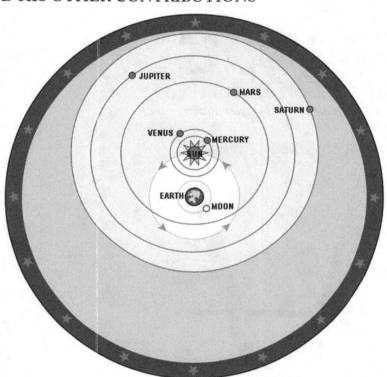

Credit: Courtesy Fastfission / Wikimedia Commons.
Website: http://en.wikipedia.org/wiki/File:Tychonian_system.svg
4.11 Tychonian System. [Public Domain]

the laws of gravitation. Thus Tycho's work had far-reaching effects. Apart from this, he also made several other contributions to science.

Tycho determined the parallax of the new star of 1572 and the comet of 1577. His observations established that these objects, without doubt, were situated beyond the Moon and it was concluded that the celestial sphere was not made of crystal, which we look at from the inside. Until that time some had thought, there were seven concentric celestial spheres, one for each known planet (except the Earth, which was not considered a planet), the Sun, the Moon and the stars. They were conceived to be hollow crystal balls. Tycho's observations showed, if such crystal spheres did exist then the comet would have to smash through the crystal spheres, as they lay in its path. No such a thing had occurred when the comet passed by. Thus proving conclusively, such spheres did not exist. So the idea of crystal spheres eventually died out.

TYCHO'S SELF-BANISHMENT

It is not hard to imagine that Tycho could be very difficult and overbearing, especially with inferiors, who came in contact with him. In the past, his vassals at Hven had complained to King Fredrick about how he made them work too hard when he was building his observatory. What was tolerated by the old king out of love for Tycho and his family would now not be tolerated by the new king. On Fredrick's death, Christian IV, ascended the throne. However, at the time he was still a minor. When he came of age, soon a disagreement arose with Tycho. As a result, in 1597 Tycho packed up his instruments and left his country never to return. After two years, we find him at the court of Rudolf II, King of Bohemia and the Holy Roman Emperor, who had invited him to Prague with an appointment as Imperial Mathematician. He was given a castle in Benátky nad Jizerou, 50 km from the city, where he built a new observatory. After working there for a year, the emperor had him recalled to Prague. He needed financial support for both his work and his extravagant life style of living. This was provided by the emperor himself, but for Tycho even this was not enough. He thus obtained monetary help from several other nobles, including Oldrich Desiderius Pruskowsky von Pruskow, to whom he dedicated his famous volume, the *Mechanica*.

JOHANNES KEPLER

We shall now leave this lively character and join, Johannes Kepler, the third player in the act. We will come back again to Tycho when circumstances and fate join the two together again, albeit briefly. This brief meeting was enough for Kepler to make Tycho's labours bear fruit later.

Kepler's life was a stark contrast to the other two. Copernicus' carrier was what every mother would have liked her child to have in those days. He had the right connections, the right education and the right position in society. His life considering the period in which he lived was smooth and without controversy. Until he died, he had not upset the apple cart, even with his great discovery. He quietly pointed out the path in his book and left the world. For the next sixty years, there would be no storm, but only ripples. The Church noted it all and held her peace. It was Bruno and Galileo, who would to stir up the hornet's nest.

Tycho Brahe in contrast was a character out of some melodrama, looming larger than life, colourful and dynamic. He was a man who "drank life to the lees". Unlike Kepler, he did not have that little drop of mysticism in him, which makes so many men of science great. However he did have vital energy, which allowed him to peruse a quest for something that only a very few would have been able to achieve during or before his time. He was a materialist through and through. Both Copernicus and Tycho had money and the resources, along with the standing in society, which enabled them to pursue their goals without hindrance. Kepler's life was different. He comes out as more human than either of the other two. He believed deeply in God. It did not appear to stem from religion; rather it was something, which came from the depths of his mind. He saw the universe as being created by God's geometry. This is the mysticism, we sense in Kepler. It was to be expressed in his idea of nested regular solids of Greek antiquity. At the same time, like the other two, he was also a man of science. When faced with facts he acquiesced to reality. We can associate with him more when we see his illnesses, struggles, poverty and the agony, which he had to face in life. Our hearts go out to him, because we feel that he is one of us.

Johannes Kepler (1571-1630) was born on 27th December 1571 at the Imperial Free City of Weil der Stadt in Baden-Wurttemberg. He was the eldest son of Heinrich Kepler and Katharina. Johannes father was a petty officer in the Duke of Wurttemberg's army. His grandfather had been the mayor of the town and his mother was descended from nobility. However, it seemed that both the families had fallen on hard times. They owned an inn that had belonged to his grandfather. Johannes Kepler was a sickly child, probably, because he was born two months prematurely. At the age of four, he contacted smallpox, which left him with an impaired eye sight. The following year his father left for the wars in Netherlands, Katharina had also accompanied him. Johannes was left with his grandmother at Weil der Stadt. It is said even as a child he showed a remarkable aptitude for arithmetic and the story goes that he was able to amaze the patrons of their inn with his mathematical skills. As a child, he had observed the Great Comet of 1577 with his mother. In 1580, he watched a lunar eclipse; he was nine years old at the time. All this made a great impression on him, so much so that he was to mention them in his writings later.

Source: Kopie eines verlorengegangenen Originals von 1610 im Benediktinerkloster in Krems (As far as I can make out this is a copy of the original in the Benadictine cloister in Krems, Austria)
http://en.wikipedia.org/wiki/File:Johannes_Kepler_1610.jpg
4.12 Portrait of Kepler 1610. [Public Domain]

While he stayed with his grandmother, he went to a local school. On his father's return from the wars, he was sent to attend another school at Leonberg. He then went to a lower seminary school at Adelberg in 1584 and moved on to a higher seminary school at Maulbronn in 1586. He graduated from here at the age of fourteen with a bachelor's degree. This enabled him to enter the University of Tubengen in the year 1587 as a theology student. Here it was probably because of his mastery of mathematics he came to the notice of Michael Maestlin, who taught him, both the Ptolemaic system and the Copernican system. Michael Maestlin himself was a Copernican. It was this leaning, which had influenced Johannes' young mind. Latter Kepler was known to have debated the merits of the Heliocentric Theory over the Geocentric Hypothesis with his peers. After obtaining a masters degree from Tubengen University in 1591 he was considering an ecclesiastical carrier. This might not have turned out well as he was a poor orator. This would also be an impediment as a teacher later, but it was not as a teacher Kepler would excel.

In 1594, Kepler took up the post of professor of mathematics in a protestant school at Graz, Austria. The authorities there expected Johannes to teach Greek classics, rhetoric and prepare almanacs. His proficiency in the subject of mathematics also enabled him to make predictions both in astronomy and astrology. The latter was the first inspiration for his first publication *Calendarium and Prognosticum* in 1595, the title of which speaks for itself. We should not think that Kepler dabbled in astrology. At Kepler's time, people did not distinguish between astrology and astronomy like we do today. They believed stars had influence on people's lives, as many do even now. At that period, this task fell upon astronomers. It was they, who had the knowledge to predict the movement of the planets. In later years, he would supplement his living from casting horoscopes and making almanacs, like so many astronomers of his day.

Kepler was a protestant and like many educated people, he read the Bible. He knew all there was to know about religious matters, because in those days most education was imparted through schools and seminaries conducted by the men of the cloth. Moreover, during his time religious influence pervaded all aspects of people's lives. So many people today think Kepler was deeply religious, but this observation need not be taken in the literal sense. Reading between the lines, I do not think he put much weight on the material aspects of religion. He did however believe in God and God's work. He believed, like Newton after him, God's work was based on mathematical symmetry, which for him was geometry. This profound belief was rooted in a sense of mysticism inside him that would manifest one day while teaching at Graz in 1595. It took the form of a sudden revelation. He came to the realization that the five perfect solids of the Greeks, the cube, the tetrahedron, dodecahedron, icosahedron and octahedron was the key to explaining the paths of the planets around the Sun. This experience made an impression on him, which was to last for the rest of his life. He later developed the idea and realized that a hollow spherical shape could both inscribe and circumscribe any Platonic solid. By nesting these solids in a suitable order, they allowed for the circular path of the then known six planets. It would be some years later he found discrepancies relating to this idea, which clashed with the formula he had proposed. Later, like a true scientist, he discarded the symbolic idea that was so near to his heart. At the moment, however, his idea of nested solids, which he thought explained the reason for Copernicus' circular paths, was to be found in the manuscript, *The Prodromus Dissertationum Mathematacarum continens Mysterium Cosmographicum* (Sacred Mystery of the Cosmos).

Before this work saw the light of day it had to pass through the censors, which meant the senate at the University of Tübingen. In this, Kepler took the aid of his old teacher and friend Michael Maestlin when presenting his manuscript to them. After reviewing the work, they gave him permission, but added a rider. Kepler was told that he should tone down his all too many allusions to the scriptures and simplify the description of the Copernican system, along with Kepler's new ideas about the nested perfect solids. This was his earliest major work. It was published in 1596. Even though he believed firmly in the heliocentric theory, this book contained an extensive chapter reconciling heliocentrism with biblical passages that seemed to support geocentrism. The clue to the fact that Kepler hid his true feelings lies in the effusive dedication to his patrons in his book. In the past, as even now, people have always found it necessary to appease the egos of men who controlled their destinies.

The work was an aforetaste of Kepler's genius that would show up one day, but for the moment, his book appeared to have been gone unread. When Kepler received his copies in 1597, he began sending them to prominent astronomers and to his patrons. Apart from his peers knowing about his interest in astronomy, he did not receive much attention. It however helped him to establish contact with other astronomers, which included Tycho Brahe and Galileo Galilei. The destinies of these two men would in time become linked with his own, in relation to the heliocentric idea. It was the formers efforts, which would make Kepler's genius flower. While Galileo's defense of the heliocentric theory would bring about his own humiliation in the winter of his life, but Galileo would die without ever appreciating Kepler's true worth.

One of the other astronomers Kepler had written to was Reimarus Ursus (Nicolaus Reimers Bär). He was the imperial mathematician to Rudolph II and a bitter rival of Tycho Brahe. Ursus however did not reply to Kepler's letter. Instead, he forwarded Kepler's work to Tycho Brahe as an example against the Tychonian System, because this happened to be one of his disputes with Tycho. The latter however did not hold this against his younger colleague when they met later. In fact, he was to write to Kepler about it. Tycho had the advantage over Kepler. He possessed a vast store of accurate observational data. He gave a rational criticism regarding Kepler's work, whose the data had been borrowed from Copernicus, which were much less accurate.

At the time, Kepler had been courting Barbra von Muhleck, a twice-widowed miller's daughter, who inherited some money from her previous husbands. He sought her hand in marriage. Not unnaturally, his suite was frowned upon by his future father–in-law, as he was deemed poor. Finally, after publication of Kepler's book, the objections were overcome and they were married on 15th of April 1597.

During this period, Kepler's fertile mind was planning other works. One of his ideas was a static universe with only the planets going around a fixed Sun in the background of stars. A second book would describe the motion of the planets. Another book, which would show the planets were physically like the Earth. Lastly, he wanted to write about his speculations on the effects of the celestial bodies on the Earth's atmosphere. It was about this time, he had a premonition of gravitational force also, about which he would write later.

Very soon after their marriage in September, Kepler and his wife had to leave their home, as Protestants were ordered out of Graz. Kepler, however, was able to return after a few months through the influence of his wife. Still, things remained uneasy; religious tension had not cleared. In spite of all he continued his work. By 1599, it became clear to him that his efforts to understand the motion of the planets were restricted by inadequate data. Most of he realized were inaccurate anyway. At this period, he was in correspondence with Henstart von Hohenberg a catholic astronomer. Through his good offices with the Emperor Rudolph II, von Hohenberg, advised him to contact Tycho Brahe, who had now succeeded his rival, Reimarus Ursus, as astronomer to the Imperial court. Kepler fearing, the developing tensions at Graz would prevent him from making the journey, he preempted an invitation from Tycho and went to meet him at Prague. Finally they met on 4th February, 1600 at Benátky nad Jizerou castle, where Tycho was constructing his new observatory. This meeting was to seal their destinies to fame. However, the path would not be an easy one.

Kepler remained as a guest with Tycho for two months and on his host's advice paid attention to Mars' orbit. Understandably, at first Tycho did not give the newcomer all his data. Kepler's ideas and dedication soon prevailed. Tycho gave him some more access, but on condition he would not copy the data. This was a great disappointment for Kepler, because he knew that lacking all the data he could not possibly test his ideas on the paths of the planets. Kepler rightly estimated, even with all of Tycho's data, it would take some years to complete his goal. So Kepler took assistance of the ill fated, Jan Jesenius (1566-1621), a physician, politician and philosopher who had predicted his own execution. When Kepler approached him, he was consultant in anatomy to Rudolf II, Holy Roman Emperor at the time. Jesenius helped Kepler and tried to workout a proper arrangement with Tycho to enable him to earn his keep, but negotiations failed. It ended in an angry argument between them. Thus, in just over two months after they had met, a disappointed Kepler left for Prague.

In the mean time, Kepler in the hope of improving his financial condition wrote an article, *In Terra inest virtus, quae Lunam ciet*, which means, "There is a force in the Earth which causes the Moon to move". It was a far-reaching idea for the day! It also detailed a new way to measure lunar eclipses. He dedicated it to Ferdinand II, Archduke of Styria (Inner Austria), who ruled from Graz, He was later to become Holy Roman Emperor in 1619. Kepler hoped it would secure a place for him in his court. He again failed. His scientific efforts however had not been wasted, because he was able to apply his method successfully during the eclipse of 10th July that was seen from Graz the same year. The *Astronomiae Pars Optica* or The Optical Part of Astronomy, which he was to write later and dedicate to the Emperor Rudolf II on 1st January, 1604, had its roots in this work. Today it is widely agreed that this work embodies the foundation of optics, even though Kepler seemed to have missed out on the laws of refraction. Nevertheless, he described many important things, such as the inverse square law of light, reflection by flat and curved mirrors, significance of parallax and about the apparent size of celestial bodies in the context of astronomical observations. He also outlined the principles of the pinhole camera. In this, some of his ideas predated those of Newton.

In spite of their earlier altercation, Kepler and Tycho soon made up. They eventually reached an agreement in June 1600. Kepler once again returned home to Graz, this time in order to collect his family. There he was told to convert to Catholicism by the authorities, Kepler refused. So he and his family were banished. Kepler left on 2nd August, 1600 with his family He remained with Tycho until the time of his death the following year. He worked for Tycho who financially supported him and his family. Also to some extent Kepler was assisted by other noblemen. Kepler tried to convince his patron about the Heliocentric Theory, but Tycho was working on his own ideas and did not pay heed him. Kepler badly wanted Tycho Brahe's observational records on planetary orbits, but the latter was still not ready to give them up. Thus Kepler had to be satisfied with working under Tycho's directions.

Predicting Mars' orbit had been an old problem for astronomers. This was because today we know Mars' orbit is slightly more elliptical than the other known planets of the time. The astronomers of the past did not know this, so the prediction of Mars' position was of great interest to them. This would prove to be a boon in disguise for Kepler, as we shall come to see later. Apart from Kepler's routine work, Tycho set Kepler the task of writing a criticism against Tycho's rival, Ursus. This was intended against the latter's earlier affront of his passing on Kepler's work to him, which supported the heliocentric theory. I consider this to be a classical example of Tychonian humour, as Ursus had past away. Tycho was now holding his late rival's post.

Soon after this, Tycho died. It is said he had died of mercury poisoning, because when his hair, which was analyzed in 1996, had a high amount of mercury. Some have even said Kepler was the person who poisoned him. The descriptions of his final days tell another tale. The story goes that after attending a feast on 13th of October 1601 at Baron von Rosenbergs, where not surprisingly he consumed much wine, so much so that he wanted to relieve himself, but it is said Tycho did not leave the table out of respect for his host. Later he found difficulty in passing urine. He was in pain for the next few days. During this time, he developed fever (urinary infection with initial bacterimea). Later he had high fever and delirium (septicaemia). Tycho Brahe died (of septicaemic shock) on 24th October, 1601. In the end he is said to have repeatedly uttered the words "Ne frustra vixisse videar", which means "May I not seemed to have lived in vain". As death closed in the fever seemed to have resolved (the final stage when the body failed to react).[20]

After his death, his instruments were kept stored, but eventually lost, which was a great pity. Tycho was buried in a tomb in the Church of Our Lady, in front of Týn near Old Town Square close to the Astronomical Clock in Prague.

What ever may have been the gossip, we may be sure the symptoms that led to his death were typical of untreated acute retention of urine. Either this may have been due to a urinary stone blocking the urethra or the result of an enlarged prostate gland. Both of which would lead to acute retention of urine. Though recent studies have revealed high concentrations of mercury in his hair, it does not imply that he was poisoned as some claim. Medications containing mercury was not uncommon in those days and Tycho may have been suffering from urinary problems before he died as indicated by the manner of his death. Mercury compounds were used as diuretics up to the middle of the last century in some countries of the world. This may explain why Tycho was using medications containing mercury. It must be remembered he was a follower of Paracelcus and a practicing alchemist. So it is quite possible he may have been medicating himself with mercury compounds. Thus there is no call to get suspicious about his being poisoned by Kepler. He may have wanted Tycho's data very badly, but he was no murderer.

In this context, it is interesting to note that high amounts of mercury was also found in Isaac Newton's body many years later when it was examined. He also dabbled in alchemy. So he may have either been given medicines or been under self-medication. This element therefore might have got into his body in either of these ways. It would be of interest to us to know whether other people, who had lived long enough those days to have urinary problems, also had a high content of mercury in their bodies.

Two days after Tycho's death, Kepler was asked to fill the position of Imperial Mathematician at the court of the Emperor. In the mean time Kepler continued to pursue his aim, but first he would have to acquire Tycho's store of data. Tycho, believing only the data would be his road to fame, had enjoined his family not to part with it after he died. Thus Kepler's repeated attempts to get the data he wanted so much failed. However, in the end, Kepler succeeded, possibly through the intersession of the Emperor.

THE DISCOVERY OF THE FIRST TWO LAWS OF PLANETARY MOTION

The next 11 years, as imperial mathematician, would be the most productive period in Kepler's life. With the data in his possession, he was now able to work on the orbits of the planets. He first selected the orbit of Mars partly due to Tycho's advice and partly because of the peculiarity in its path. It took him few years and much labour. In the year 1605, Kepler would stumble on the truth. He was able to deduce that the orbital path of Mars was elliptical and not a circle as had been assumed previously. By extrapolating his findings to the other planets he was able to show, from Tycho's data, that *"the paths all of all the planets were elliptical with the Sun's centre of mass lying at one focus"*. This was the first law of planetary motion! This had been a crucial discovery. Once he had established this fact, by further studying the data, he realized the progress of the planets in their orbital paths could not be uniform; some times, they travelled faster sometimes slower. From this, he figured out *"that a line joining a planet to the Sun swept equal areas in equal time"*. This was his second law of planetary motion. The discovery of the second law explained the phenomenon of unequal rates in their movement and this was consistent with observed findings. It was this that had led to the discrepancies between predictions through calculations of circular motion and observations before. These findings culminated in Kepler's work *Astronomia Nova* (1609).

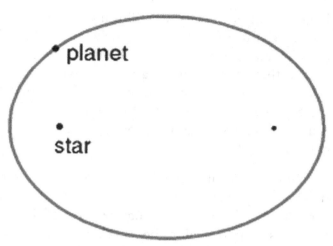

Credit: Courtesy Arpad Horvath.
Website: http://en.wikipedia.org/wiki/File:Kepler-first-law.svg

4.13 The first law of planetary motion. [CC-BY-SA-3.0]

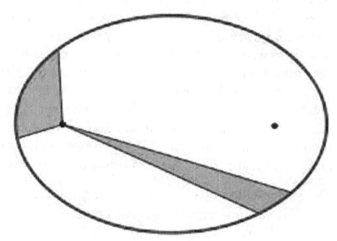

Credit: Courtesy Arpad Horvath.
Website: http://en.wikipedia.org/wiki/File:Kepler-second-law.svg

4.14 The second law of planetary motion. [CC-BY-SA-3.0]

Here Kepler opens with a discussion of the three existing models of planetary paths, namely, that of Ptolemy, Copernicus and Tyconian model. He went on to point out these three models could not be distinguished by observation alone. None of them could predict the positions of planets with accuracy, even though the values were near enough. Here Kepler had realized that a good scientific model should not be beset with such uncertainty. He went on to show, these older models would never do, so a new model was required. He also showed if Mars' orbit were considered circular, it would not fit with Tycho's observational data. At first he considered this orbit to be an oval. An oval is egg-shaped figure, which if folded along its long axis it is symmetrical. Still things would not fit. Lastly, he came to the realization that the orbit of Mars was elliptical. It was only then observations fitted Tycho's data. He went on to extrapolate his discovery to other planets with success. What is of significance in this work is that he shows an insight into gravitational force as well. In doing so, he foreshadowed Isaac Newton. Lastly, he makes a speculation that reaches out centuries ahead. He says that the Sun is not at the centre of the universe but probably follows a certain orbit of its own. This far-reaching insight was only possible, because of the mysticism in him, as he had no data to this effect.

Through all this, it became obvious to his astute mind some sort of force was involved in moving the planets around the Sun. He thought this force was emanating from the Sun, as it revolved on its axis and along the plane of the orbit of the planets. He also concluded this force was proportional to the length of the radius vector. Even though Kepler may have had a premonition of the idea of gravity, he never found the real answer.

We all know Isaac Newton discovered gravity, but it may surprise many to find how close Kepler had come to understanding this still unknown force. This is what he had to say in his *Astronomia Nova*. *"If two stones were placed anywhere in space near to each other, and outside the reach of the force (of other bodies), then they could come together ... at an intermediate point, each approaching each other in proportion to the other's mass."* As further proof of this understanding he was to write, "If the earth ceased to attract the waters of the sea, the seas would rise and flow into the moon – *"... then he goes on to add, "If the attractive force of the moon reaches down to the earth, it follows that the attractive force of the earth, all the more, extends to the moon and even further..."* However, here Kepler was not talking about the same thing as Newton. He misunderstood the nature of the gravitational force. He thought this attractive force was the result of magnetic force. It explains what he was saying. He never correlated mass with gravitational attraction. He had reached out far, but not quite touched the answer.

The reader will remember that Kepler had once sent a copy of his first major work to Galileo. We will also remember it had been ignored. It was now Galileo's turn to seek Kepler's support and seek Kepler's opinion on his book. He sent a copy of *Sidereus Nuncius* (Starry Messenger), which gave his discoveries of his early telescopic observations. Kepler was not one to hold a grudge. He enthusiastically replied to this in his *Dessertatio cum Nuncio Sidereo* (Conversation with the Starry Messenger), which he published in 1610. He not only agreed with what Galileo said, which in essence was what Galileo wanted, he also went on to suggest the significances of Galileo's observations. It seems Galileo did not make any reply.

Kepler saw the invention of the Galilean telescope. He used one borrowed from Duke of Cologne and viewed the Galilean moons for himself. He was the first to use the word *satellite*. We have used the word ever since. The results of his observations were he published the *Dioptrice* 1611. Here Kepler mentions how the instrument could be improved, if a convex lens was substituted for the concave eyepiece, in the Galilean telescope.

After nearly eleven years of relative peace in Kepler's life, his troubles started again. In 1611 Emperor Rudolf II (1576-1612), who was a patron of arts and sciences, was suffering from ill health. He was under pressure to abdicate in favour of his brother, Matthias (1612-1619), who unlike his brother was not so liberal either to the Protestants or towards the sciences.

Kepler for a time was unsure as to whether his post would be renewed or not. At about this time Kepler's wife, Barbra, had contacted Hungarian spotted fever. His children had fallen ill with small pox and one died. He had been away from home at this time, because he was looking for a new opportunity due to the uncertain future. He tried to obtain a post at the University of Tübingen in Württemberg, but because of religious scruples there, he was not given the position. On Galileo's recommendation, before his death, Kepler was asked to take his seat as the professor of mathematics. However, Kepler declined to go to Italy with his family, as he wanted them to stay within protestant territory. Instead, he went to Austria to seek a position as teacher in Linz.

Kepler's wife became ill once again. She died on Kepler's return. His relationship with her had not been so good. She had never been able to appreciate his greatness or the merit of his work; like so many wives of other great people. Maybe she was right from her point of view as a wife. Kepler had failed her as he was often in financial difficulties and was too much involved with his work to attend to the day-to-day problems of the family. Kepler stayed in Prague until Rudolph's death in 1612. His now became involved in a legal dispute regarding his wife's estate. Kepler's work came to a standstill for a time. During this time, he was not entirely idle. He took time off to piece together a chronology manuscript, which resulted in his *Eclogae Chronicae*. Matthias on becoming the emperor confirmed Kepler's position again as Imperial Mathematician and allowed him to move to Linz.

In the years at Linz, Kepler's life was much quieter, even though the Lutheran Church had excommunicated him; he was free to do as he pleased. He was now able to take up his work on the *Rudolphine Tablets* again. This time he was able to complete the work, but did not publish. At Linz, he held a teaching post, which was not a heavy responsibility for him. He wrote a book, *De vero Anno*, in which he discussed the probable year of Christ's birth and got it published. Realizing the advantages of the Gregorian calendar, he proposed its adoption by the Protestants, but they would not even hear of it.

Kepler was 41 and now looking for a second wife. After the experience of his first marriage, Kepler had become circumspect in selecting a wife. It was said he had looked at eleven different prospective brides. Today we would call it "market survey". Kepler finally selected Susanna Reuttinger, a twenty-four-year-old orphan, who had taken the trouble to find out Kepler's requirements and his children. On 30th October, 1613 they got married. At this time, he happened to note that he could work out the exact volume of the contents in the barrels mathematically. This he set out in his book *Nova Steriometria Doliorum vinariorum* (New Stereometry of Wine Barrels). This in a way paved the way to infinitesimal calculus. This work was however published two years later. Kepler would have six children by his second wife, three of whom survived. He was happy in this marriage.

THE THIRD LAW OF PLANETARY MOTION
Kepler in his *De Harmonice Mundi* (The Harmony of the World) was in essence trying to find symmetry in the world he knew, just as physicists do today. It opened with a chapter, which explained the symmetry of regular polygons. The second was about congruence. The third is about musical harmony. While in the forth he discusses about harmonic configurations in astrology. In the fifth chapter, he enunciated his third and last law of planetary motion

PLANET	ORBITAL RADIUS	ORBITAL PERIOD
MERCURY	0.39	0.24
VENUS	0.72	0.59
MARS	1.00	1.00
EARTH	1.51	1.88
JUPITER	5.20	11.86
SATURN	9.53	29.46
URANUS	19.17	84.40
NEPTUNE	30.00	164.00
PLUTO	39.40	247.00

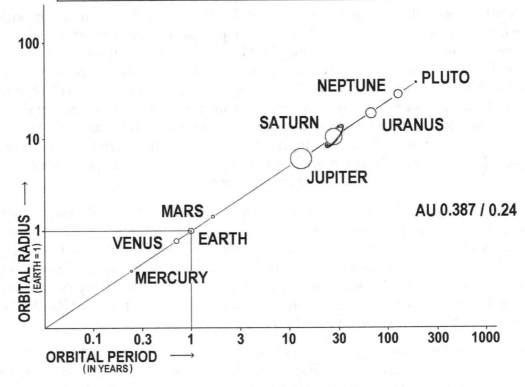

4.15 The third law of planetary motion.

It was published in 1621. The law stated, *"The square of the period of a planet is equal to the cube of its average distance from the Sun"*. Kepler did not know the exact distance of the Earth from the Sun, but he knew the exact ratios of their distances with respect to the Earth and the orbital periods of all the then known planets. Here Kepler takes the Earth as the reference planet; the distance of the Earth from the Sun is taken as one (or 1AU) and the period that represented the revolution of the Earth around the Sun, which is one Earth year, is taken as one unit. In this way, he confirmed his third law.

At the time when Kepler was preparing his work De Harmonice Mundi, his mother, Katharina, was accused of witchcraft. A woman who had some dispute with Kepler's brother regarding financial matters accused Kepler's mother, Katharina, of having poisoned her in 1615. As a result, things came to a head when Katharina was accused of witchcraft two years later. Things dragged on till in August 1620, when she was imprisoned. She was finally released, as there was no evidence against her. Her accusers had brought up trumped up evidence by distorting the story in Kepler's *Somnium* (The Dream), in which the mother of the author seeks the aid of a demon to learn how to travel through space. There was no rocket travel in those days. So naturally, Kepler's story took this form, in order to explain how the author could reach another world through space. Somnium was a blend of fairy tale and science. In fact, it was a work of science fiction, probably the first of its kind. In his book, Kepler described how the Earth could be viewed from Moon, in order to explain the heliocentric system through his story. The manuscript of the Somnium had been written earlier in 1611 and circulated. Many had read it. It was this allusion to the woman consulting the demon in the story, which was used as distorted evidence against Kepler's mother during her trial. During her imprisonment, she was subject to *territio verbalis*. This implied she would be given a vivid description of the physical torture, which she would have to undergo, if she did not confess. All this must have been a terrifying experience for his mother and very unsettling for Kepler.

During all this period even though much of Kepler's time had been taken up to defend his mother, he continued to work with his usual zeal in spite of his personal distractions. He knew the value of time. He is known to have completed the reading of a work on the theory of music by Vincenzo Galilei on his way to defend his mother. Incidentally, Vincenzo Galilei was Galileo Galilei's father. Thankfully, after 14 months of incarceration in prison Kepler's mother was eventually freed in the end. Although we might not appreciate it today, even in those days there were the right and wrong way in going about legal torture. So due to Kepler's legal efforts and the failure of the authorities to follow proper procedure in using torture, Katharina was released.

Later, following the trail, Kepler was to modify the manuscript of the Somnium by insert more than two hundred commentaries, which only reflects how cautious people had to be in those times. In the end, the result was that the "tail" became longer than the tale!

It had always been Kepler's wish to write a book on astronomy after completing Astronomia nova. In 1615, he had completed the first three volumes of this book, *Epitome Astronomia Copernicanae* or Epitome of Copernican Astronomy, which consisted of seven volumes. The first were printed in 1617. The subsequent volume was completed by 1620. The last volumes 5 – 7 were published in 1621. This proved to be Kepler's most influential work. It contained all his laws on planetary motion and his concept that the motion could be attributed to physical processes. After his death it would be the most widely read amongst all of Kepler's works.

He was the first to apply the three laws in his work in computing an astronomical table using logarithms. This was published under the name of *Chilias Logarithmorum*. In 1627, the overdue *Tabulae Rudolphinae* or Rudolphine Tables ordered by Emperor Rudolf II was finally published. It superseded the earlier tables that were available at the period. The reason for this delay in publishing was that Kepler's life had not been easy. Added to all his problems, he was plagued by Tycho Brahe's family, as they wanted the share of the profits. These star tables would become the standard for the next hundred years. In the following year, Kepler moved to Silesia, where he continued to work on astronomical tables and in his *Admonitio ad astronomos* in 1629, where he drew attention to several impending transits.

He died in Regensburg (Ratisbon) on 15th November 1629 an unsung hero of science during his lifetime. Even his grave has been lost in the turmoil of the Thirty Years War.

Kepler's work Somniun, which he wrote in 1611, was published posthumously in 1634.

THE DFENDERS OF HELIEOCENTRIC THEORY

Kepler's work was not appreciated in his lifetime. When theories are born, they can be like ugly ducklings; rejected when young; only to be appreciated when they mature.

GALILEO GALILEI

Galileo Galilei, a contemporary of Kepler, ignoring the latter's epoch making contribution to science, went on to take up, almost eighty years later, the brief for Copernicus' circular orbits. Galileo was a remarkable mind, an excellent orator and teacher. When he taught, students packed his classes. At the time, we are speaking of; he had already made a name for himself. He had friends and patrons in high places and was at a high point in his carrier. All the while, he was enjoying the lime light; Galileo did not see the gathering clouds. Many of his fellow teachers who taught were jealous of his popularity. As we have already mentioned Galileo, unlike Kepler, was a very eloquent teacher, so his colleagues found their classes empty. This raised their ire against him. This was one of the factors, which would lead to his eventual fall.

Secret opposition was mounting against Galileo. This was an unstable period, when religion was an instrument used freely against one's opponents. Already petitions to the Inquisition had been made against him. Ominously in 1614, Tomasso Caccini had denounced Galileo publicly. Galileo being a man of science, naively thought himself secure in his logical reasoning. He failed to grasp the politics going on behind his back. By 1616, only when things came to a head, he realized what was going on. Galileo now quickly decided to travel to Rome and personally place his views to the Pope. It was too late; he had been preempted by circumstances. Cardinal Bellarmine at the behest of the Inquisition delivered an order to Galileo. It stated that he should neither hold nor defend the idea of heliocentrism.

After the warning Galileo remained quite for seven years. This however did not diminish the flame within him. He saw his chance in 1623, when Cardinal Barberini, with who he was in favour, was elected Pope. He took the name Urban VIII. As a cardinal, he had opposed the stricture on Galileo in 1616. Galileo met him six times after he became the Pope and was much encouraged as a result of their discussions. He set out to write his book *Dialogue Concerning the Two Chief World Systems*. It was written as a discussion between Simplicius, who supported the Ptolemaic hypothesis and two others, one of whom was Salviati, who tried to convince him about the Copernican system. In doing so Galileo forgot on which side his bread was buttered. The Pope had enjoined him to present the work in such a way that Galileo should in no way appear to favour either hypothesis and that he should explain the two different views only. In a way, the Pope was doing Galileo a great favour, because he himself believed in the geocentric hypothesis. Urban VIII had his own pet ideas also and he requested Galileo to mention them in the book. When the book came out, it was sold out like any best seller of today. Little did Galileo realize that his great success would very soon herald an even greater catastrophe for him! In the book, Simplicius came out not only for the worse, but positively looked foolish. What was even more devastating, Galileo had unwittingly put Pope Urban's views in Simplicius' mouth! Pope Urban VIII was not amused as one might guess and quite naturally took umbrage to Galileo's book.

Galileo was now seventy years old and ailing. He received a letter. The letter was from the Inquisition asking him to present himself at Rome. He tried his very best to excuse himself to avoid the Inquisition, but it was of no avail. Under duress, he was compelled to travel to Rome where he had to face his judges. Amongst them were the Pope's brother and nephew and all of them favoured the geocentric hypothesis.

After the trial Galileo was imprisoned. Due to his failing health, he was allowed to go back to his house, but remained there under house arrest. His younger daughter, Virginia who had taken the veil as Sister Maria Celeste, came to look after him until he died. After his arrest, he was grief stricken. He could not work for many months. In the end, he became blind, but was able to complete *Discourses and Mathematical Demonstrations Concerning the Two New Sciences*. This was his last work. Here he was to write, *"assume that the speed acquired by the same movable object over different inclinations of a plane are equal, whenever the heights of those planes are equal"*. He died on 8th January, 1642, while still a prisoner at his house in Arcetri, Tuscany, Italy.

FILIPPO BRUNO

Filippo Bruno (1548-1600) was the other prominent figure who supported the heliocentric theory. He was born at Nola. On becoming a Dominican friar, he assumed the name Giordano. So he is known to us as Giordano Bruno. He conditionally accepted the Copernican Theory and went on to propose there were other worlds in an infinite number of universes that had life. He was a rebel. His ideas, even though conjectural, were far advanced even for the foremost thinkers of his day.

Whether there is life out there in the universe, other than ours, is not known even today. However, in the order of things, life could not be something unique to this planet. There must be other places in the universe that can harbour life. To consider such a thought would be called heresy, by some, even today. I would say to them, we would be then making the same mistake, as what we have always made in the past. We should not think that man holds some unique status in the universe. Bruno was a philosopher, what he envisaged many years ago might not have been proven until now, yet today we have taken his ideas seriously in our search for life in the form of SETI (Search for Extra Terrestrial Intelligence). Bruno was an intellectual and the Church saw his ideas as a threat, so they lured him to his doom. He was invited by a Venetian nobleman, who later treacherously gave him up to the Inquisition. He was taken to Rome where he remained imprisoned for seven years. Finally, after his trial he was burned at the stake on 17th February, 1600. The exact allegations against him are not known, but one of them was for his support for the heliocentric theory.

ISLAND UNIVERSES

After the death of Bruno, in the new century, the invention of the telescope came to peoples' notice about 1608. No one knows exactly who devised this instrument first. There are three contenders, all of whom we shall come across later. It seems the most likely candidate must have been one of the spectacle-makers, who handled lenses. Whatever is the real story, the use of telescopes soon spread across Europe. The designs improved and with the solving of chromatic aberration by Isaac Newton and replacing the objective with a concave mirror, telescopes became more and more powerful. By the 18th century William Fredrick Herschel, who discovered Uranus, realized there were other galaxies beyond ours. He called them *"island universes"*. It also became clear the Sun was a part of our Milky Way Galaxy.

THE FALL OF THE GEOCENTRIC HYPOTHESIS

Copernicus' explanations solved a major problem of the Eudoxan, Aristotelian and Ptolemaic models and simplified the mathematics. Tycho demolished the Aristotelian idea of crystal spheres. Kepler discovered the true paths traced out by the planets. Galileo observed the moons of Jupiter giving him a model for the solar system. He also observed the phases of Venus, which proved two things. One, planets did not have any light of their own and secondly, which was even more important, they revolved round the Sun and not the Earth. It was a triumph of power of human intellect to be able to build a model of the solar system from the Earth, only through naked eye observations and without sophisticated modern technology, like powerful telescopes or going into space.

The astronomers by this time realized the geocentric hypothesis had become untenable. Yet they did not ready to embrace the Copernican system immediately. This was probably due to the fact there was no direct proof, until that time, the Earth went around the Sun as Cardinal Bellarmine had said, which I have quoted at the end of the last chapter. Thus, Europe took the middle line and they embraced the Tychonian system for a while. Even with Kepler's Laws of Planetary motion, things did not change. It was only with the publication of the Newton's *Principia* and the understanding of the concept of inertia, as embodied in Newton's three laws of motion and his laws of gravitation, people came to accept the heliocentric theory.

The last nail in the coffin of the geocentric theory came when in 1838 Friedrich Wilhelm Bessel (1784-1846), a German astronomer, measured the parallax of Cygnus 61, a star. This manifestation of parallax proved that the Earth went around the Sun. (See Fig. 3.8)

THE FINAL VICTORY OF THE GEOCENTRIC THEORY

After all the struggle that the pioneers went through to establish the heliocentric theory, it appears strange that we still use the geocentric reference frame in most of our everyday activities to this day. But that is how things work. All our activities are in reference to a stationary Earth. We leave the heliocentric reference frame for space travel and other scientific activities. So after all the final victory goes to the geocentric hypothesis.

5 GRAPPLING WITH TIME

We have to live by our allotted measure of time
Within its compass we have to finish our earthly tasks
For neither time nor events will await any man.

AN INTRODUCTION TO THE NATURE OF TIME

We all know what we mean by time. Yet, if asked to elaborate, one would be hard put to explain what it is. This was also what St. Augustine (354-430) had to say many centuries ago. Things have not changed much to this day. When we learn how the nature of time was explained by different people, including great minds such as Newton and Einstein, we would like to think, things would become lucid. However, this did not prove to be the case. Looking back into the history of science we find, whereas most other things become clear as we progress, paradoxically the nature of time has remained an enigma. Today we are still far from understanding the true nature of time, as we were before.

Time has various aspects. Over the years, people have debated these many aspects. Men have asked, whether time had a beginning or not. To some the obvious answer seems, like everything else it began with the universe. But how could the universe begin, if time had not already been there before? The ancients thought that time had always existed. Newton proposed time was independent of the universe. This meant it was absolute. His rival, Gottfried Wilhelm Leibniz (1646-1716), asked a very subtle question. Though it did not challenge Newton's idea, it nevertheless did pose a moot philosophical point. The question he asked was, "If time was absolute then why did God create the universe when He did?" What he implied was if time had indeed been always present, then God could have created the universe at any time, why at that particular time? So, if there was a reason, then time could not be absolute as there was a preferred instance of creation. Since we do not know God's intention for choosing that particular time, we do not know the answer to Leibniz's question. Aristotle (384BC-322BC) had an entirely different approach to the problem. He thought time was the measure of motion. This was an excellent idea and difficult to disprove. Every body in this universe is in constant motion. Whenever we measure time, it is always measured against movement. All clocks are based of this fundamental principle, whether it is celestial bodies acting as clocks, mechanical clocks or even the atomic clock. The latter uses the vibrations of the cesium atom to measure time. Today we define a "second" by 9,192,631,770 pulses of the cesium atom. It is the most accurate clock that we have in the world. It loses a second only after about every 20,000,000 years.

Scientists like Henri Poincaré (1854-1912) put the spanner in Aristotle's argument by saying time was relative. So the question arose, if time was indeed relative, then which clock shows the actual time? Poincaré answered, each was correct! Of course, this was subject to such things as slowing down by friction, low batteries or because of malfunction. So with time being relative is the universe out of gear? We do not have any such evidence. As far as we know, everything seems to be in proper order. Here Einstein came to the rescue. He showed if a clock was subjected to different speeds or different gravitation fields, in each case, the clock would show different time. This effect has been shown to be real! Moreover, he went on to give a new interpretation. He united space with time and called it space-time.

By now, people had come to understand there is no privileged position in space, so with space and time blended into one, many came to think there was no privileged position in time also. Thus the question of past, present and future from this point of view has no real meaning. If this seems confusing, let me explain. If we say that Tokyo is east of New York, then we can equally say that Tokyo is west of New York. That is if we do not consider the artificial grid, which is made up of longitudes and latitudes. After all, the 0° meridian at Greenwich is man-made. Similar to the analogy of space, an event has no preferred position in time. We can only give it a position, if we describe it in relation with another event. Only then can we say whether it is before or after. This is what Leibniz was trying to tell Newton when he said time was absolute.

Yet we know that time has a direction, which is real. We call it the arrow of time. If a cup falls from the table and breaks, there is no way the pieces can come together on their own again, no matter how long you may happen to wait. Again, there is the psychological aspect of time, for various reasons it appears to "flow" or to "pass". This is understandable, because events around us are continually unfolding and like the broken cup, they do not revert. Thus, we get the arrow of time! However, the rate at which it appears to flow may differ. Imagine a criminal waiting in front of the judge, who is about to hear his sentence pronounced within the next five minutes. Not knowing whether it will be a death sentence or not, it may seem like an eternity to him. On the other hand, to a young man who meets his girl for five minutes, time seems to pass only too quickly. Such is the complexity of time.

We will all however agreed, time is one of the fundamental ingredients of our universe; without it, all events would grind to a halt. If such an event were to occur, then as far as we are concerned the universe would be no more. In our present universe for time to work it needs space and like the three dimensions of space, time constitutes another dimension. Thus, appropriately, it is known as the fourth dimension.

Everything in this universe is subject to time. Our quest to understand the nature of time is important from not only the scientific point, but also the philosophical as well. However, we will keep such understanding for the second part of the book. Here we will only deal with our biological and practical relationship with time. All living things, to exist, must possess an innate perception of time. We will thus discuss the importance of this relation and show how our bodies abide by time. We will then see how this eventually led man to measure time and in doing so, we may think we have taken some control over it. Such, however, has not proved to be the case. Today time has taken over our lives. Probably more than what is desirable for our well-being.

MAN'S INNATE SENSE OF TIME

Since my school days, I have been intrigued with man's ability to sense time. We once had a maid, she had no sense of time and always came late. We would consider ourselves lucky if she was late by an hour. She was only tolerated because she was quite meticulous and did her work well. Our other maid was exactly the opposite; she was always punctual. You could almost tell the time to the minute by when she came and went. This was remarkable in a country where people do not have much respect for time. She did not have a watch nor did she have a clock at home. She always finished each part of her total work within a fixed period. The latter could be explained because there is a wall clock in every room in the house; but her coming to work in time always intrigued us. One day when asked, she said that the shadow cast by a certain post near her house was her indication of time. She also appeared to know the difference between the shadows cast during winter and summer. This was like telling time by a sundial, which is quite understandable. When we asked her again how she managed during the rainy season, when the skies were overcast. She could not explain, but only said she "knew". By this she probably implied that she had an instinct for time.

All this would have passed my mind had I not already been aware through an earlier observation that the attribute of sensing time is present in all humans, only some have it to a greater or lesser degree. This variation is like the differences seen in any other attribute, which humans possess. Just like, there are persons with a well-developed mechanical sense, persons who can sing well, persons who are good at sports or persons who have mathematical skills. These attributes are always inherent. In some, it may be so well developed it is obvious. While in others, such attributes may not be present at all. In case of the latter no matter how hard they may try, they can never attain these skills, unless of course they happen to lie dormant. In my case, familiar tunes play in my mind, but I could not get one out, even if I tried for a thousand years. The ability to sense time is also a variable attribute. My wife has quite a well-developed sense of time. Compared to her sense of time, her sense of space is less well developed. She sometimes bumps into the furniture in the house; whereas in a similar situation I would not have done so. My sense of space is good, but the sense of time is low. My father was the same. So is my daughter. This diminished sense of time in my case does not imply that I am habitually late for appointments; in fact, I am very particular. What I lack is the ability to tell the passage of time. If I am working, I will not be able to say how long it has been without consulting a watch. My wife on the other hand will be able to tell the time almost to within a minute or two, without consulting a watch.

People may rightly point out that my examples are only subjective impressions. This may well be true. I have come across an even more remarkable instance that was to create my first impression about humans sensing time. I once came across a person, who had what can be only described as a very acute sense of time. I recall that he used to be able to tell the time to the exact minute without the help of a watch. We were schoolchildren then. We used to call him Mister Time. I never knew his real name. He appeared to be a Marwari, who are a business caste from Rajasthan, which is a western province in India. He was a thin built, middle aged with graying hair. His forehead was adorned with a large red "tilak" made from red sandalwood paste. It was considered an auspicious mark, which was not uncommon in his community. It is still used by some even today. He was always clad in a spotless white shirt and a white dhoti. He never carried a watch.

Those days we used to go to school in a public bus and it was here that we came across him often. As school boys will, someone on seeing that he had no watch must have asked him the time. He gave him the correct time to the minute. Since then it had become a game with us, asking him the time. As a precaution against his looking at our watches, we would keep our watches in our pockets. I also remember testing him in various other ways. One of the ways we would do so was by surrounding him as he sat in one corner of the bus and then engage him in conversation. After a while, we would all of a sudden ask him to tell us the time. Another technique we used was to synchronize our watches with All India Radio before leaving our houses. There was no television in those days. We made sure we set our watches a minute or two in advance or late by prior arrangement over the phone, so that even if he happed to have a glimpse at our watches he would not know the correct time. To all our amazement, he never failed any of these tests, which we put to him. Looking straight at us, he always gave the right answer. Thinking it was some form of magic that he was performing, I once asked him if he could show me some magic. He answered in the negative. This convinced me what he was doing was not any form of magic.

All this may seem rather strange and distant. I will now describe something that may be more familiar to many of us. Say, we want to wake up at 4.00am and get ready to catch an early train. Naturally, we would set the alarm for 4.00am the night before. Many will have experienced that under such circumstances they have woken up a minute or even seconds before the alarm goes off. Though some may, point out this has never happen to them. I can only repeat what I have pointed out earlier; such attributes are not present to the same degree amongst all. They vary widely from person to person. So we must allow for this fact.

BIOLOGICAL TIME

Apart from the cycles of life and death of many plants, which are approximated with the seasons, common people, especially farmers, knew that the activities of certain plants and insects were finely attuned to time of the day or the seasons. Before machines were invented to tell us time, people came to accept the telling of time through such behaviour of living organisms. The first biological rhythm recorded was by Androsthenes during Alexander the Great's progress east. He had observed the regular opening and closing of the tamarind leaf. Then for a long, long time there is no other record. No one appeared to take further interest in such matters. After many centuries Jean Jacques d'Ortous de Mairan (1678-1771), a French geophysicist, documented a similar diurnal movement in the leaves of *Mimosa pudica*. People were already aware certain flowers open only at certain specific hours of the day. Carl Linnaeus (1707-1778) was able to use this phenomenon to make a "flower clock" by using a variety of plants whose flowers opened or closed at different, but specific hours of the day.

> **OPENING TIMES FOR FLOWERS USED TO CREATE A FLOWER CLOCK**
>
> | Time | Flower |
> |---|---|
> | 6 am | Spotted cat's ear, catmint (both open) |
> | 7am | African marigold, dandelions (both open) |
> | 8am | Mouse-ear hawkweed (opens) |
> | 9am | Field marigold, gentians (both open) |
> | 10am | Californium poppy (opens) |
> | 11am | Star of Bethlehem (opens) |
> | 12Noon | Passion flower, goatsbeard (both open) |
> | 1pm | Childing pink (closes) |
> | 2pm | Scarlet pimpernel (closes) |
> | 3pm | Hawbit (closes) |
> | 4pm | Four o'clock plant, Bindweed (closes) |
> | 5pm | White water lily (closes) |
> | 6pm | Evening primrose, moon flower (both open) |

Source: BBC Home page » Flower Clock Garden
Website: http://www.bbc.co.uk/dna/h2g2/A5170024

Box 5.1 Time of opening and closing of some flowers.

Whatever may be the capacity of either plants or animals to sense time; all other forms of life have been endowed with this ability also. The commonest manifestation of this phenomenon is the *circadian rhythm*. This term was introduced by Franz Halberg (1919-) the founder of modern chronobiology (chrono = time; -bio = life and –logy = study). The word circadian comes from the two Latin words "circa" meaning "about" and "diem or dies" meaning "day". This rhythm approximates a 24 hours cycle and triggered by sunlight, which is manifested by awaking, opening of flowers and such other diurnal activities of organisms. We do not know the details of how this mechanism works in many life forms, but we know something of how it works in humans. In man, the external rhythm of the daily morning light triggers the melanopsin ganglion cells, which are found in the retina. These cells are distinct from the rods and cones, which are related to vision. Melanopsin ganglion cells constitute about 1%

of the total number of cells of the retina. When light falls on the retina. Unlike the signals from the rods and cones, the signals from the melanopsin ganglion cells are not relayed to the visual cortex at the back of the brain. Their signals pass down the optic nerve to reach the optic chiasma. Above the optic chiasma, there is a collection of about 20,000 cells known as the suprachiasmatic nucleus. It is here the impulses from the melanopsin ganglion cells pass. The optic chiasma is like a railway junction, where the visual fibers of optic nerve fibers from each eye are sorted out in such a manner, so that the image from the right half from each eye will pass to the left half of the brain and vice versa. We will discuss this in a later chapter in the second part of the book in another context.

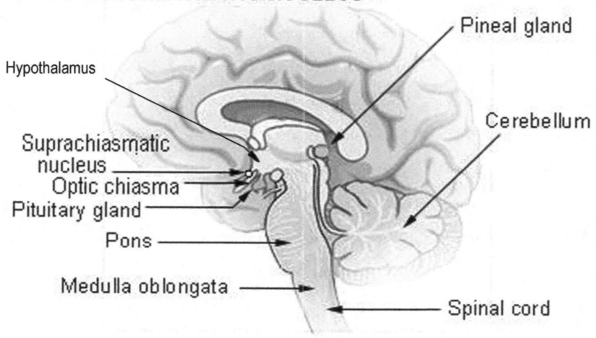

Credit: United States Federal Government.
Website: http://en.wikipedia.org/wiki/File:Illu_pituitary_pineal_glands.jpg
5.1 The suprachiasmatic nucleus. (Labeling added to suit text) [Public Domain]

Apart from the circadian rhythm, there are rhythms in our bodies, which correspond to other intervals of time. There are those rhythms, which are shorter than the day, like our heartbeat, respiration rate and maintaining tone of our muscles. There are many other such examples in the body. These functions are usually dependant on the various metabolic requirements of the body. They are known as ultradian rhythm. Then there is the infradian rhythm, where cycles are longer than the day. The behavior of the human immune system is one such an example, which appears to have a weekly rhythm. The lunar rhythm, like the menstrual cycle in women, is also another example.

Some rhythms appear to be independent of external indicators of time. There is an insect known as the cicada, which in Latin means the "hummer". One variety, the periodical cicada (*Magicicada septendecim*) or 17-year locust remains secreted underground for 17 years. After this time, they emerge, all at the same time, chirping in unison. Their sound has been described by some, as a railway whistle. After emerging, they climb certain trees, mate and then soon die. The nymphs that hatch from their eggs soon drop to the ground and bury themselves. Once underground they seek roots of suitable trees and live by sucking their juices. Eventually they form pupae. After which the adult form emerges and their 17-year cycle begins, when a new generation appears again.

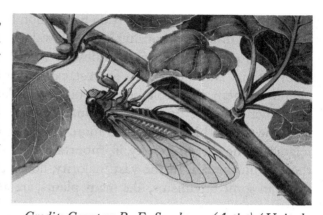

Credit: Courtesy R. E. Snodgrass (Artist) / United States Department of Agriculture.
Website: http://en.wikipedia.org/wiki/ File:Snodgrass_Magicicada_septendecim.jpg

5.2 Magicicada septendecim or the 17-year locust. [Public Domain]

All the various biological rhythms represent our link with external phenomenon of time. This link is so fundamental to life that it is present in all cells; and cells constitute the basic unit of life. Even primitive cells, like bacteria, have been shown to posses the sense of time. There is a very ingenious experiment, which reveals this property in the in a spectacular way. Here the gene for luciferase, which is an enzyme, was incorporated into the genome of certain bacteria. Enzymes are a special class of proteins produced by the genes. Each is capable of facilitating a particular step in the biochemical process. Once the gene for luciferase is incorporated, it causes the organism to posses the capability to light up. Surprising these bacteria did not give off light continuously, as would be expected. Instead, they only light up periodically in synchronization with their biochemical activity. It revealed not only that these bacteria had an internal clock, but also this biological clock had a genetic basis, which was geared their metabolism, thus confirming cells are subject to time. Not only this, what is even more significant, it highlights the fact that whether an organism has a brain or not, it still possesses an innate sense time.

The work on the fruit fly, Drosophila melanogaster, has shown that it possesses genes that are clearly responsible for its "understanding" the time of day. These insects in their natural state always emerge out of their pupae stage in the morning. By manipulating their genes, they can be made to emerge at other times of the day as well. This experiment again reinforces the fact that biological time has its roots in the genes. Through other experiments and from various observations we know today that all living things, including fungi, algae and bacteria possess similar genes. Life's link with time was thus laid down very early in the evolutionary process. It constitutes one of the fundamental properties of the simple cell, without it, life could never have evolved.

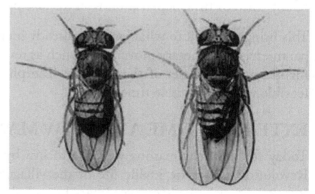

Credit: Courtesy NASA.
Website: http://en.wikipedia.org/wiki/ File:55542main_maflies_med.jpg

5.3 Fruit fly, Drosophila melanogaster. [Public Domain]

Our association with time rests on a common basis. The web of life is synchronized through the sense of time. The morning light will wake animals, hunting instincts will take over, leaves will open for photosynthesis and flowers will open to attract insects for pollinations. The day is for activity and the night is for rest, except for those creatures who have evolved nocturnal habits. In them, their role between night and day is reversed. Some sea creatures have their reproductive cycles tuned to the lunar cycle, as their reproduction is necessarily associated with the tides. This makes them more attuned to the lunar cycle. The tidal rhythm, which is related to the moon's gravity, is important for the life cycles and activity of many organisms that live in the sea. In the vast majority, however, reproduction, growth and maturity; even death in some organisms, like many plants, are geared to the seasons.

It is of interest to note, all other senses such as light, sound and temperature are all manifestation of energy transfer. Even touch and pressure are a result of mechanical energy exchange. Only time is an exception. It is not a form of energy. Yet we can sense it. This paradox can be explained by the fact that all living organisms have the ability to detect different forms of energy that surround us. In some, these organs may be prominent, such as the eye, while others do not appear to have any sense organs. Even though, not all organisms have a sense organ *par se*, yet they can sense various forms of energy. The amoeba is a single cell and does not have an eye, yet it can sense light. Through this link to the environment living organisms can perceive changes in the environment, especially light. The variation in the levels of light are detected and translated as time. Without the ability to sense light, neither they nor we would have the ability to appreciate changes in time.

The importance of the link between our senses and time is underscored when we are subject to time deprivation. This becomes apparent under conditions such as in the laboratory, in the case of freak accidents and in case of torture where the victim is cut off from perception light, sound etc, which gives us our sense of time. Apart from the psychological stress it creates, even short periods of such condition result in disorientation and inability to function properly. The phenomenon of jet lag is but a mild example. Any prolonged period of exposure of any form of time deprivation would cause permanent psychological malfunction.

This brings us back to what we have already mentioned in the first chapter as to how certain parameters of planetary movements, such as revolution and rotation are important in looking for life in other parts of the universe. The phenomena of life and evolution can never be feasible without suitable time cycles.

EXTERNAL TIME AND HOW MAN CAME TO ABIDE BY IT

Today for different reasons we have to live by calendars and clocks. When the Industrial Revolution came, the idyllic life of the villagers changed forever. Now in the newly born industrial towns, the lives in ordinary households gradually came to revolve around the clock. The more accurate of such clocks were used in factories and offices. Appropriately they came to be known as *regulators*. Clocks now started to regulate our lives. We know what days are for work and the days for rest from our calendars. The clocks tell us the time to wake, time to eat, time to go to work, time for rest and time to go to bed. They synchronize our lives to that of others, so we may keep appointments, catch trains and so on. This form of control of our lives by time happens, not through internal regulation by our biological clock, but through external agency of artificial timekeepers. So we may call it external time to distinguish it from internal time, which is our biological clock.

Even though external time keeping does not have any genetic or evolutionary basis, it has taken over much of our lives today. By the help of clocks, we have stepped beyond the control of biological time. We work late into the night, we do night shifts, we party till morning and sleep during the day. Those who regularly work at night have their biological clock reversed. Their day's metabolism shifts to the night. So body processes like skin turnover and hair growth, which takes place mostly at nights, now shifts to the day. While other hormones that are normally active during the day will become active at night.

DIVIDING TIME

The attribute of time has taken on a new dimension in man. As man evolved technologically, he had to learn how to divide his time more and more accurately. To do this he had to device various ways of measuring the interval of time, which would cater for his required needs. The testimony for this can be seen in edifices, like the Stonehenge to highly sophisticated instruments like the atomic clock. What follows gives us a short survey of this relation of man's needs and the various ways by which he measured time.

Credit: Photo: Courtesy Wikipedia User: Retro00064.
Website: http://commons.wikimedia.org/wiki/File:Schoolhouse_Regulator_Pendulum_Clock.jpg
5.4 A schoolhouse regulator clock. [Public Domain]

Man's reason to tell time differed at each stage in his history. During the earliest stage when man was primarily occupied with hunting and gathering, his need to tell time did not have the same purpose as it has today. At that stage, his life was quite different. Just after the ice age receded, there was no agriculture at this period. He occupied himself with making garments out of animal skin; rope out of creepers; weave baskets; make snares for birds and small game; clubs, spears, arrows to hunt or for war; tamed animals and even use colours to paint on the wall of his caves. It was only necessary for him to know the seasons, so he would know when and where to get his food. During the warmer seasons, he hunted and in autumn, he gathered fruit. He had to depend on stored food during the winter months and glean from the woods what ever he could get. So for them the understanding of time lay in their ability to understand the seasons, because it related to availability of food and predicting the changes of the seasons. This was enough for him.

To those early men, the subdivision of the day was significant only for their daily concerns. It meant such things as setting out to hunt early morning; gathering firewood, make fires and cook food during the day. In the evening, he would sit by the fire recounting their experiences and by nightfall, he would go to sleep. Any period shorter than the day had no great significance for them, except for perhaps waiting for the family or tribal members to return from a hunt before the closing of the day. Without any minute divisions of time, like ours, their lives must certainly have been tension free. Impending birth or a person at the point of death would make him aware of the psychological effect of time. This only meant anxiousness or hurrying to take quicker action. Minutes or hours would have no meaning for them. If necessary, at the most, the day could always be divided into dawn, morning, afternoon and evening. Distances could be described in terms of so many days or nights. Any longer length of time, such as describing age or a memorable event could always be made in terms of so many moons or so many summers or winters.

Whereas, in those days, the daily changes in the phases of the moon made it easy to track time, it was not of much use when man changed to an agrarian life. Growing crops is a much more precise affair than most would imagine. If sown too early the cold of early spring would stunt the growth. If late, the heat of the summer could make young seedlings wither and the frosts of an early winter could even kill the crops. Growth and maturity of crops had to be synchronized to the seasons and the seasons in turn were linked to the Sun and not to the Moon. So the sowing had to be done in the right time in relation to the Sun. Thus to be able to predict this "right" time accurately was of utmost importance. It could make all the difference between a successful harvest and starvation. At first, man did this empirically from his observations of the heavens. Once man knew when the year began, he could judge the time for sowing fairly accurately by calculating back from when he expected the harvest. He knew this from how long the corn takes to ripen.

Though about twelve lunar cycles approximately added up to a year, they did not coincide exactly with the solar year. Therefore, the lunar cycles kept shifting and they do not fall in with the seasons. If only the lunar cycles were used to calculate the time to sow crops, it would mean many tedious adjustments to the seasons. So no agricultural community took up this method. Some cultures still use lunar reckoning. They are people who have their origins in the harsher life of deserts where there was little or no agriculture and their lives were more or less nomadic. Others like the Chinese and Jews still clung to the lunar calendar for traditional reasons. While others like the Hindus and Christians maintained a combined solar and lunar calendar, otherwise known as soli-lunar calendar. The lunar portion of the calendar was essentially used for religious purposes, whereas the solar calendar was used for agriculture and secular work.

Some however took to other methods. People were aware that stars behaved in a predictable manner. The Egyptians chose the heliacal rising of Sirius on the horizon, which means the first rising of this star after a period of invisibility due to its conjunction with the Sun. Their priests kept watch for its first appearance each year to let the nation know of this auspicious event. The reason for so much importance being given to this celestial event lay in the fact that this particular phenomenon coincided with the annual flooding of the Nile. It brought in the rich soil from the upper reaches of this river, which renewed the fertility of the surrounding plane each year.

Most cultures used the daily excursions of the Sun to calculate the solar year. Apart from the east to west movement of the solar disc, man had observed, the Sun moved in another way also. In what I have to say next, I will speak from the point of view of people of the northern hemisphere. This is because for most of man's history, all the significant progress came from this hemisphere. After a certain day in the summer, the days become shorter and shorter. During this period, the Sun rises on the horizon at a more southern point each successive day to reach a southern most point on a particular day. This happens about 22nd December of the calendar. From the next day, the Sun rises on the horizon from a more northern point each successive day. This continues, until after six months in summer the Sun reached its northern most point on the horizon. After which it turn back again. This is repeated faithfully each year. (See Fig. 3.1) The two days, one when the Sun rises from the northernmost point on the horizon and the other when the Sun rises from the southern most point on the horizon, is known as the summer and winter solstices respectively. The day light lasts the longest on the summer solstice and on the winter solstice, the night lasts the longest. Thus to know the number of days in a year one had to calculate the number of days from either summer solstice to the next summer solstice or winter solstice to the next winter solstice. For his purpose, man chose the latter, because sowing time came in spring and spring follows winter.

Behind all this there lay a greater ethos that we could hardly appreciate today. There is the story of Sir Gawain and the Green Knight, where the latter comes to King Arthur's court on Christmas Day. The Green Knight lays a wager. He proposes that one of the Knights of the Round Table should cut off his head. If he lives through this macabre experience, then he will come back the following year to do the same for the knight, who agrees to his challenge. Sir Gawain, Arthur's nephew, came forward to take up the challenge. He cut off the Green Knight's head. To everyone's amazement, the Green Knight took up his fallen head and placed it on his shoulder. Then he rode off into the night. This is the story in its basic form, which in the Middle Ages was further expanded and embellished. The details of which are not relevant to our point here. In the mean time after many adventures, Sir Gawain returns to Arthur's court in time for the Green Knight to claim his forfeit. The story ends happily with Sir Gawain being spared in the end. Though many other interpretations have been suggested, I personally feel the people of that time saw it as a great renewal of life after winter. Not only is the timing of the events in the story significant, but also the green colour of the knight. The beheading represents the end of the annual cycle of plants when life becomes dormant with winter settling in, whereas the Green Knight getting up to replace his head indicates return of life that would follow with the coming of spring.

Website: http://en.wikipedia.org/wiki/File:Gawain_and_the_Green_Knight.jpg

5.5 Sir Gawain and the Green Knight. A medieval illuminated manuscript by unknown artist.

Though the shortest day fell on or about 22nd December the warm weather would not come until another three months. It did however indicate that spring would soon come, because from the next day, the days would start lengthening again. To determine the day of the winter solstice man had to observe the rising of the Sun and determine the southern most point of the Sun's rising on the horizon. In order to do so he must have initially taken natural landmarks to fix the winter solstice. Later man built his own landmarks in the shape of stone circles, which were probably the earliest observatories. These stone circles can still be seen scattered all over northern Europe. The most famous being the Stonehenge.

All was very well, but nature could be unreliable. The rising of the Sun could be masked by weather conditions such as mist, fog or clouds. So it became obvious it would be best to make one good observation and then working from there to create a calendar, which would give the days of the year. Then the next would follow from the first and so on. All this was easier said than done.

After few years, discrepancies arose. This was because the solar year did not round off to a whole day, but man was not aware of this fact at that stage. Thus, the ancients who followed the solar calendar must have used some form of correction. Maybe they adjusted the dates with one of the solstices each year. This however did not solve the inherent problems mentioned earlier.

Some early farmers followed what to them was a better method. In many parts of the world, they did not take the help from the skies to learn the time for sowing crops. Instead, they took the help of certain predictable natural phenomenon in the living world. Such examples may be found in different cultures all over the world. Hesiod was an ancient Greek poet and rhapsodist, who lived around 700BC. He wrote, "the cries of the migrating swans indicate the time for plowing followed by sowing; when the snail climbs up the plants then there should be no more digging in the vineyards; and when the thistle blossoms the summer has arrived".[21] Even today, farmers in many parts of the world rely on such familiar observations, which have been found to be reliable over the years.

Often such ways of predicting the weather in local areas can be fairly accurate, but they can by no means be regarded as a universally applicable. This is why it makes the solar clock so important. When the earth completes its journey round the sun, starting from one point in its path and returning to the same point, it completes the cycles of the seasons. The year is broadly divided into four seasons; spring, summer, autumn and winter. Seasons are weather conditions only. Their coming and going can not only differ at any particular place, but also differ from place to place. In some parts of the world, there are finer distinctions to the usual four seasons. In Bengal, there are six seasons. Seet meaning winter; Basanta, the season that heralds the hot dry summer; Griswa meaning summer; Barsha implying the rainy season; Sarat equivalent to autumn; Hemanta a mild season between autumn and winter. In the Mediterranean, there are roughly two phases of the year a hot dry summer and a cold wet winter.

Thus seasons are arbitrary divisions of the year describing average weather conditions, which are dictated by the tilt of the Earth's axis and its position on its orbit around the Sun. However, these are not the only things that determine the weather conditions. We have to take into account other things, like distance from the sea, whether it is on the east or west of a continent or if the place is on the windward of leeward side of a mountain range. Height of the place also matters. Of course, as we all know, the latitude is very important also. Though the seasons have a regular cycle, nonetheless at any one time, the seasons and their effects at any one place may differ from year to year. The effects of the seasons are more apparent in the extremes of latitudes and less so on the equator, where the spring and autumn can be warmer as the Sun lies overhead. Again, when it is summer in the northern hemisphere it is winter in the southern hemisphere and vice versa; so it is with spring and autumn.

There is another great problem with the solar cycle however. It is not synchronous with the rotation of the Earth around its axis. The latter causes the day to occur. The day is important, because in many ways it forms the basic unit of time, especially from the biological point. Therefore, to make a calendar, the days have to be fitted to the year. This turns out to be an insoluble problem. If you count the year by the number of days in a year, you will find there are 365 days and an extra part of a day is left over. This extra part amounts to about a quarter of a day. This small difference may not appear to be significant initially, because in those early days the life of society revolved around agriculture with one main harvest in the year. Once the harvest was over there was enough to feed all the members of the society. So after the harvest people went about their business and did not pay much attention to the intervening months. This fact is underscored in cultures such as the Romans. They had no names for some of these intervening months in their earlier calendar.

As societies became more and more complex, there arose other needs. Transactions and trade started taking place throughout the year. Armies were on continuous move. They had to know when and where they had to be on a particular day. Wages had to be paid in proper time. Interests on loans required to be paid on stipulated dates. The government and the offices were now open throughout the year. It now became important to have a calendar, which gave the days and dates all round the year.

THE IMPORTANCE OF SEASONS AND THE CALENDAR

All solar calendars have evolved from the lunar calendar, as societies turned to agriculture. Initially, these societies clung to the lunar cycle, which was predictable and so they tried to fit it with the solar calendar. Here we will only trace the Roman experience, because it was from their calendar that the calendars of other European countries were adopted. Now it is used all over the world. In fact, the very word calendar comes from the Roman word Kalends, which denoted the first day of any month in the Roman calendar.

Initially the Romans had ten months in a year. The system was inherited from the Greek lunar calendar, which had preceded it. The Greek calendar In turn came from an earlier Babylonian calendar. The earliest Roman calendar is believed to have been introduced by Romulus. He was the first king of Rome, who ruled about 738 BC. According to their calendar, the year started with the month of Martius (March) and was followed by Aprilis, Maius, Junius. All named after the gods and goddesses. Then came the months that were not important to agriculture, so they bore numerical names only - Quintilis, Sextilis, September, October, November and December, thus indicating their lesser importance in the Roman world. The named months had 31 days each, while the others had 30 days only. The Roman calendar thus consisted of 304 days only. There remained a gap of 61¼ days in their year during winter. The winter lay fallow, so to speak. To compensate for this gap, Numa Pompilius, who followed Romulus on the throne in 716 BC, added January to the beginning of the year and February to the end of the year. In this way, February became the last month of the year. This was not surprising, because the last month was dedicated to the Februus the Etruscan god of the underworld, who was also the god of purification. The last few days of this month was thus marked by religious rites of purification and festivals. The first month, January, was dedicated to Janus, the god who presided over doorways. He therefore symbolized the connection between the old year and the new. Janus was represented with two faces, which looked in opposite directions. It was many years later in 452 BC February came to take its present place. In spite of the adjustments, their year still fell short by 10–11 days. This was because their month of February had 23 or 24 days only. To compensate for this error another month, known as Mercedinus was added. The name Mersedinus comes from the word "merces" meaning wages. It was about this time the wages of workers were paid in the Roman world. Again the month of Mersedinus had 22 or 23 days. So it was added to every alternate year to keep the seasons in order. Since the Roman calendar was still based on the lunar cycle, it became fraught with discrepancies even after these attempts at correction.

When the Romans tried to convert their existing lunar calendar to a solar one, the lunar year fell short of the solar year by 11.24 days. This was due to the fact that the interval from one new moon to the next, known as *lunation*, is equivalent to 29½ days. In this context it is to be noted, lunation is not same as the period of lunar revolution. The latter is equal to 27.3 days only. Thus, 12 lunations come to 12 * 29.5 = 354 days. If the lunar calendar was converted directly to a solar, it would fall back by 11¼ days each year. Thus in 3 years a little over one month would be lost, while a whole year would be lost in 32years 5months and 17days! Long before this happened, it would create havoc in peoples lives. Sowing would be disrupted and winter months would recede into autumn and so on. The Romans corrected this by adding another lunar month to the already present 12 months, once every three years. It was not a very bright move on their part, because their calendar still remained out of step. This time, instead of falling behind, it started to advance about 4 days every 3 years. To add to the confusion, the officials concerned, who were in charge supervising the changes in the calendar were corrupt. They used their powers to change the calendar, so that their favourites might benefit by holding office longer and those they did not like could be turned out earlier! Such misuse of power distorted the calendar further.

In 46 BC, Julius Caesar consulted Sosigenes of Alexandria, the astronomer, regarding the confusion in the calendar. Nothing much is known about Sosigenes except from Pliny the Elder's reference to him in his *Natural History*. Sosigenes took the step to reorganize the calendar to take recognizable form. It now had 365 days, with 12 months for each year. Days thus had to be added to some of the months. February now had 28 days, April, June, September and November had 30 days and the rest had 31 days each. This left the calendar only about ¼ day out of step with the solar cycle. This was solved by adding one day to the last day of February every four years. Not only was February the shortest month, but for traditional reasons as well. February had always been the last month in the Roman calendar. It was thus appropriate that the extra day should be added to the last month even though it had been now shifted forward. Thus the calendar appeared finally adjusted with the solar cycle, but the month now became disrupted with respect to the lunar cycle.

Though it was a great improvement on the previous calendars, the assumption that the solar year contained an extra quarter of a day or six whole hours had been a great mistake, so it was still remained a few minutes off; the value being 11 minutes 14 seconds short for each year. Initially as this error was comparatively small, people failed to notice it for some years. Over the centuries, however, the discrepancies became apparent. The effect of this assumption was to have serious consequences on the calendar and soon came to reflect on the lives of people again. All this affected not only the farmers and the Church, but society in general as well. Since Julius Caesar's time in every century there was a gain of 18.72 hours or 0.77 days. By the time of Pope Gregory XIII, this discrepancy had reached 10 days.

The solar calendar now in use all over the world for secular purposes was designed by Aloysius Lilius (1510 -1576), an Italian philosopher, physician, astronomer and chronologist. He was also known as Luigi Lilio Ghiraldi. However, he was fated to die before his new calendar came into use. According to his advice, Pope Gregory removed 10 days from the calendar, which happened to be 5thOctober to 14thOctober inclusive. Thus the next day after 4thOctober, 1582 was

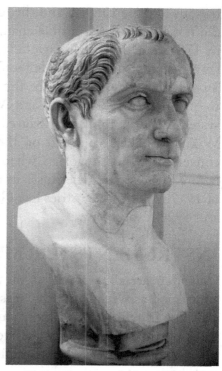

Photo: Courtesy Andreas Wahra. Website: http://commons.wikimedia.org/wiki/File:Giulio-cesare-enhanced_1-800x1450.jpg

5.6 Bust of Gaius Julius Caesar (100BC - 44 BC) at National Archaeological Museum of Naples, Italy. [Public Domain]

Photo: By Wikipwdia user GKD Source: Unknown Website: http://commons.wikimedia.org/wiki/ File:Gregory_XIII.jpg

5.7 Pope Gregory XIII, portrait by Lavinia Fontana.

15thOctober, 1582. He kept the leap year. However, as an added precaution against further problems of discrepancies accumulating, he decreed the centuries, which were not divisible by four hundred would not be considered as leap years. This brought the error down to 1 day in 3322 years. So it has been proposed that any year divisible by 4000 was not to be considered as a leap years. It is hoped that this will come into practice in the year 4000. This would bring down the error to one day in 20000 years.

Though the Gregorian calendar was very sensible, all the European countries would not come to adopt it. Why? Because it was a Catholic innovation! So the Protestant nations like the English and the followers of the Greek Orthodox Church like the Russians did not accept these sensible changes for many years to come. It was only seventy years later in 1752 that the English would come to accepted it. So many of the dates that we have from records such as say Newton's birth on Christmas Day of the year 1642 actually falls on 4th January 1643. Therefore, the letters OS and NS are often seen to be appended to dates to qualify *Old Style* (Julian) and *New Style* (Gregorian) to avoid confusion between the two systems. People think that Saavedra Miguel de Cervantes, the creator of Don Quixote and William Shakespeare died on the same day, 23rd April, 1616. Spain, however, was already converted to the Gregorian, whereas England had not. So Cervantes actually died ten days before Shakespeare. The Russians only accepted the Gregorian calendar after the Revolution of 1917. Whereas countries like Greece and Turkey adopted the new calendar as late as 1923 and 1926 respectively.

INSTRUMENTS FOR MEASURING TIME

Calendars are all very well to measure intervals of time in days or any larger increments of time, such as weeks, months, seasons and years or even decades and centuries. They are of no value in telling time of any smaller than the day. As civilization and science progressed, there was pressure in many ways to tell time in hours, minutes, and seconds. Sometimes even in smaller measures, such as milliseconds and nanoseconds. Clocks were invented for this reason.

WATER CLOCKS

With trade came money. A successful journey meant the gods had to be thanked. Thus, temples sprang up in the towns and cities along the routes where trade took place. They brought rich revenues through offering to the gods, especially after a successful venture. Added to this some times after long journeys of abstinence the traders were glad to see faces of women. The temples organized this service for them so that the income would not go to others. In many places, it was the custom for parents to dedicate young girls to the gods. Some of them who were too poor to provide their daughters with a dowry or at times they would dedicate them for some material gain, which they expected from the gods. These poor girls were then used by the priests or some powerful patron and then were relegated to a life of prostitution to gratify the lust of sailors, travellers, traders and any other who would come to give offerings to the gods of these temples. All of which was organized under the aegis of the priests under the garb of religion to fulfill their greed for money.

Like all professional people, time was money for these girls. Whether they worked for the temples or they worked in the prostitute quarters of the towns, it did not matter. They had a stipulated time for their clients. There were no watches or clocks in those days. Instead, they had a simple device, which told them the time. A metal bowl with a small hole or holes at the bottom was placed on water in a metal container. After a time the water would seep in increasing the weight of the bowl. This made it sink to the bottom. The bowl would make a metallic sound on touching the metal container. This signaled the end of the session for the client.

Overtime more and more complex water clocks were developed. It was a common method for time keeping used allover the world. Starting from Korea and China in the east, across India,

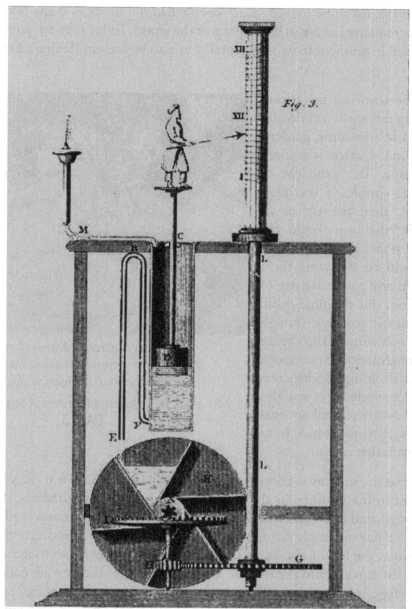

Water enters and raises the figure, which points to the time of the day. The spillover water operates a set of gears to show the appropriate hour for any particular date. The ancient Greeks and Romans divided the day into twelve hours, so the summer had hours that are marked at longer intervals than winter.

Arabia, Babylon, Egypt, Greece, Rome and Europe in the west, water clocks were developed during various periods of history. They were able to give time with reasonable accuracy. Moreover they were also amenable to development in various ways. In China a complicated water clock was devised, which is said to have had an escapement mechanism. This allowed movement to be quantified. In spite of all, such clocks had many disadvantages.

Not only were these devices difficult to maintain and cumbersome to use, they were also not accurate in the long run. Water provided the driving power and as time passed, the level of water would fall. This in turn would reflect on the time shown by the clock. Each time the water ran out, it needed to be refilled. Since people had not developed any better alternatives, the water clocks remained in use in many parts of the world. Today they are very rare. One such beautifully crafted instrument in working condition is to be seen in Beijing's Drum Tower.

SUNDIALS

Sundials had been in use for a long time. Today they remain as a curiosity; only to be seen in museums, gardens of old houses and squares of ancient towns of Europe. The principle of the sundial was simple. If you kept a stick upright, then the tip of its shadow cast by the Sun would fall at a particular point on the ground. This point would be the same for a specific time on any particular day of the year. So from this the time could be deduced with fair accuracy. Though in principle it was simple, the theory behind its construction was much more complex. Building sundials was not only an ancient science it was also an art with its own terminology that carries with it a mystique, which to us today may seem rather quaint.

Photo: Courtesy Hannes Grobe.
Website: http://commons.wikimedia.org/wiki/File:Sundial_berggarten_hg.jpg

5.9 Sundial, Berggarten, Hannover, Germany. [CC-BY-SA-2.5]

The stick that cast the shadow is known as the *gnomon* and the shadow it casts is known as the *style*. Some refer to the gnomon as the style, which is wrong. Nevertheless, this changeover has become entrenched in common language. The shadow of the gnomon is allowed to fall on *a dial plate* or *dial face* or simply *face*. The face is marked by lines to indicate the hours and is therefore known as the hour lines. There may be other lines in addition to indicate the month or the horizon, the equator and the tropics. These markings on the face are collectively known as the *dial furniture*. It is traditional for a sundial to have a *motto* on its dial face, like "Come along and grow old with me; the best is yet to be" or "An hour passes slowly, but the years go by quickly". There is a particular point on the gnomon known as the *nodus*, which indicates the time of the day and some times even the day and the month on the dial face. This point is represented usually by the tip of the gnomon. The gnomon is aligned north south and its angle with the face plate is known as the *substyle* height; a strange way to describe an angle.

Though the original sundials were constructed empirically, they became more and more sophisticated as people came to learn more about the movement of the Earth and the Sun. The gnomon was placed so that it was aligned to the geographical poles. This ensured against distortions of the shadow of the gnomon that fell between the corresponding hours, before and after noon, say 9am and 3pm or 8am and 4pm. Moreover, the gnomon had to be also parallel to the axis of the Earth. Otherwise, there would be distortion again in the shadow, as the seasons changed. Thus a particular sundial can only work for particular latitude.

The sundials do not show the standard clock time. This is because of the Earth's motion around the sun is elliptical and like all other planets it goes faster at perihelion and slower at aphelion. Therefore, the translation from the apparent time that the sundial shows may vary from zero to as much as 15 minutes before and after the standard clock time. This correction is independent of the latitude.

Thus in the pursuit of perfection some dial faces have incorporated curved hour-lines and curved gnomons to allow for this fact. The designs of sundials have been taken to such heights that they can compensate for time zones or even for summer time and winter time. Today there are heliochronometers that are accurate to a minute or even less. It is noteworthy that during the Middle Ages portable sundials were developed, which could be used by navigators. In spite of all these ingenious innovations, the sundial was useless without the Sun. Thus it could not be used in bad weather or at night.

THE CANDLE CLOCK

To overcome this problem, people who worked late into the night in the past used candles. They were marked equally, at alternate intervals. In this way as the candlestick burnt down, it would indicate the time. There were no windowpanes in those days and the candle could be blown out by the wind. So a covering was made. Lacking glass, initially this was made from a thin strip of translucent material such as horn, thus the name lanthorn or lantern.

THE INCENSE CLOCK

Another exotic variation of the candle clock was the incense clock. Various incenses were incorporated in such away that each variety would burn at a particular hour in a consecutive fashion. From this people could tell the time by smell.

THE HOUR GLASS

The hourglass came into vogue with the advances in glass making. It consisted of two conical reservoirs of glass, which stood, one inverted over the other and connected at their apices. This resulted in a narrow waist. One of the glass reservoirs was partially filled with a measured amount of fine grains of dry sand and sealed. The whole was then mounted on a wooden frame with a flat base at each end and the two were linked with three of four narrow columns for support. Now, if the conical end at the bottom containing the sand were inverted, it would allow the sand to run down through the narrow connection in the middle to the empty reservoir now at the bottom. The rate at which the sand would run depended on the narrow waste between the two containers. The amount of sand and the size of the narrow opening in the middle would dictate how long the instrument was designed to run.

It was from the sandglass we get the phrase "the sands of time are running out". The hourglass got its name from the fact its emptying was adjusted to one hour for each cycle. It could be adjusted to any given length of time, such as three minutes in case of the egg timer, which is used in our kitchens even to this day. Though it often does not contain sand any more, but the principle is the same.

Some believe that the hourglass was used by the Venetians first. Others think it could have been used as early as the 11th century, because of its close connection with use of the knotted line in navigation. Though we do not know for sure, the hourglass probably came later for reasons relating to its technological development. The first evidence of an hourglass may be seen in the painting by Ambrogio Lorenzitti (1290-1348) in 1338. It is called, *"Allegory of Good Government"*. By the next century, the hourglasses came to be used widely. They found their way into churches, courts, offices, ships and even households.

The hourglass was utilized in other ways also. At one time in England, it was the measure of the parson's preaching abilities. If the sermon ran out before the hourglass did, then the parson was not considered a good preacher. Whereas, if the parson over stepped the hourglass people would leave. It was the same for the courts, where the pleader was given only a certain time to present his case. This was also measured against the hourglass.

THE PENDULUM CLOCKS

The idea of the use of the pendulum as a harmonic oscillator for a clock is attributed to Galileo. Some time about 1581 or 1582 while still a medical student at the University of Pisa Galileo observed that a lamp hanging from the ceiling and swaying in the wind. It appeared to him that it always took the same time no matter how big or small the swing. Being a medical student, he used his pulse to establish the interval of the time of each

Source: Wikimedia commons / The Yorck Project: 10.000 Meisterwerke der Malerei. DVD-ROM, 2002. ISBN 3936122202. Website: http://en.wikipedia.org/wiki/File:Ambrogio_Lorenzetti_002-detail-Temperance.jpg

5.10 Temperance holding an hourglass. Artist: Ambrogio Lorenzetti. [Public Domain]

swing. Galileo must have been a very cool person, many would have been so excited by such a discovery that their pulse would have raced! Lacking a stopwatch Galileo thought the amplitudes of the swings of various pendulums to be exact. In reality, it was only an approximation, but was near enough for small amplitudes. Thus using shorter amplitudes had the potential for making clocks. In 1637, Galileo attempted to devise a clock on this principle, but it did not work satisfactorily. Later his son was to take up the project again in 1649, but he died, so nothing came of it.

It was through Christiaan Huygens that Galileo's idea of the pendulum would become a reality. After hearing about Galileo's observation on the pendulum, in the year 1656, he was successful in making the first pendulum clock. The introduction of the pendulum increased the accuracy of clocks. It was a great step towards modern time keeping.

The principle of the pendulum clock rests on the fact it had to have a power source. Before the advent of the spring, this power had been supplied by weights on a pulley. It had to be pulled up every time one of the weights reached the bottom. If one tried to make a clock just on this principle, he or she would find that the weights would run down very quickly and the clock would run fast. So some form of control was needed to prevent this from happening.

Since the power was transferred to the clock face to drive the hour hand by a set of gears, known as the *gear train*, some resistance was offered. However this was not enough to modulate the speed, to allow the hour hand to show proper time. There were no minute or second hand in those days. The trick was to modulate the power, so that the power was supplied in small quanta. In this way, the power derived from the weights would not expend itself too rapidly. In order to do this a very ingenious mechanism was invented. It was known as the *escapement mechanism*. No one knows who designed this mechanism, but it was certainly present in clocks as early as the 13th century, when tower clocks could be seen in medieval towns. This design used at the period was known as the verge escapement. Interestingly it used a primitive balance wheel, which was known as *foliot*. It was now hoped with this clocks would keep better time. However, such hopes were belied. The foliot did not do its job well. Thus without an oscillating resonator like the pendulum, these early clocks were left without the essential element for keeping time accurately. They would soon fall out of time by about 15 minutes per day; some times even more. With the introduction of the pendulum, this discrepancy was narrowed down to around 15 seconds.

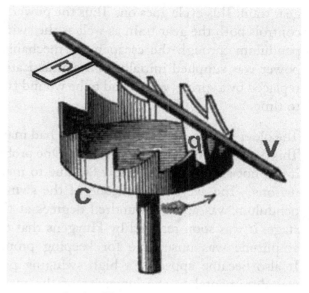

*Credit: Wikipedia user Chetvorno.
Website: http://en.wikipedia.org/wiki/File:Verge_Escapement_Labelled.png*

5.11 Verge escapement mechanism from A Handbook of Applied Mechanics, William Collins & Sons, London, fig.58, p.153. Drawing by Henry Evers. [Public Domain]

Labeling: c = crown wheel; v = verge; p&q= pallets.

Note: In the drawing there are 12 teeth on the crown wheel; this is an error. A verge mechanism must have an odd number of teeth, say 11 or 13, to function as pointed out by Wikipedia user Chetvorno.

The pendulum is a weighted rod, which beats equal time with each swing. This is also known as isochronism (iso=equal and khronos = time). Thus the pendulum clock utilizes this swing to keep correct time. In theory, once the pendulum starts to swing, it should continue to do so without interruption. In practice, however, over time it will eventually slow down and then stop altogether. The cause can be laid down to friction. Firstly in the mechanism and secondly, due to resistance offered by the air to the swinging pendulum. In order to prevent this from happening, a part of the power supplied is transferred to the pendulum. This shift of power is achieved by the *escapement mechanism*, which converts circular motion into oscillatory movement. Thus, the power it receives is changed into small impulses that go to supply the oscillating pendulum. This allows the pendulum to keep swinging. At the same time, the escapement mechanism ensures that the gear train does not speed up or slow down. The familiar "tick-tock" sound of the clock arises, because continuous delivery of power is interrupted at each step by the escapement mechanism, which stops and then releases the gear train. This cycle goes on. Thus the power supplied controls both the gear train as well as the swing of the pendulum through the escapement mechanism. This power was supplied initially by weights. Later it was replaced by a spring, which had to be wound from time to time.

The older *verge escapement* mechanism had many flaws. This became apparent early in its use. One problem was it did not allow a swing below 80° due to mechanical reasons. The average amplitude of the swing of the pendulum was about a hundred degrees at this early stage. It was soon realized by Huygens that this high amplitude was unsuitable for keeping proper time. It also became apparent a high swinging pendulum was detrimental to the longevity of the mechanism. Huygens thus redesigned the escapement mechanism. In 1670, he introduced the *anchor escapement* to replace the earlier verge escapement. The swing of the pendulum now could be reduced to about 5° only. Now the function of his clocks improved. With the swings becoming less the pendulum needed less power and they could be made longer. This enabled the "seconds" pendulum to be developed. It was 39.1 inches long and took exactly one second for each swing. Due to the pendulum, becoming longer clocks became taller and narrower. William Clement in 1690 designed a clock shape around such a mechanism, which came to be known as the grandfather clock.

The early pendulum clocks still needed frequent adjustments to maintain time. This was in a large part due to changes in temperature, which resulted in expansion and contraction of the pendulum. In winter, the pendulum contracted and thus its period was shorter, so that the clock ran faster. In summer it ran slower, because of the expansion in length of the pendulum, its period was longer. John Harrison invented the *compensated gridiron pendulum* about the year 1726, which guarded against such changes in length, due to fluctuations in temperature.

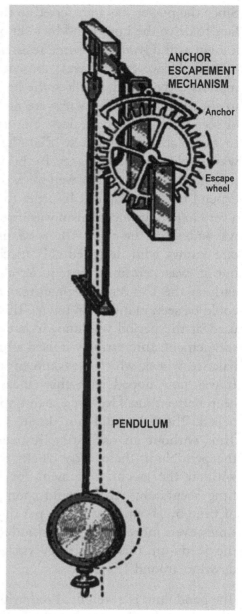

Website: http://en.wikipedia.org/wiki/File:Pendulum-with-Escapement.png

5.12 Pendulum and the anchor escapement mechanism from Silas Ellsworth Coleman (1906) The Elements of Physics, D.C. Heath & Co., Boston, p.109, fig.87. [Public Domain]

Huygens replaced the weights by a spring, which had to be wound daily. Later the "seven day" clocks appeared; where the clock had to be wound once every seven days. As time went on many refinements and embellishments were added, making these clocks into works of art. The wooden cases could be inlaid and the clock face made very ornate. Some of them were even designed to show the phases of the moon. The pendulum was eventually replaced by the balance wheel, so clocks could be made smaller. Now the clocks, like the French carriage clock, could be taken on journeys.

WATCHES

There soon emerged a new elite amongst the artisans, they were the watchmakers. They were the master craftsmen of their age. The invention of reliable springs to replace the pendulum, along with a suitable escapement mechanism paved the way to development of the watches. As early as the 16th century, Peter Henlein (1479-1542) a locksmith turned watchmaker from Nuremburg is recorded to have made the first pocket watch in 1524. These earliest watches were not of sufficient accuracy as they only had an hour hand. One such instrument was presented to the Mogul Emperor Jahangir by Thomas Roe a representative of the East India Company. The Emperor found it very useful, as he could know his prayer times in winter and summer with equal certainty.

Gradually these instruments became more and more refined. Taqui al Din Muhammad ibn Ma'ruf al-Shami al-Asadi (1526–1585), who was a judge, inventor and also appointed astronomer to the Ottoman emperor, Sultan Selim II (1566-1574) is reputed to have made the first watch with a minute hand.

By the seventeen hundreds, the watches were not only quite accurate, but also became very ornamental. They started catering for diverse requirements of their period till it became a rage at the time. A great many changes would take place over the following centuries. These exquisitely hand crafted pocket watches would eventually give way to cheap mass produced watches, which could be worn on the wrist.

THE NEW CHALLENGE

Our story would not be complete without mention of the development of the marine chronometer. It was the last link in the final achievement of point-to-point navigation, which we have spoken of in the discussion on navigation. With the sea routes already opened, it became necessary to find an accurate way to determine the longitude. The sextant could only determine the latitude. Determining the longitude accurately, remained an unsolved problem.

To solve this problem of determining the longitude Charles II of Great Britain founded the Royal Observatory at Greenwich in 1675. The astronomers of the time came up with a solution, which came to be known as the "Lunar Distance Method". In essence this method was based on the fact, the Moon moves at a certain rate, which is 33" arc across the sky to every minute in time. This makes the Moon a natural celestial clock. The mariners could now determine the longitude of the ship by consulting charts of the skies for the Moon's position at different times of the year and calculate the time by a system of references to Greenwich Time. This proved to be a complicated system. Moreover, the Moon was not always visible in the sky.

An easier method would be to use a clock, which had to be very accurate and sea worthy as well. This method was possible, because the Earth rotates 360° in 24 hours, so for each hour's difference in time would amount to 15° longitude. Thus, a 4 minutes difference would be 1° longitude. If the ship had one such accurate clock, which was set to the time of noon, when the Sun was at its highest overhead at the port where the journey started. This could be taken as a reference point in time relating to the longitude of that port. Today the Greenwich meridian, which is 0° longitude, is taken to be the standard, from which all other longitudes are measured. Thus by knowing the difference between the ship's local time and the standard, the longitude can be established.

The local time could be known for certain, if the ship carried a number of accurate clocks primed to certain fixed meridians through which it would pass on its voyage. One would have to be primed to the homeport, another set with the port of destination, others could be matched with the salient meridians. They then could be then compared to the ships own clock adjusted to the local time, which could also be got accurately from the Sun's position when it was at its highest or by knowing the position of a known star overhead at night from a star table.

Once this was known, the only thing then remained was to subtract or add this time to the time of the home port, depending on whether the ship was going west or east, for the longitude to be known. This is why there was an urgent need to develop an accurate clock, which came to be known as a marine chronometer.

The British had become established as the most powerful seafaring nation by the 18th century and to maintain their supremacy they took up the problem on a serious footing. At the time, Queen Anne (1702-1714) was on the throne. The British Government announced a prize of £20,000 in July 1714. In today's terms, this value would probably be more than £6,000,000. They stipulated that the prize would go to the person or persons, who could determine the longitude of a ship at sea to an accuracy of ½° of a great circle. This value translated into time was equal to 2 minutes or as distances at sea to about 33 nautical miles. A committee known as the Board of Longitude was set up, whose responsibility would be to test the authenticity of any claim and see that it is verified by testing it in actual practice at sea.

Many entered the fray for the prize, which amounted to a princely sum in those days. There were astronomers, clockmakers, scientists like Newton and even cranks. Many thought the problem insoluble. The watchmakers of those days represented the apex of the then technological world. Many tried, but failed to meet the exacting standards that were required to build this instrument. It was from an unlikely quarter that the problem was solved. John Harrison a carpenter devised a clock, which met all the required specifications of a chronometer that could be used at sea.

JOHN HARRISON

John Harrison was born in Yorkshire in Foulby village near Wakefield on 24th March 1693. The family subsequently moved to Barrow in Lincolnshire. His father was a carpenter and he grew up learning the family trade. Though he came from humble origins, but he showed remarkable abilities and intelligence that rivaled him as one the best minds of the day. This was the age when people took great interest in developing clocks; for many it was a hobby. So Harrison even though a joiner had developed an interest in making clocks from an early age.

By the time he was twenty, he had built a long case clock. What was remarkable, it was made entirely of wood including its mechanism! This instrument is to be still seen today at the Worshipful Company of Clockmakers' Guildhall, London. The reason for his building a wooden mechanism was that it did not need oiling. He had an innovative mind and could see problems clearly. He had realized one of the problems with mechanical clocks was that they needed to be oiled. The oil used during that period was not of suitable quality and could often clog the mechanism. Later he took up work for a clock that was placed on the turret of the stables at Brocklesby Park, North Lincolnshire in England, which would also not require any form of lubrication. This experience would help him to develop the marine chronometer one day.

Website: http://en.wikipedia.org/wiki/File:John_Harrison_Uhrmacher.jpg

5.13 John Harrison. (P L Tassaert's mezzotint (1768) of Thomas King's original portrait of John Harrison in 1767, located at the Science and Society Picture Library, London)[Public Domain]

A few years later while still in his twenties he designed what came to be known as the Harrison's compensated pendulum. In those days, the change in length of the pendulum, due the expansion and contraction as a result of variation in temperature, caused a pendulum clock to loose or gain time, making it necessary to carry out constant adjustments. Harrison ingeniously introduced alternate wires of brass and iron into the pendulum. The different rates of contractions of brass and iron at different temperatures would offset the problem, by keeping the length of the pendulum same in all weathers. Through this invention the pendulum clocks neither gained nor lost more than one second a month; an accuracy unheard of in those days. Later the pendulum came to be mounted on a frame consisting of alternate iron and zinc rods of suitable lengths. Soon the zinc was replaced with brass, which was made from zinc and copper.

In 1727, Harrison took up the challenge of designing a marine chronometer, which would fulfill the stipulation of the Board of Longitude. He put his idea on paper. The design of his new clock was in essence based on the plan of his earlier wooden clocks, but this time it was made sturdy by adding certain metal parts. Moreover while constructing it Harrison sensibly replaced the pendulum with a spring. The moving parts were counterpoised by springs to make it immune to the movements of the ship at sea, even in severe weather conditions. In 1730 he set out to London with the intension of expounding his ideas to the Board of Longitude and probably also in the hope of getting a loan from them. To make an instrument of that sophistication needed lot of money. We will see later in the second part of the book, a ship's chronometer would come to be priced at about a third of a ship's value! For Harrison to design and develop such a device from scratch would naturally cost more. Such a sum was quite beyond his means. At London Harrison was unable to locate the office of the Board of Longitude. He then went on and approached the Astronomer Royal, Edmond Halley at Greenwich. Halley directed him to George Graham who was not only a renowned watchmaker, but also a member of the Royal Society as well. Graham on hearing Harrison's idea was so impressed that he encouraged him to make the instrument. He even loaned Harrison some money to start on his work.

John Harrison now set out to transform his idea into a reality. After five years of work, he was able to complete his first marine chronometer. He called it the H1, which stood for Harrison - (mark) 1, as one might guess. In 1735, Harrison showed his marine chronometer at the Royal Society. The members were impressed. It also created a great interest in London society as well. The following year the H1 was put to test on the HMS Centurion. The ship reached Lisbon in Portugal. His instrument had performed well. In fact so well that a miscalculation by the ship's master (navigator) whose calculations had predicted they would see land earlier, was proved wrong. His calculations had put it 60 miles east of the actual position of landfall. Harrison, by using the H1, had predicted the actual distance and the time when they would see land. This earned him praise from both the captain and the master. In spite of the excellent performance, Harrison did not qualify for the Longitude Prize, because he did not demonstrate his instrument by sea trials to the West Indies as stipulated by the Longitude Act. Harrison wanted to improve on the design of the H1. He therefore did not want any further tests by going to the West Indies. This time the Board of Longitude was sufficiently impressed to grant him £500 to enable him to carry out further work on his design. He requested the Board for two more years and another advance of £500 to improve on H1 and reduce its size also. They however advanced him £250 only and told Harrison to submit his improved version for sea trials and then only he would receive the balance of £250.

Harrison was able to begin work on his next model and by 1740 he had developed the H2, which was a sturdier and a more compact version of the H1. He presented this to the Board. By this time, the War of Austrian Succession was in progress and the British Government and the navy considered it unwise to carry out sea trials. They could not risk such an important invention falling into the enemy's hands. So now Harrison had to be satisfied with testing it on land. After the trials, he realized that the performance of the H2 was not to his satisfaction.

He now started work on developing the H3. It would take him another nineteen years to perfect his chronometer. He requested the Board for further time and an advance of another £500 from the Board of Longitude. In the mean time, he kept on working on the H3 and took further advances of £500 in stages, totaling £2500 in order to complete the work. The H3 had a bimetallic strip incorporated in the balance spring to compensate for temperature changes. The other innovation was a caged roller bearing that was to become the perfect anti-friction device. Both these inventions of Harrison were to be used widely for watches and other devices later. In spite of all his efforts, the H3 did not perform well. This was 1753 that Harrison realized he could not improve the H3, if he depended on his previous designs. He started looking for new ideas.

The same year with this in mind Harrison arrangements with John Jeffrys to make a watch according to his design. He had intended it to be for his own use, which could be carried on his person. No one at that period could imagine that a time keeper could be carried on one's person.

In 1755, he went back to the Board and asked for further support for his work at this stage. It took Harrison another six years to perfect his H4. Its design was different from the others. It was just 5¼ inches in diameter and weighed 3 lbs and 3 ounces, small enough to be carried by any one. John Harrison was now 68 years and too old now to take a sea voyage. He thus entrusted his son with the testing of his latest model. His son set off on 18th November, 1761 on the HMS Deptford for West Indies. When they reached Port Royal, Jamaica on 19th January, 1762 the Boards representative there checked the H4's time. It was found that the H4 was only 5.1 seconds behind in terms of degrees, which meant an error of 2.75 miles only! This was a remarkable feat, considering the fact they had been 63 days at sea. When the news came, John Harrison now hoped for his £ 20,000 reward. His expectations however proved wrong. Even after such an excellent performance, the Board behaved in a niggardly fashion to one who deserved much more from his nation. They said that the performance was the result of a chance and not due to the merit of Harrison's instrument! Harrison was exasperated. This was a natural response to such an outrageous comment by the Board. Harrison had solved a problem, which was though to be insoluble by many. The matter was eventually referred to the Parliament. They offered Harrison only £5,000, which he rightly refused to accept. He was obliged to submit to a second trial. This new trial was made on the HMS Tartar, which set out on 28th March, 1764. His son was again on board with the H4. On arriving at Bridgetown in Barbados, the H4 was out by only 39.2 seconds in a voyage that took 47 days.

On this second voyage, a Reverend Nevil Maskelyne had been asked to be present. He was there to test the method of Lunar Distances so that the two systems may be compared. The H4 was clearly superior, being only out by 10 miles, whereas the Lunar Distance method used at the same time had missed by 30 miles. The later system moreover had proved much more complicated as explained earlier. The H4 had won hands down. The board of Longitude convened once again in 1765. This time however, there was our friend, Reverend Nevil Maskelyne, on the board. He had succeeded Edmond Halley as Astronomer Royal and this also explained the reason for his presence both during the second voyage and on the board.

The Board once again did not believe the results. To add salt to his wounds the Board further implied that the success of Harrison's chronometer was due to chance rather that due to the quality of performance of his clock. This time there are no prizes for guessing under whose influence the Board gave such an opinion. Maskelyne had given a negative report on the H4, which did not comply with the Board's criteria for accuracy.

Once again, the matter was referred to the Parliament. They told Harrison that they would only give him half the prize money. On receiving this money, he would have to hand over all four models to the Board. He would also be required to show them how his chronometer worked. Only then would he get the rest of the prize!

Harrison did not like the Board's attitude and refused to accept the Board's proposals. The Board remained adamant. Weeks dragged on and Harrison realized that he would not be given his due, so he had little alternative left. Eventually, he accepted the Board's deal reluctantly.

In 1765, six experts, appointed by the Board, went to Harrison's house in Lion's Square to look at the inner workings of his instruments. A week later, after reviewing details of the workings, the experts were satisfied. The Board now asked Harrison to hand over the four chronometers and asked him to name someone, who they could trust, to make a copy of the H4. Harrison felt that the Board was taking things out of his hands, but at the same time, he realized that he would not get the sum promised unless he complied. He named one Larcum Kendal. Finally, he received £10,000, a handsome sum in those days, but it was only half of what had been promised by the Board

Harrison was now left face to face with Nevil Maskelyne, who virtually controlled what the Board of Longitude had to say. Maskelyne was still unconvinced about the efficiency of Harrison's watches. He dogmatically persisted that the Lunar Distance was superior. Harrison was now put to another ordeal. The Board asked him to make copies of the H4 himself. Harrison was now 72. He and his son, William, constructed the H5 model. Kendal in the mean time was progressing. By 1769, he made a copy of the H4. William Harrison saw and testified to its excellence in the presence of the experts, who had earlier assessed the H4. This copy came to be known as K1.

Photo: Courtesy Racklever at en.wikipedia
Website: http://en.wikipedia.org/wiki/ File:Harrison's_Chronometer_H5.JPG

5.14 John Harrison's chronometer, the H5. [CC-BY-SA-3.0]

Still the Board was not satisfied. Now they changed their tactics and told Harrison that if he expected to get the rest of the money he had to make two copies of the H4 and these had to be also made by the Harrisons themselves! Harrison was now not in possession of the H4, it was with Maskelyne. Three years had gone by and Harrison was now 75 However he had managed to make a new model the H5.

All this proved too much for Harrison. John Harrison considered himself "extremely ill used" by these people, whom he out of politeness called "gentlemen", who in fact were mean minded. They were nothing but pen pushing bureaucrats, who were occupied with their pompous self-importance. Like a bunch of crows, they were trying to judge an eagle. Harrison now approached King George III (1760-1820), who granted him an audience on 31st January, 1772. William also went with his father. When the king heard his story, he expressed his anger over the treatment meted out to Harrison. He tested the H5 himself and on finding that it had lost 4½ seconds in ten days. The king was very impressed and at the same time very annoyed at the Board of Longitude's attitude. He is reported to have exclaimed, "By God Harrison, I will see you righted!" He advised Harrison to petition the Parliament and expressed his wish to appear on Harrison's behalf. It was in 1773 that Harrison received an amount of £8,750 from the Parliament for his work. However, he never got the full award, which he so much deserved. He was 80 years then. He died three years later on his birthday 24th March, 1776. He was buried at St. John's Church, Hampstead.

Thanks to John Harrison, the British could now place their ships on any point on the waters of the world. His achievement was one of the reasons the British were able to maintain their supremacy at sea. It is only today over the last few decades or so we have changed to using the global positioning system (GPS) to find positions of ships on the seas through satellites.

It was during John Harrison's lifetime in 1740's that England saw the Industrial Revolution and keeping time became important. Now there was a time for clocking in and a time for clocking out. The management saw to it that the workers came in time. The workers saw to it that they left in time. Now a great crime was born for which there is no adequate law - the crime of stealing time! The strict idea of time gradually pervaded into man's life as never before. It had literally eaten in to our lives, so much so that to keep to our schedules we are stressed. This affects our blood pressure, which in turn also reflects on the heart. In the long run, it ultimately cuts down on the quality of our lives. To solve this we have made great strides in medical sciences. That however is another story!

6 ARISTOTLE, GALILEO AND NEWTON

*"To follow knowledge like a sinking star,
Beyond the utmost bound of human thought."*
Ulysses by Lord Alfred Tennyson
(1809-1892)

ARISTOTLE'S IDEAS

Aristotle's influence has been felt across the centuries. We still talk of him today, because he was a man of many ideas and interests. Even though most of Aristotle's ideas eventually proved to be wrong, we still respect his efforts to understand the world around him. His interests were varied and he wanted to know about all that came to his notice. We have already had a taste of his interest in lofty matters, such as motion of celestial bodies. This did not however preclude him from being curious about mundane things. I will relate a story here, which I had heard from a friend of mine when I was still at school. Aristotle was interested in whether women had the same number of teeth as men. Since for obvious reasons Aristotle could not go looking into the mouths of other peoples' wives, he was reduced to looking at his wife's teeth only. The story goes that he found she had twenty-eight teeth and he had thirty-two. He therefore concluded, men possessed 32 teeth and women possessed 28 teeth. Today we know better. We all have a set of 32 teeth. It is only when the wisdom teeth are impacted they are often not visible. No doubt, this was the reason for his mistake. Though Aristotle's sweeping conclusions were erroneous, it showed the vivacity of his mind.

We have seen before that at one time, some of Aristotle's ideas had become dogma, but paradoxically it was one

Credit: Courtesy Ludovisi Collection (Inv. 8575)/Photo: Courtesy Wikipedia user Jastrow (2006)
Website: http://en.wikipedia.org/wiki/File:Aristotle_Altemps_Inv8575.jpg

6.1 Bust of Aristotle. Marble, Roman copy after a Greek bronze original by Lysippos from 330 BC; the alabaster mantle is a modern addition. [Public Domain]

of the factors responsible for bringing human thoughts back into the main stream of science. People had come to challenge his ideas from time to time and in doing so found the right path to a better understanding. Aristotle's great disadvantage had been time. The world was not ready for the answers to the various questions he had sought. All of that would come much later, not just by observations or through metaphysical thoughts, but by investigation and reason.

INERTIA

Aristotle believed a stone fell to the ground, because the ground and the stone were similar in nature. Both, according to his view, were composed of the element earth. He also believed smoke rose into the air, because smoke contained the element air. He envisaged that the heavenly bodies were pure in some way, made from some sort of quintessential substance, which allowed them to have continuous uniform motion in a circular fashion. Conversely, objects on Earth did not have the same degree of purity. Therefore, they had to be pushed, if they were to be moved. He concluded from this that it was the nature of the substance, which determines resistance to movement.

His ideas were reinforced by the fact that anything heavy, say a huge rock, would lie stationary under normal circumstances. It could only be moved when some great force was applied to it. In those times inertia meant resistance to movement. He went on to conclude, greater the object greater is its inertia and thus greater would be the force required to move it. He also noticed, if the force was no longer applied, the movement stopped. Thus, he came to believe, any movement was always associated with application of force. All this appeared quite reasonable to people at the time. However, this did not clarify everything. Aristotle found it difficult to explain, why arrows once released from the bow continued to fly through the air and land at a distance. Since, once the arrow was released from the bow, no force was being applied to it any longer! So how could the arrow continue to fly without any force to help it along? Aristotle's explanation was that after the arrow was sped, it pushed the air out of its way, leaving a partial vacuum in its wake. He then went on to say that, air rushed in again into this partial vacuum and pushed the arrow forward. The question remained, if this was so, then why was it that the arrow did not keep on flying? Here Aristotle applied one of his earlier ideas. He reasoned the material of the arrow was composed of the element earth and therefore it was drawn to the ground. This sums up Aristotle's ideas of force and motion. No one was there to question Aristotle for many centuries. We shall now see why.

FILLING IN THE HISTORICAL BACKGROUND

The Romans, who had inherited the Greek legacy, were great empire builders, but did not contribute any further thoughts to science. Their aristocracy was happy to leave the care of their children's education to Greek teachers. In doing so, they were to hold the Greek legacy for the future generations.

After the fall of the Roman Empire came the Dark Ages. As the Germanic tribes from the north overwhelmed the Roman Empire, the empire was split into two parts and the centre of gravity now shifted east to Constantinople or Byzantium, as it was then known. It was strategically placed between two continents and was one of the greatest cities of the time. Byzantium now became the seat of the Eastern Empire. The Western part of the empire was eventually overrun by the Visigoths. The ethos of the empire however remained in the minds of those who had been under its influence, including the conquerors. The Church took advantage of this fact. As the Christian Church's influence grew, it espoused Aristotle's ideas about force, motion and inertia. At this time most learning was centered on the Church. It had no intension of questioning Aristotle's ideas; it did not need to do so. It was a religious institution. As long as they could give the answers to any question, they looked no further. In the meantime, the Church was too preoccupied in expanding its influence across Europe and consolidating her position at the same time.

After the fall of Rome, a new power arose. They were the Frankish tribes in what is now Belgium. Clovis I (d.511), who till now had been a tribal leader, went on to gain dominance over the Germanic tribes to build what can be loosely termed as a kingdom - the Merovingian Kingdom (481–751). Though it was not a kingdom in the true sense, it brought a semblance of order in those turbulent times. The tribes that the Merovingian kings ruled had their own laws and did not always obey the king's orders. There was no centralized army to guard its boarders; no central bureaucracy; no tax collection by its kings; or any legal system, which was common to all. The kingdom was thus only nominally based on the allegiance of the tribal leaders.

There came a time when the military power of the Merovingian kings came to rest on the *mayors of the palace*. Initially they had been advisors to the kings and led their armies into battle. It was one of them, Pepin the Short (ruled 751–768), who finally displaced Childeric III (ruled 743–751), the last of the Merovingian kings in 751 with the Pope's blessings. The Carolingian power thus came into being. This name was derived from Carolus Magnus or Charlemagne

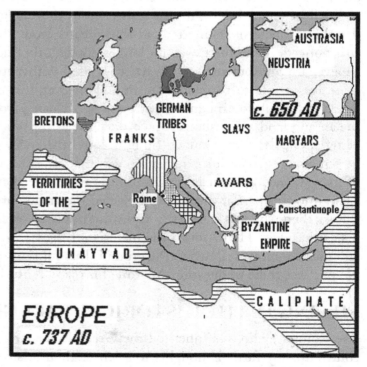

6.2 RIGHT: Map of Europe during the Dark Ages during transition period between the Merovingian and the Carolingian dynasties. Also shows the Arab expansion into the Iberian Peninsula. Inset: Frankish Kingdoms 737AD.

Redrawn by author from The Penguin Atlas of Medieval History by Colin McEvedy. Publishers – Penguin Books. Reprint 1986. ISBN 0 14 051152

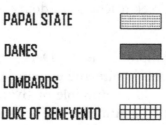

(742-814), who was the most prominent member of their dynasty. His ancestor Pepin of Héristal (d. 714) had already become the de facto ruler of the Franks as early as the year 687. He, however, refrained from actually taking over in spite of being offered the position as consul. Charles Martel (714-741), the son of Pepin by Chalpaïda, his concubine, took over power after his father. It was through him Austrasia gained supremacy over Neustria and Burgundy and united the Franks. He was also able to establish Frankish hegemony over western Germany. Later, it was he who would check the Moslem advance into Europe at the Battle of Tours in 732. This earned him the epithet "The Hammerer". Many have credited him as being the role model of our idea of knighthood. It is also said he encouraged the spread of chivalry in an age when barbarism ruled. His son Pepin III or Pepin the Short (741-768) was the father of Charlemagne (ruled 768 – 814).

The ancestors of Carolingian kings, in the person of Pepin of Héristal and Charles Martel were already inclined to the Church, now as kings they became a great support to the Church. It was with their help the Church gradually infiltrated the German tribes to obtain their conversion. This proved to be laborious process. They furthered their objective by making Charlemagne the Holy Roman Emperor in 800; an honour Charlemagne could not refuse. Nonetheless, it was a distraction for him. Charlemagne had his sights on the Byzantine crown rather than being the Emperor of some half-savage and unruly German tribes. However, the Church got its way. This takes us to the ninth century. It was later during the tenth century with the fading of power of the Frankish Emperors that the position of the Holy Roman Emperors came to be elected by the Princes of Germany. They were nothing but the glorified chiefs of the earlier German tribes after they had been brought under the influence of the Church. From here onwards, the Middle Ages emerged in Europe.

As a result of stories brought back by Christian pilgrims, who were harshly treated by the Seljuk Turks in the Holy lands, all Europe plunged into a Holy War, which begun in 1099 with the Pope's blessings. It came to be known as the Crusades. There were seven Crusades, all for the control of Jerusalem. Many fought with valor on both sides, but even more died; many of them not on the field of battle, but from disease and starvation. The Crusades ended by 1291. As if this was not enough, the Black Death that followed in 1348 left decimated Europe's population.

The feudal era had its origins in the last days of the Roman Empire and spanned well into the Middle Ages, until its back was broken by the Crusades and the Plague. The Middle Age itself would end with the beginning of Renaissance, which was marked by the use of the *moveable type* for printing by Guttenberg (c.1398-1468) in 1454.

The *feudal system* got its name from the word *fief*. It came about in this way. So long as the Roman Empire remained undivided, its eastern portion was profitable and therefore compensated the economic short fall of the west. When the empire split, the west had to support it self through heavier taxation and much of this fell on the peasants. To escape the burden of excessive taxes, a part of which no doubt went into the tax collectors pockets, the peasants used to take refuge under the most powerful landlord in their area. In doing so, the peasants gave up many of their rights, including their right to possess land. They now became tenants, who could be evicted any time on any pretext. In such deteriorating conditions, trade and industry also suffered. Europe was now poorer and had to fend for itself. Thus, the middle ages came to be notoriously short of money. Instead, they relied on free labor for everything, from building their castles to the labour in the fields. This is how things worked.

Initially the Germanic tribes who had settled on Roman land had chiefs. The most powerful amongst these chiefs, became their leader. In time, the leader's position gradually evolved to be king. The lesser chiefs, who were next in order, became barons. They owed allegiance to their king. A hierarchy gradually formed. The king reigned over the country, but did not necessarily rule. He might also own some land in a baronial capacity. Though he was entitled to receive taxes from the barons, he had no direct control over their lands. The barons, who were next in order of rank, owned vast tracts of land. They had either inherited the land from their ancestors or were given it by the king. They were thus vassals of the king. They had to swear allegiance to him in a ritual. This had to be done in a public ceremony in church or a gathering, where the vassal paid homage and pledged his fealty to his king. Below the barons ranked the knights. This form of expressing fealty went right down to any common man, who had been given a fief. A fief could be any thing from, say, land, a position at court or a military appointment for nobles. For the common man it could be a tollgate, a toll bridge, a ferryboat or even a piece of land. In return for the fief he held, the person had to make a pledge of loyalty. This oath was central to the integrity of the feudal system's structure. In times of peace, the vassal was expected to look after the interest of his lord. While in times of war, the vassal had to provide and equipped one person from his family to follow his lord. In doing this, the tenant would hold allegiance to the baron and not the king. This maintained the integrity between the king and his barons, the barons and their vassals; like knights, small landowners, such as franklins, who were free, but not of noble birth and any lesser people, if they happened to be the owners of a fief. This oath was sacrosanct and any deviation was considered a felony, thus placing the offender beyond the pale of the law. It could not only cost his life, but lose the fief for his heirs as well.

The plague and the crusades left the manpower of Europe reduced to nearly a third. This brought the feudal system to its knees, as the system depended on labour. Now with shortage of men, labour came to have a price. The kings and the nobility of European countries began looking desperately for money. Many nobles, who had sold much of their lands in order to pay for the crusades, now became impoverished. The Church in turn became rich and powerful. In the absence of the nobility, who went to the crusades, they took over many of their feudal rights. They held court and meted out punishment just like the lay nobility had done before them. Their bishops became more and more rapacious and drew envy as they flaunted their wealth and arrogance. With the unquestioning supremacy of the Church, there was no one to challenge Aristotle's ideas. Times however were changing. It became obvious to all that the clergy had strayed from the path of religion, except the very naïve or those blinded by religion.

In the mean time a new class was emerging - the merchants. Trade was now expanding. Under Prince Henry the Navigator, Europe saw changes in ship construction and the beginnings of exploration of oceans. Some who had started as bankers or merchants, like the Medici's of Florence, would become the neo-ruling class. Great changes also took place in the common man's life in Europe. The village black smith or his modified version the armourer, the carpenter, weavers, lace workers, silversmiths, goldsmiths, cobblers, stonemason and such others gradually came to be organized into, what came to be known as, *guilds*.

Each industry was controlled by a council formed by its members. The guilds became more and more important to the life of the cities. They even came to be more important than the earlier merchant guilds, which had been established in the 10th century. In time, they came to demand a hand in the civic administration. They were powerful in their own sphere like the unions of today. No outsider was allowed in to practice their trade. One had to become their member of which there was a stipulated number only. After becoming a member, they trained to become an apprentice. They learnt the rudiments of the trade under a master craftsman. If they attained the satisfactory skills needed, then they graduated to being a *journeyman*. This meant they were given a letter or certificate from the master or the guild itself, which allowed them to earn daily wages for their work. The word journeyman comes from either the French word "jour" or "journée", which meant a day. The English came to call it "journey", when it took on a new meaning, denoting day labourer. After a few years of this form of experience, they were allowed to produce an item of their trade, as stipulated by the guild. This had to be made at their own cost and in their own time. It is only after the work was scrutinized and its quality approved by the guild, a journeyman became a *master craftsman*.

With the coming of industrial revolution, the guilds ultimately fell into disfavour. The industrialists found they obstructed free trade, resisted technological innovations and did not want to share their secrets. They were a hindrance to their new system. However, the contribution to quality of work and society by the guild workers before the Industrial Revolution should not be underestimated.

There was much good in the system. For example, all prices were regulated. There could be no undercutting of prices by members nor could they charge too much. Plating was forbidden, as it could easily give rise to fakes. It was seen to that the quality of work was maintained. The market could not be cornered by any one member through buying up all the raw material. They looked after the needs of their poorer members or the families of those who died. They performed social services like giving dowry to poor girls, donated towards the building of churches and embellishing them by supplying such things as stained glass windows, which still can be seen to this day. They policed city streets at night. They also helped to raise armies during war. Above all, they contributed to two things. One to give a man the dignity of labour, which no slave, villein or serf had ever experienced before. The other was to provide lay education. These and other things, like the invention of the moveable type ushered in the renaissance.

Now freed from the constraints of the intense labour that went to create illuminated manuscripts, printed books were now able to spread knowledge as never before. People were now ready to ask questions. All this culminated in Martin Luther questioning the ways of the Church in 1517 and Copernicus' work in 1543. What had started off as a trickle before the Middle Ages would become a flood by the 16th century. It was just at this time that Galileo stepped in.

GALILEO

Galileo Galilei was born on 15th February, 1564 in the town of Pisa, which was then a part of the Grand Duchy of Tuscany. He was the son of Vincenzo Galilei, a musician, a famous lutenist and a master of musical theory and Giulia Ammannati. They had five other children, but Galileo was their first-born. In 1574, his family moved to Florence. He received his early education at the monastery of Vallombrosa, which was situated near their new home. He was considering entering the monastery, when his father induced him to take up the study of medicine. He enrolled himself in the University of Pisa in the year 1581. At the end of the first year at Pisa, Galileo is said to have noticed a lamp that was swinging in the wind. He noted that the period of the swing was the same by noting his pulse no matter how great or small the range of the oscillation. He was later to remember this fact and suggest this principle could be applied to the measurement of time in the form of a pendulum in clocks. Soon after this, Galileo chanced to hear a lecture in geometry, which made a profound impression upon him. This caused him to take up the study of science and mathematics under Ostilio Ricci. However before he could complete his degree he had to leave the university in 1585, because of financial constraints.

Source: Author – Kelson / Wikipedia user Pseudomoi Website: http://en.wikipedia.org/wiki/File:Galilee.jpg

6.3 Galileo Galilei. Portrait in crayon by Leoni. [Public Domain]

He now went back to Florence and lectured at the Florentine Academy. The following year he wrote a description of a hydrostatic balance he had invented. This invention was to make his name familiar amongst the scientific community of Italy. In 1588, he wrote a paper about the centre of gravity in solids, which assured his position as a scientist. It was to procure for him the position of lecturer in mathematics at the University of Pisa. Though this post was prestigious, it did not get him out of his financial difficulties. With this, he entered the second phase of his life. He was now 24 years of age.

It was here at Pisa that Galileo took to studying the motion of bodies. The story goes that he dropped a coin and a cannon ball from the leaning Tower of Pisa and demonstrated for all to see that both the objects fell on the ground below at the same time. This oft-repeated story has gained a mythical status in science. However much we might want to believe the story, it is not true! The story was suggested through the writings of Vincenzo Viviani, a student of Galileo's. The probability of the story being true or not has been investigated by many people, including science historians. They found that it was unlikely to have happened. It may have been quite possible that the student had misinterpreted what Galileo was trying to say in trying to illustrate his point, while teaching his students at a much later date.

In 1591, his father died leaving the care of the family to Galileo. The following year he moved to the University of Padua as professor of mathematics, though he taught mechanics and astronomy as well. Now his financial difficulties were somewhat allayed. He was to remain here for the next eighteen years. There Galileo continued his studies on the motion of bodies. This period of his life had been very fruitful. He was able to make fundamental discoveries regarding this subject, which no one before had understood much. People until now had been only repeating what Aristotle had said earlier.

In studying motion of falling bodies, instead of dropping them from a tower, what Galileo actually had done was to allow bodies of various masses to roll along an inclined plane. This was akin to a body falling, but only more slowly. This would make it more amenable to measurement, in order to judge their rate of progress, at a time when stopwatches were not available. It was very likely that Galileo used some type of water clock in doing his experiments.

6.4 A map of places associated with Galileo.

Galileo's work disproved what Aristotle had taught. By 1604, he was able to show that all free falling bodies travel at the same rate, revealing the important fact that acceleration of a falling body was independent of its mass. It is known to us today as the law of uniform acceleration. Though Giambattista Benededitti, a mathematician, had come to similar conclusions 51 years earlier in 1553, Galileo's achievement lay in the fact that he demonstrated this through experimental proof.

Galileo's work on the motion of bodies along with Kepler's work paved the way for Newton's laws of motion and his theory of gravitation. Though Galileo never enunciated the laws of motion in his writings, it is clear from his works on the subject that he understood their significance. Most importantly, Galileo realized that force was the reason behind motion.

Many may wonder, after gaining so much insight about motion of bodies, why a great mind like Galileo missed out on the idea of gravitation. It seems when it came to falling bodies, he did not extend his idea to the Earth as being a source of the force, which we now know as gravitation. Even great minds have their limitations. He visualized the falling of bodies not as the result of some force acting on them, but rather as an "occult" quality of the Earth. He thus failed to grasp the significance that it was force, which was this "occult" quality that made all bodies fall to Earth. Though Galileo missed out on one of the greatest discoveries of science, he struck out in a new direction, which was to have far reaching consequences for science and particularly for astronomy.

THE INVENTION OF THE TELESCOPE

In the spring of 1609, while in Venice, Galileo had heard of a new invention in Netherlands, which was made about that time. It was said it could make distant objects seem to appear nearer. There were three claimants to this invention, all of whom were Dutch. However, even today, the origin of the telescope remains shrouded in mystery. It was rather by chance than design, as Charles Hutton mentions in his Mathematical and Philosophical Dictionary

(1795). He goes on to say that according to Wolfius, one John Baptisa Porta invented the telescope in 1560. It seems, however, he did not understand its implication well enough to turn it to the heavens. Later, when Kepler_was asked to review Porta's description of his instrument by Emperor Rudolf, because of its vagueness Kepler was unable to make out what Porta had written. Kepler was no fool. Though such crude instruments may have been invented and even used before, it is certain that the modern telescopes originated from Netherlands. About thirty years later, we hear of a telescope, which was presented to Prince Maurice of Nassau by a spectacle maker from Middleburg. But his identity remains obscure. In his treatise on telescopes, Sirturus in 1618 credited Hans Lippersheim or Hans Lippershay. While another author Borelli in 1655, who solely discussed this invention, credits Zacharias Jansen.

The first contender was Jacob (or James) Metius (?1580-1628). He was an instrument maker and was also an expert in grinding lens. His claim to having invented the telescope lay in the fact that he had applied for a patent. However, his claim had been preempted by Hans Lippersheim by a few weeks. Thus, many have doubted his claims. Moreover, he was known to have bought a telescope. The question is why did he buy one when he could make one himself? This is what Charles Hutton again had to say in his afore mentioned work. "In 1620, James Metius of Alcmaer, brought telescopes from Jansen's children, who had made them public; and yet Adrian Metius has given his brother the honour of the invention, in which too he is mistakenly followed by René Descartes."

The remaining contenders were both spectacle-makers, Hans Lippershey (1570–1619) and Zacharias Janssen (c. 1585–c. 1632). The story goes, Lippershey's children, while playing with his lens discovered distant objects could be magnified by placing two lenses in the line of sight. This was to inspire Lippershey's creation of the telescope. This story is quite plausible. His application of a patent is first mentioned, strangely enough, in a diplomatic document from the King of Siam in 1608. Although he failed to receive his patent, he was rewarded for letting them have his design for his "Dutch perspective glass". His telescope was the first recorded practical device that gave a clear image and had a magnification of x3.

The last contender was Zacharias Janssen or Sacharias Jansen, as he was known, lived in Middelburg in Netherlands. The story now takes on a new twist. This was the very town where Lippershey also lived. Not only did they live in the same town, but they lived on the same street as well! Their houses were just a few steps away from each other. It seems he and his father had made a compound microscope in 1595. This shows he probably had a better understanding of optics than Lippershey. Many years later when Janssen died, his son, Johannes, would testify under oath that Lippershey had stolen his father's invention. In all probability, Jansen has a better claim.

Galileo was able to acquire the description and promptly set about making one after he returned to Padua that year. His first telescope had a magnification of x3. He was quick to realize that the quality of the telescope depended not only on its magnification, but also on quality of the image produced by the lens. The less the distortion of the image produced by the lens, the better would be the image. The distortion depended on the smoothness of the curvature of the surface of the lens. Galileo ingeniously devised a method of determining the curvature while shaping his lenses. In this way, he was able to obtain a higher standard for his telescopes, which would make it possible for them to be used for astronomical observations. It was only appropriate that he was the first person to use it for this purpose. Galileo's telescopes came to be in great demand in all parts of Europe, so much so that he made it into a side business. He subsequently made other models and was able to achieve a magnification of x32. This was certainly an achievement for the time. On 25th August, 1609 he demonstrated his first telescope to the elite of Venetian society.

The greatness of Venice was due to its sea trade. To know which ships were returning from sea formed an important factor in their lives. The merchants keenly scanned the seas everyday. Returning ships usually meant a successful voyage. There were many who looked forlornly at the sea for days, only to realize that they had lost their fortune. The news from the seafront of Venice was somewhat like today's stock market, where a merchant may become rich or poor overnight. His credit worthiness depended upon whether his ship returned or not. The earlier they came to know about it the better. Now the telescope became an important devise by which such information could be obtained. They were now able to identify their ships, when they were just a speck on the horizon to the naked eye.

THE NEW ASTRONOMY

On the night of 7th January 1610 Galileo turned his telescope towards Jupiter and saw three of the larger moons of Jupiter, which we now know as, Io, Europa, and Callisto. At first he thought they were fixed stars in Jupiter's background. He went on to record that by chance he turned his telescope to the same region of the sky the next night. He found that these bodies had all shifted to the west of Jupiter. On the 13th a fourth body had appeared. This was Ganymede. Studying them over the next few weeks, he noted these bodies appeared and disappeared periodically. At the same time all of them appeared to move with Jupiter. This made him rightly conclude they were revolving around the planet. Galileo named them the "Medicean Stars" in honour of his patron Cosimo II, the Grand Duke of Tuscany and his three brothers. Today we call them the Galilean satellites. He recorded all these observations in his short, but important work called *Sidereus Nuncius* (*Starry Messenger*), which he published in March 1610.

Credit: Courtesy NASA/JPL/Malin Space Science Systems.
Website: http://photojournal.jpl.nasa.gov/catalog/PIA04532

6.5 The photograph resembles what Galileo saw on the night of 7th January, 1610 as seen through a modern telescope. (Note the order of the moons are not the same as Galileo had seen.) [Public Domain]

Later Simon Marius (1573-1624), claimed this great discovery for himself in his book *Mundus Jovialis* (1614). His claim that he made his discovery in November 1609 is difficult to substantiate. It seems that the earliest picture of his observation he records of Jupiter and its moons in his book, appears to have matched Galileo's diagram of 7th January, which could have been possible but not probable. if he had made an earlier drawing the chances were it would look different. His case is not helped by the fact that he had helped one Balthazar Capra to plagiarize Galileo's notes for a military compass. This is not to be confused with a magnetic compass. It was more like a geometric compass, which had useful scales at the side to be used for military purpose, including determining the elevation of the barrel of a canon in order to fire a cannon ball at a particular distance. Moreover, unlike Galileo's conclusions, which were to break the backbone of Aristotelian and the Ptolemaic system, Simon Marius ideas supported the Tychonian system, which Galileo's observations had proved false. However, it was he who on the suggestion of Kepler, named the moons after Jupiter's loves, as shown in his writing, "This fancy, and the particular names given, were suggested to me by Kepler, Imperial Astronomer, when we met at Ratisbon fair in October 1613. So if, as a jest, and in memory of our friendship then begun, I hail him as joint father of these four stars, again I shall not be doing wrong." He also observed the Andromeda galaxy and described it as being like "a candle shining through horn".

On the other hand, Galileo's discovery of Jupiter's moons leaves us with no doubt at all. On the side shows a photograph of a draft letter by Galileo to Leonardo Donato, the Dodge of Venice, about the year 1609. It translates as follows, "Most Serene Prince, Galileo Galilei most humbly prostrates himself before Your Highness, watching carefully, and with all spirit of willingness, not only to satisfy what concerns the reading of mathematics in the study of Padua, but to write of having decided to present to Your Highness a telescope that will be a great help in maritime and land enterprises. I assure you I shall keep this new invention a great secret and show it to Your Highness. The telescope was made for the most accurate study of distances. This telescope has the advantage of discovering the ships of the enemy two hours before they can be seen with the natural vision and to distinguish the number and quality of ships and to judge their strength and be ready to chase them, to fight them, or to flee from them; or, in the open country to see all the

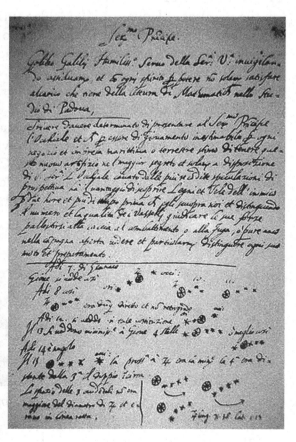

Credit: Courtesy NASA/JPL/Malin Space Science Systems. Prepared by Adrian Pingstone - NASA/JPL Website: http://en.wikipedia.org/wiki/File:Galileo.script.arp.600pix.jpg.jpg

6.6 A draft of a letter and the doodles below. [Public Domain]

details and to distinguish every movement and preparation." Below the draft, the reader will note that there are some doodles. They lay ignored for over three and a half centuries. The document passed many hands. In the end it was gifted to the University of Michigan by Tracy W. McGregor in 1938. It now lies at the Harlan Graduate Library's Special Collection. It was not until the turn of the eight decade of the last century that someone realized that they recorded Galileo's discovery of the satellites that go by his name.[22]

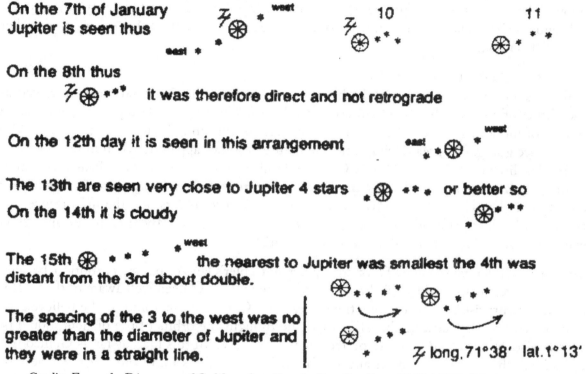

Credit: From the Discovery of Galilean Satellites by Ron Baalke - NASA/JPL (Under diagram translation).
Website: http://www2.jpl.nasa.gov/galileo/ganymede/discovery.html
6.7 Interpretation of Galileo's doodles. [Public Domain]

When Galileo also turned his telescope to the Moon, he showed the Moon was not plane as previously thought. It had mountains. He looked at the phases of Venus. He also realized that the Milky Way consisted of a collection of stars. He thus opened a new chapter in astronomy with the use of the telescope. More importantly, he realized that the speed of light, though very great, was finite. This he concluded from the behaviour of Jupiter's moons. Later Ole Rømer was to calculate the speed of light from this in 1676 from similar observations.

The Venetian senate gave Galileo the honour of a lifetime appointment at Padua University, but he left the position in 1610. Instead he became the "first philosopher and mathematician" to Cosimo II. The new appointment would now give him plenty of opportunity to pursue his researches unhindered.

THE CLOSING PHASE OF GALILEO'S LIFE
It was the year 1611 Galileo was now 46 and entering the last phase of his life, little did he know

that it would lead to his undoing 22 years later. That year he went to Rome to demonstrate his new instrument to the dignitaries of the Papal court. He got a flattering reception there. In 1613, he wrote *Letters on Sunspots*, which was published in Rome. This made his leanings towards the Copernican theory obvious. In this work, he deduced by observing sunspots, the Sun rotates on its axis and also the Earth revolves around the Sun. This, along with his earlier observation that the moons of Jupiter orbited the planet, made the geocentric model untenable. He made his views clear to students in his lectures. Galileo's lectures were very popular, because of his exceptional gift of oratory. This aroused great jealousy amongst his colleagues, whose classes went empty. We have already seen what all this led to in an earlier part of this book.

AN APPRAISAL OF GALILEO

Galileo is an important figure in the history of science. He has been variously described as the "father of modern observational astronomy" or the "father of modern physics" or even the "father of modern science". Whatever one chooses to call him, Galileo was the person who got science going. His great insight into the nature becomes apparent, when in response to a pamphlet directed at him by Orazio Grassi, a Jesuit priest, Galileo made a brilliant counter in his *Saggiatore* (Assyer). He said the *"Book of Nature is written in mathematical characters"*; a profound observation far in advance of his time. He also understood the importance and the necessity of testing ideas, through what he called *"ordeals"*. By this, he meant doing experiments. This revealed Galileo as a true scientist.

One may wonder why Galileo failed to appreciate Kepler's laws of planetary motion, because these laws were discovered in his lifetime and so he must have been well aware of them. This should not surprise us, because like the Greeks and Copernicus before him, he believed the cosmic order to be faultless. Thus, like them, he thought the orbits of the planets should be perfect also. What could be more perfect than a circle? With this idea, as far as Galileo was concerned, the answer had been already found. Thus Galileo's mind did not take the next step, which would be the question "if the paths of the planets were circular, why was it that their actual positions did not coincide with their predicted positions?" If a question does not arise in the mind, then one cannot go looking for solutions. Thus, he failed to grasp the significance of Kepler's laws of planetary motion.

Regarding gravitation, again Galileo thought he already knew the explanation. Thus he felt there was no need to look any further. He considered that objects fell to earth because of an occult quality, which the Earth possessed. It is true gravity can be considered as a hidden quality. Even Newton, who gave us the idea of gravity and the mathematics of how it worked, could not explain its nature. Neither can we do any better today. However Galileo thought of it in terms of a mystic quality, rather than a force, so he failed to enquire further into the matter. This lapse into the realm of metaphysical led to his failure.

During his lifetime, Galileo had never been able to make the clock designed on his idea of the pendulum. After Galileo's death, his son had tried to make a pendulum clock according to Galileo's conception, but he was never able to make a working model. His other contribution was to make improvements on the microscope. This was after he had seen one in Rome, which was made by Dutch inventor, Cornelis Drebbel (1572-1633). Incidentally, it maybe of interest to the reader to know, Drebbel was the first to make a working submarine in 1620.

Later in 1624 he made an improved version and took King James I of Great Britain on a short journey underwater in the Thames.

Printing saw to it that throughout Europe, amongst whoever knew how to read, had heard what Galileo had to say. His support for the Copernican theory was an important step in swaying the people of Europe away from the geocentric idea and this proved to be an important step in the history of science.

GALILEO'S CHILDREN

Galileo was not married though he had fathered three children by Marina Gamba. They had two daughters and a son. The eldest daughter, Virginia, was born in 1600 and his second daughter, Livia, was born the following year. Being illegitimate, they were not considered marriageable in those days. Their only option lay in entering a convent, which they did. His son, Vincenzo was born in 1606 and was legitimized later.

HIS LAST DAYS

Galileo's final years were sad. He had become blind, but he kept on working until his last day. This was only possible, because his disciple, Vincenzo Viviani (1622-1703) came to help him. Viviani would become a scientist and mathematician in his own right and would one day celebrate the works of his master by recording them in stone on the façade of his house. Evangelista Torricelli (1608-1647), who was already an established mathematician, also helped Galileo during the last three months of his life. In the end Galileo caught fever while dictating his ideas on impact. to them. Galileo died on the 8th January, 1642, while still under house arrest at his home in Arcetri.

NEWTON

Isaac Newton was born on Christmas day in the year 1642 at Woolsthorpe Manor in Woolsthorpe-by-Colsterworth, a small village near Grantham in Lincolnshire County, England. There is a discrepancy regarding Newton's date of birth, because England had not yet adopted the Gregorian calendar due to religious scruples. His actual date of birth was 4th January, 1643. He was born prematurely and was a posthumous child. His father had died the previous year, a few months before his birth. His mother, Hannah Newton, née Ayscough, married a preacher named Reverend Barnabus Smith three years later; leaving Isaac under the care of his maternal grandmother.

Isaac Newton started his education at the village school. When Newton was about twelve, he went on to attend King's School at Grantham. He was an indifferent student at first. It is said the turning point came, when he became involved in a fight at school with a fellow student from which he emerged as the victor. It is thought that this raised his spirits and from then onwards he became the best student in his

Credit: Courtesy Portsmouth Estate. Website: http://en.wikipedia.org/wiki/File:GodfreyKneller-IsaacNewton-1689.jpg

6.8 Isaac Newton by Godfrey Kneller (1689). Owned by 10th Earl of Portsmouth. [Public Domain]

class. Though there may not be any reason to doubt this story, but the cause of his becoming a successful student could not be that simple. Newton must have been often subject to bullying, which is not unusual amongst peers at school. Boys of that age can be cruel, especially as he did not have a father. Thus for this reason Newton must have found school depressing. Newton showed his mettle by winning the fight and therefore earned the esteem of his fellow students. The bullying stopped and Newton's genius took over.

In 1656 his mother became a widow again for the second time and came back to live in Woolsthorpe. Newton's mother brought him back home to look after their farm. Newton was 14 years at that time and it is not surprising that a boy of his intellectual abilities had distaste for farm work. Instead, he occupied himself with mathematics. In 1660, his maternal uncle William Ayscough, who was rector of Burton Coggles and a member of Trinity College, Cambridge, prevailed on his sister to send her son back to school in preparation for a future academic life.

He matriculated as a subsizar at Trinity College in June of 1661. A subsizar was a student who did not have to pay tuition fees, but instead had to make it up by helping out in the kitchen and doing other chores for his institution. By 1664, he was elected scholar and in January 1665, he took the Batchelor of Arts degree. Two years later, he was elected fellow of the college. That year due to the plague, the university was closed and Newton returned to Woolsthorpe.

RETURN TO WOOLSTHORPE

The months that he spent at Woolsthorpe during the Great Plague were very fruitful for Newton. During the first three months of his stay, he discovered what we know as the binomial theorem. He then went on to discover differential calculus, which he termed "*fluxions*". It was a landmark discovery. Soon he devised a form of calculation that would give areas under curves and volumes of solid shapes whose sides or bottoms were curved. This would come to be known as integral calculus. It was also here at Woolsthorpe that Newton got his first insight into gravitation on observing an apple fall from a tree. He was later to workout this idea and express them in mathematical form in his book, *Principia*. These discoveries would have been enough to assure his fame. His genius however would not stop here.

He was to write later "I was in the prime of my age for invention, and minded mathematics and philosophy (read science) more than any time since." He returned to the university when it opened again in the spring, the following year.

BACK TO CAMBRIDGE

Newton went back to Cambridge in 1667, but did not publish any of his discoveries. His teacher, Isaac Barrow, who was also interested in the study of light, came to know of Newton's work. Recognizing Newton's genius, Barrow stepped down from his chair in his favour. Isaac Newton now became the Lucasian Professor at the young age of 26 years.

At this period, Newton was involved with the study of light. To do this he had to grind his own lenses. He also worked with prisms and other glass surfaces to see how light behaved. Newton was not the first to show, when white light passed through a prism it would produce many colours of the rainbow. People were aware about this phenomenon, long before Newton came into the picture. At the time people thought when white light passed through a prism, the light became "stained", thus producing the effect of different colours. This was not very surprising, as they associated this phenomenon with colours produced by stained glass. This type of glass had been in use since the Middle Ages to produce the beautiful stained glass windows seen in churches.

THE RIDDLE OF THE COLOURS

Newton studied the emergent colours from a prism. He placed a board with a narrow longitudinal slit in the path of each of the colours, so that only one colour would be allowed to pass the slit. He then placed another prism along the path of this single coloured beam. No further colours were produced by the second prism. He did the same for all the other six colours with the same result. This proved conclusively, the prism did not have any inherent property of "staining" light.

He then went on to his next experiment, where he placed a second prism in the path of the spectrum formed by the first prism. This second prism was placed in reverse with respect to the first. The emergent light beam from the second prism was white! This indicated that the first prism had split white light into seven colours, but because the second prism was held in the reverse way, it had an opposite effect. It recombined the colours of the spectrum to produce white light again. Thus, Newton proved white light was made up of seven different colours.

Today we know that when light strikes a glass surface perpendicularly it passes straight on. Whereas, when light strikes a surface at an angle, it is bends. Newton knew this fact, but what he did not know is that the individual components of white light, which is to say the individual colours, have different energies levels, which depends on their wavelengths. Thus, when a beam of white light strikes the glass surface of a prism at an angle, the colour with highest energy, violet, is deflected slightly in its passage. The colour with a little less energy will be deflected slightly more and so on. Till we come to the last colour red, which posses the least energy and thus deflected most. Thus the different constituent colours of white light are sorted out according to the energy they posses. Newton's theory of the composition of white light could explain many other phenomena, which were not clearly understood till that time.

One such example was the contribution to the understanding as to why lenses of telescopes had aberrations. Such aberrations not only reduced the quality of the image, but also caused colours to appear at the edges of the image. Aberrations are of two types. First, there is spherical aberration, which happened due to the curvature of the lens. The more the magnification one tries to achieve, greater the curvature that will be required by a lens. In doing this the curvature at the margins will become relatively more compared to the central part. The rays coming through the periphery will thus be bent more than those passing through the centre. So the image will become blurred, as all the rays will fail to coincide at the point of focus. So in a powerful lens the image will be distorted unless some form of correction is provided.

Secondly, there was the problem of colours appearing in an image seen through the telescope was related to the edges of the lens also. This is known as chromatic aberration. The edge of the lens when seen in cross section looked like a prism. It is therefore not surprising that the edge behaved like one. These lenses naturally tended to produce a spectrum at the periphery of the image. At the time people thought such defects could not be overcome in anyway.

Later, the problems were solved, by using a convex and concave lens made from crown glass and flint glass glued to each other. This is known as a achromatic doublet. These two contrasting lenses bent light, each to a different degree, so they complemented each other. This breakthrough would come in 1729 from an unlikely source. Chester Moore Hall (1703 – 1771), an English barrister and amateur lens grinder, stumbled on the fact that if two lenses one convex and the other concave made of flint glass and the other from crown glass if glued together would resolve the problem. In order to keep his discovery secret he gave each type of lens to be made professionally by two different opticians. However it turned out that they in their turn subcontracted their work to the same person, one George Bass. He realized that the work had come from the same source and by putting the two lenses together he came to know the secret. This goes to show you can never be too careful.[23]

In the mean time, Newton solved the problem by converting the primary lens, used for collecting light in a telescope, known as the objective, into a concave mirror. The concave mirror instead of being placed at the front end of

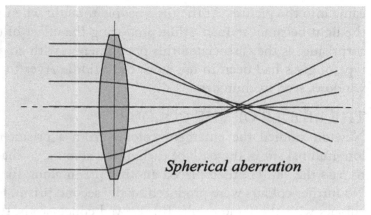

Credit: Courtesy Wikipedia User: DrBob (talk | contribs)
Website: http://en.wikipedia.org/wiki/File:Lens5.svg
6.9 Spherical aberration in a convex lens. [CC-BY-SA-3.0]

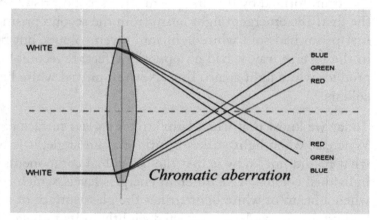

Credit: Courtesy Wikipedia User: DrBob (talk | contribs)
Website: http://en.wikipedia.org/wiki/File:Lens6a.svg
6.10 Chromatic aberration in a convex lens. [CC-BY-SA-3.0]

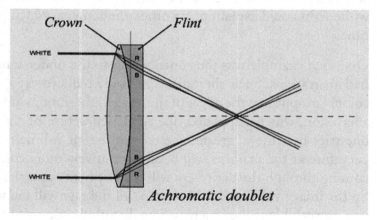

Credit: Courtesy Wikipedia User: DrBob (talk | contribs)
Website: http://en.wikipedia.org/wiki/File:Lens6b.svg
6.11 Correction of chromatic aberration by using by using one lens made of flint glass and the other from crown glass. [CC-BY-SA-3.0]

the telescope was placed at the back and the image was then reflected on to a flat mirror and back to the viewing lens, known as the eyepiece. In effect, this put the eyepiece of the telescope at the side of the front end. Whatever the advantages of Newton's telescope from the scientific point of view, it was certainly easier on the neck for those who viewed through it. The design was a new discovery and people came to know about it. He sent a model he had made to the Royal Society. It was six inches long with a two-inch mirror. Newton must have realized that though his use of a concave mirror solved the question of chromatic aberration, he would be still left with the problem of spherical aberration, if the mirror happened to be large enough. Christiaan Huygens was to point out this problem would only be solved, when parabolic mirrors could be made. It makes us wonder if this was the reason why Newton sent a telescope with a smaller mirror to the Royal Society. Newton could certainly have made a bigger model, which would be much more powerful and therefore more impressive. Today the largest telescopes are reflecting telescopes like those that Newton made, except that they all have parabolic mirrors.

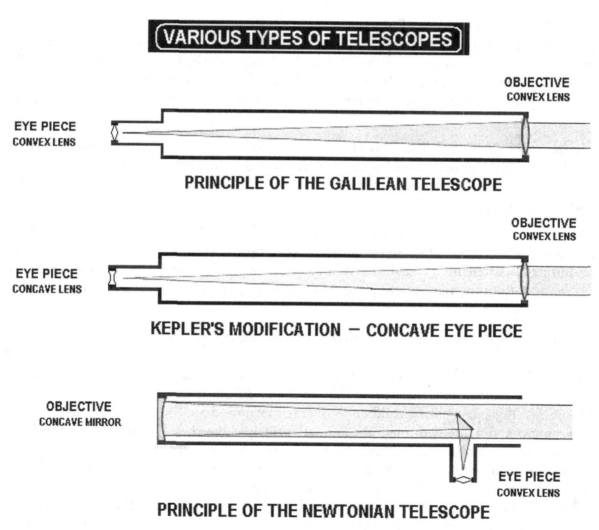

6.12 Various types of telescopes.

Newton sent the findings on his studies on colour and light to the Royal Society. His theories were misunderstood. Understandably Newton could not see why? His feelings regarding the matter were summed up "… the Theory, which I propounded, was envinced by me, not inferring 'tis thus because not otherwise, that is, not by deducing it only from a confutation of contrary suppositions, but by deriving it from Experiments concluding positive and directly." Then he goes on to say, "For the best and the safest method of philosophising seems to be, first to enquire diligently into the properties of things, and of establishing these properties by experiment, and then to proceed more slowly to hypotheses for the explanation of them." He goes further to say to his critics, "certain properties of light, which, now discovered, I think easy to be proved, … which if I had not considered them true, I would rather have them rejected as vain and empty speculation, than acknowledged even as an hypothesis." In the end, Newton was tired of all the criticism directed at him, which he felt was unwarranted. It left him bitter. In December 1675, he was to write, "I was so persecuted with discussions arising out of my theory of light that I blamed my own imprudence for parting with so substantial a blessing as my quiet to run after a shadow."

CORPUSCULES OR WAVES?

Newton was also involved in another controversy. This time it was about the nature of light and the way it propagates. One school of thought championed by Christiaan Huygens (1629-1695), believed that light travelled in waves. While Newton argued light was particulate in nature. The two theories came to be known as the corpuscular theory of Light as opposed to Huygens' wave theory.

Huygens had already published his ideas in his work, *"Treatise on Light"*, in 1690. However, much of his thoughts on the matter had been conceived by him as early as 1678. He explained that light was waves. Robert Hooke had also suggested this idea earlier in 1664, but it seems that he had not gone into the matter any further.

Newton on the other hand thought of light in terms of motion of bodies, a concept with which he was already familiar. He based this idea on reflection of light from surface of mirrors and on refraction, where light is deflected as it passes through medium of one density to another. However both properties of light, reflection and refraction, were already known long before Newton was even born.

Leaving aside the metaphysical thoughts of the ancient philosophers concerning light, the only person whose ideas are worth considering during those times is that of Hero of Alexandria. It is recorded he was the first man to have noted that the angle of incidence was equal to the angle of reflection. From this, he deduced light takes the shortest path between two points. This is something children learn at school nowadays. Today modern physics states this somewhat differently through the Fermat's principle. Pierre de Fermat (?1601-1665) was a French lawyer and mathematician. Fermat's principle translates Hero's "least distance" to "least time" and states, *"the path taken between two points by a ray of light is the path that can be traversed in the least time"*. Today it is often taken to be the definition a ray of light. This is because if a ray of light travels through a number of different media, say through a composite lens made of flint glass and crown glass, then light is refracted many times. First from air to flint glass, then from the flint glass to the crown glass and again from the crown glass to the air. Here, the path taken is not the shortest path, which would be a straight line, but it is the path that takes the shortest time for the ray to travel.

The phenomenon of refraction of light was investigated by the Dutch astronomer and mathematician, Willebrord Snellius van Roigen (1591-1626) or Willebrord Snell who succeeded his father as professor of mathematics at the Leiden University in 1613. In 1621, he discovered the two laws that bear his name. His work however had been preempted by Ibn Sahl (c. 940-1000), who was an Arab mathematician and physicist. He lived in Bagdad and wrote a book, *On Burning Mirrors and Lens* in 984. Around 1602, Thomas Harriot (1560-1621), an English astronomer, mathematician, translator and ethnographer and a contemporary of Snell, rediscovered these laws independently. He had even communicated his results to Kepler, but for some reason did not publish his work. René Descartes (1596–1650) came to the same conclusion independently in 1637, which he set out in his *Discourse on Method*. However not agreeing with Descartes, Fermat proved this phenomenon using his principle of "least time" in 1657. By the time Newton came into the picture these properties of light, reflection and refraction, were already established on a sound basis.

Again, Newton's idea of the corpuscular nature of light was reinforced by his observation that light could not get round obstructions. This property of light appeared to be in contrast to sound, which could be heard even if one stood behind a pillar and everyone knew sound was transmitted as waves. Light on the other hand, he said, does not reach the observer under similar circumstances, but casts a sharp shadow. So Newton argued light could not be transmitted as waves. Newton's prestige went a long way to give support to his corpuscular theory of Light, which was to hold sway for the next hundred years. Round one went in Newton's favour. Nevertheless he knew well what he was saying did not explain everything.

At the time people already knew about a phenomenon related to light, which showed that light indeed could bend around corners. This property could be demonstrated if light was allowed to pass a very narrow slit. This contradicted Newton's idea that light threw sharp shadows. In fact, though we usually think of the shadow cast by bright sunlight as having sharp edges, nevertheless its edges are actually slightly blurred. We cannot distinguish this because the effect is only very slight. So it passes unnoticed by the casual observer. Francesco Maria Grimaldi (1618-1663), who was an Italian mathematician and physicist, discovered this phenomenon and named it *diffraction*. Robert Hooke (1635–1703) had also discovered this independently, even though he explained it in a slightly different way. The phenomenon of diffraction thus became an embarrassment to Newton's ideas and Newton as we have already remarked was aware of it.

Controversy and Newton was not unacquainted with each other. Newton had always been guarded in propounding his theories. He had already realized that his theory had flaws. In his first edition of *Opticks* in 1704, Newton proposed a dualistic theory, which implied that light behaved like both corpuscles as well as waves. Such a contrary idea involving opposites was not even to be considered in those days. People could conceive that something could have many different properties at once, but they would not conceive of anything having opposite properties at the same time; a thing, physics readily accept today. So Newton's suggestion lay dormant, only to be confirmed much later by physicists of the twentieth century.

THE CORPUSCULAR THEORY ECLIPSED

There was silence for the next century after the general adoption of Newton's corpuscular theory, which he had proposed in 1703. This silence was broken by the unheard cries of two children who were born during this time. One was Thomas Young born in 1773 and the other was Augustin-Jean Fresnel who was born five years later. The ominous cries of these two new born babies would portend no good for the corpuscular theory. Within the century of its birth crucial experiments were devised, which declare in favour of the wave theory. Newtonian theory would thus become untenable.

THE WORK OF YOUNG AND FRESNEL

Since light was waves, the wave theory of light predicted that light waves should interfere with each other in the same manner as sound waves. This fact was to be later confirmed by Thomas Young (1773-1829). Young was born in Milverton, Somerset. He was an English doctor who held a position at St. George's Hospital, London. As an Egyptologist, he helped in deciphering a few of the hieroglyphics in the Rosetta stone. Napoleon Bonaparte had brought this stone back from his military expedition to Egypt. History records that one of Napoleon's soldiers found a portion of a black basalt slab, when he was digging a fortification near the town of Rashid on the Nile delta in the year 1799. Hence the name Rosetta stone, which is a corruption of the name "Rashid". What proved to be of utmost importance, in this find, was that it had inscriptions in three languages hieroglyphic, Demotic and Greek! The significance of which was not lost on the men of science who had accompanied the expedition. Europe for some time had been keen to know about the great Egyptian civilization.

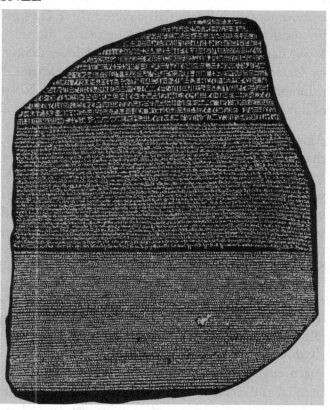

Source: *Courtesy European Space Agency (ESA) / British Museum.*
Website: *http://esamultimedia.esa.int/images/Science/rosetta_stone_50.jpg*

6.13 Rosetta (Rashid) Stone in the British Museum. [Public Domain]

Many tried to decipher the inscriptions, but failed. Young had based his researches on the work of Johan David Åkerbald (1763-1819), a Swedish diplomat and orientalist, who was a student of Baron Silvestre de Sacy (1758-1838), a French linguist and orientalist. When Sacy had tried to decipher the Rosetta Stone and failed, Åkerbald took up the challenge in 1802. His success was however only limited. He was able to decipher only one or two names, which were written in demotic. It was the brilliant Jean François Champollion (1790-1832) a French classic scholar, philologist and orientalist, who eventually was able to crack the code, would eclipse efforts of all the others. He went on to find the key to the grammar and wrote a book on it. However, Champollion had taken the help of Young's work that preceded his own.

Unlike Young, at first, he had failed to recognize that hieroglyphics were phonetic in nature and thought they were ideograms. Ideogram means a representation of ideas through pictures, whereas these hieroglyphics actually symbolized vocal sounds representing a system of spelling, which always used the same letter for the same sound. Young's work had been of crucial help to him in this regard. Later even though Champollion did not give the credit due to Young, Young on his part had applauded his success in a paper he wrote in 1822.

However, Young is better known for his outstanding contribution in the field of optics. He compared ripples in a water tank passing through two narrow slits to light that also pass through two slits. He showed both showed interference patterns. This meant when two waves met crest to crest or trough to trough, they reinforced each other, but if trough and crest or crest and trough met, they would cancel out one another. This confirmed light travelled in waves, because it behaved in a similar fashion to sound.

In his paper *Experiments and Calculations Relative to Physical Optics*, which he published in 1803, he described the phenomenon of interference of light by placing a very narrow card (less than a millimeter) in the path of a beam of light from a small opening. The edge of the shadow was fringed with colours. This behaviour was the same as light passing through a prism, where light bent to produce a spectrum. It showed that light did bend around corners. By placing another card suitably, before and after the narrow strip, these fringes could be made to disappear. This was further confirmation of the wave nature of light.

POLARIZATION OF LIGHT

However at the time, the proponents of wave nature of light did not realize that light waves and sound waves were not of the same nature. Sound waves travel longitudinally in the same direction as the sound is propagated, but the light waves travel transversely to the direction of their travel. What this means is that sound travel by creating alternate high pressure and low pressure zones that passes through a medium in the same direction as that of sound. The sound waves therefore do not have freedom to produce compression in any other direction. Thus longitudinal waves, which are compression waves, necessarily have to travel in one direction only. Light being transverse waves can travel forward, whether the waves are horizontally disposed or vertically disposed or for that matter disposed in any other way between these two options. This gives light waves the freedom to travel in more than one way along the axis of propagation. This can easily be understood by examining the picture. The various dispositions of the transverse waves are restricted by some means and only waves in one disposition are allowed through, it gives rise to the phenomenon of polarization. When we use Polaroid glasses, they allow the light waves disposed in one way to pass and reduces the glare by cutting out all the other waves. This makes Polaroid glasses soothing to the eyes.

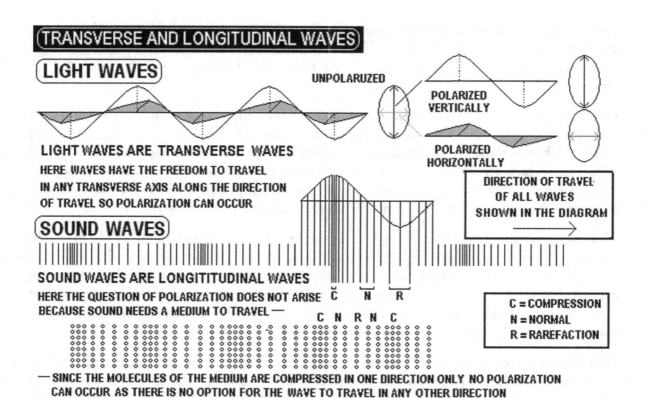

6.14 Polarization of light.

After Young there was Augustin-Jean Fresnel (1788-1827), he was a French physicist. His name today is associated with the Fresnel lens used in lighthouses. He was to develop the wave theory of light further. In 1817, he propounded his ideas at the Academie of Sciences. It was his ideas that were to shape our notion of wave motion of light and carry us into the turn of the twentieth century. Simeon Denis Poisson (1781-1840), another French mathematician and physicist further contributed by supporting Fresnel's work.

The wave theory implied light would travel more slowly in a denser medium, which contradicted Newton's corpuscular theory. Until now, the speed of light had not been determined accurately, so it was not possible to confirm whether it was true or not. Jean Bernard Léon Foucault (1819-1868), a French physicist, in 1850 was able to calculate and give a near accurate prediction of speed of light to a value of 0.6% less than the currently accepted value. This helped to confirm the predictions of the wave theory. Foucault's name is also associated with the pendulum of that name, which was used to demonstrate the Earth's rotation.

The mounting evidence for the wave theory of light dealt a body blow to Newton's corpuscular theory. Round two thus went to Hugyens' wave theory and Newton's corpuscular theory faded from peoples' minds. It was not until the beginning of the next century Einstein would come to Newton's rescue, but that is another story.

LUMINIFEROUS ÆTHER (OR ETHER)

The proponents of the wave theory of light, in order to explain the passage of light, introduced the concept of luminiferous æther or æther or ether. They assumed light like sound needed some medium to travel. Huygens was one of the first to introduce the idea of ether. He proposed there should be some medium between the eye and the object for it to be seen. According to his theory, space was filled with an extremely rare substance, ether. Huygens' ether was not to be confused with the chemical substance obtained from the action of acids on alcohol. Rather it was derived from the metaphysical ideas of the Greeks concerning the heavens. It was conceived as some sort of rarified fluid pervading all space, throughout the universe. In time, various properties would be attributed to it.

Light, he thought, travelled just like sound by causing a series of vibrations in ether, which are set in motion by the pulsations of the luminous body. In doing so, he unwittingly introduced a dogma, which would stick in the minds of scientists until Einstein came and threw it out of the window. We shall see how the story of light ends in the next article.

UNITING KEPLER'S THIRD LAW WITH NEWTON'S LAW OF GRAVITATION

When Kepler worked out the Laws of Planetary Motion, he had done so by trial and error. He, as you will remember, fitted his mathematics to the data of Tycho Brahe. He showed observations and the predicted positions of planets could only fit, if their paths were taken to be ellipses. No doubt, his was a great discovery, but for all Kepler's achievement, he did not explain the "why" of it all. This is where Kepler failed. This "why" could only be explained through Newton's laws of motion and by his the concept of gravitation. Interestingly, as we have noted before, Kepler understood that there was some force at work. However, he did not perceive its true nature nor did he attempt any mathematical interpretation, which Newton accomplished later. In failing to do so, Kepler's work on the Laws of Planetary motion was unable to gain the acceptance it should have had during his time. The world thus had to wait for Newton to explain it all. Moreover, Kepler's last law on planetary motion only worked for the Sun and the bodies that moved around it, but it did not apply for, say, the Earth-Moon system or binary stars. This point needs further elaboration.

NEWTON'S LAWS OF MOTION

Newton had expanded on Galileo's ideas in his three laws of motion. When we talk of motion here, it signifies uniform motion. It means that a body travels equal distances in equal time. In his, first law Newton states – *"A body will continue to be in a state of rest or in a state of uniform motion unless otherwise impressed upon by an external force"*. This sums up the concept of inertia. Here the word "inertia" has a different connotation from what Aristotle had originally implied. Newton says. *"Inertia resists not only the change in the state of rest, but also the state of uniform motion"*. To change either state, a force is needed. It is from the latter part of this law we can derived the definition of force. Thus force can be defined as, *"anything that tends to change or that which changes the state of rest or of uniform motion of a body"*. Thus, it may be taken the law is implicit on the fact that *no force is required to keep a body moving in uniform motion*. The reason behind this remains an unsolved problem in physics.

At the root of force lies energy. Energy supplies the force. Thus if a body at rest is to be moved or if a body is travelling at an uniform rate in a particular direction has to be speeded up in relation to the direction of the travel, then energy must be imparted to it. I always like to think of force as the transfer of energy, because it is only when energy is transferred it manifests as force.

Before we go on, I would like to make another point. I wish to ignore the part about "tends to change" for our purposes of our present discussion. This is because, here, we are only concerned with bodies in space where friction does not have a part to play. For the curious who want to know. All it means is that if you are trying to push, say, a heavy truck, you will not be able to make it budge. It is quite true that here you are applying a force, but due to friction, you cannot get it to move. If you want to move it, you must overcome the friction first. This happens wherever there is friction like we experience on Earth. Therefore, the force you are applying on the truck, though will *tend* to move it, but actually will not make it move. Thus, when we talk of Newton's laws here, it is implies that we are talking of bodies in space, which are not subject to friction.

His second law states, "The change in motion is proportional to the force that is applied". Note Newton does not mention for how long the force is applied. I have stressed this point here, because this is significant to our understanding of the matter. When Newton talks of a force being applied, we must understand what he is saying.

If a force is applied to a body for an instant only, then it will speed up the body to a new level. This new speed will be proportionate to the force applied and the body will continue to move at this new speed even after the application of the force stops. On the other hand, if the force continues to act, the body will then go on speeding up at a uniform rate proportional to the force as long as it continues to act. This is known as acceleration and is measure of distance coved in unit time per unit time. It is represented as say, 2 meters per second per second. Here the speed of the body is increasing at the rate of 2 meters each second as long as the force is

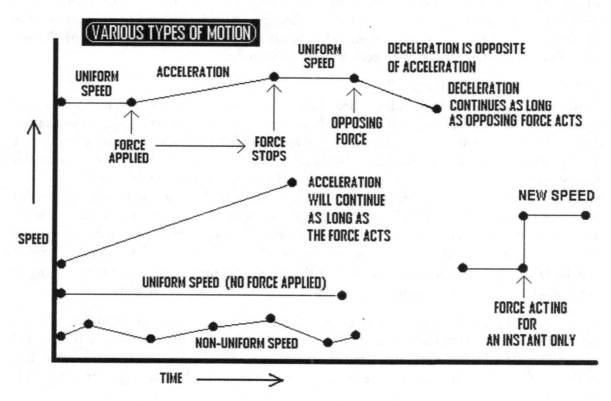

6.15 Various types of motion or the concept of inertial and accelerating frames in relation to force

acting. So if the body was travelling at, say, 3 metres per second before the force started acting, it will speed up to 5 m/ per sec. after the force acts for one second, then to 7 m/sec. after the force acts for two seconds and so on. Whereas, a body in uniform motion will travel at a constant rate, which will be represented by a uniform speed of, say, 2 meters per second.

Newton's third law of motion is the most widely known law of physics, except perhaps for Einstein's E= mc^2. It states, "every action has an equal and opposite reaction".

The force of gravity is an attractive force and is a property that all bodies possess, no matter how big or how small. Thus, the force of gravity due to a body exerts a continuous effect on whatever comes under its sphere of influence. Therefore, it is called *universal gravitational force*. As it is a continuous force, it will produce acceleration on any other body under its influence. The acceleration produced by the Earth's gravity is equal to 32 feet/sec/sec or 81cms/ sec / sec. It should however be remembered that the force of gravity decreases according to the inverse square law. Thus when a body moves further away from the Earth, the effect of the Earth's gravitational force gets lesser and lesser. The value of 32 feet/sec/sec will decreases as we recede from the earth's surface. The rate of acceleration produced by a body is therefore not only proportional to its mass, but the distance also.

The force of gravity is proportional to its mass. In a more massive body than the Earth, the acceleration will be greater than 32 feet/sec /sec., whereas, if the body is less massive the acceleration will be less.

NEWTON AND THE APPLE: HIS DISCOVERY OF GRAVITY
Newton already had a clear concept of inertia and force, when he observed an apple falling from his window, during his stay at Woolsthorpe. Thus observing the falling of the apple was not an inspiration, but acted as a catalyst to his thoughts. The result was the idea of gravitation. According to his own version, Newton stated that the idea of gravitation came to his mind when he noted an apple falling from a tree. When he observed the apple fall, he immediately realized that there was a force at work, which had caused it to fall from its position of rest on the tree. Otherwise, this fall could not have taken place at all! He knew from common knowledge that all things fell to earth. He was also aware that the Earth was round. So by extending the logic he concluded that a force must be coming form the Earth itself and that the nature if the force had to be an attractive one.

Also experience told him no matter how high the tree or where it was placed, whether at sea level or on a mountain top, the result would be the same – the apple would fall down straight to earth. So Newton realized this force was emanating from the Earth and was everywhere around it. He concluded that this force could extend far beyond the Earth and even to the Moon. Newton called this force gravity, meaning that which lends weight. Our mass is the same in what ever part of the universe we may be, but it is the presence of gravity that changes our weight. We would weigh less on the Moon, while, hypothetically, if we were on Jupiter we would weigh much more than we do on Earth.

NEWTON APPLIES HIS NEW CONCEPT TO THE MOON
The idea of gravity enabled Newton to work out how celestial bodies circled around their parent body. If the moon was travelling in a tangent to the Earth, then according to his first

law of motion, the force of gravity of the Earth was continuously acting to pull it back. Thus at each instant while the Moon was trying to go off its orbit in a tangent, the Earth was pulling it back. Therefore in the Earth-Moon system, the Moon kept on going in an orbit around the Earth, by a delicate balance. This was also, what happening to planets going around the Sun.

If the velocity of the Moon had been somewhat less, then it would gradually spiral inwards and plummet into the Earth with disastrous result. On the other hand, if the velocity of the Moon happened to be any greater, it would go off into a tangent and free itself from the Earth's gravitational influence forever. Today in fact we know that the Earth's gravity is very, very slowly releasing its hold on the Moon and the Moon is slowly moving away from the Earth. In time, after freeing itself from the Earth's gravitational hold completely, the Moon may become another planet will be going around the Sun.

Newton based his gravitational laws on Kepler's laws of planetary motion, though he did not acknowledge the fact. Kepler's first law states, *"Planets go round the Sun in elliptical paths with the Sun lying at one of its foci"*. Before Kepler, Copernicus and the savants of the past had thought this path should be perfectly circular, because of metaphysical reasons rather than reasoned speculation. The ellipse is only a slight distortion in case of planets.

The second law of Kepler states that *"equal areas, are swept in equal time"* by a line joining the planet to the Sun. This line is known as the *radius vector*, which in simple terms means it is the imaginary line, which joins the Sun to the planet moves in a particular direction. Now to understand the "why" of the second law, let us go back to the hypothetical situation, where a planet goes around the Sun in a perfect circular path. If the planet travels at a uniform speed, then it will necessarily cover equal distances in equal time as it travels along the orbit. In which case, the radius vector will also sweep equal areas in equal time. Now since an ellipse is derived from a circle, some of the properties of the circle will be inherited by the ellipse. Therefore, a planet going around the Sun in an elliptical path will also *sweep equal areas in equal time*. In the circle, this means the radius is sweeping equal distances on the circumference. Thus if the planet was going around in a circular path, there would be a ratio between the time and the area swept by the radius vector, which translates to the speed of the planet in its orbit. In case of an ellipse this relation is maintained, because, remember the ellipse is a distorted circle. Thus in an ellipse the radius vector sweeps equal areas in equal time also. This is a philosophical way of looking at the problem. Of course, it can also be derived mathematically. This is an example in science, where mathematics and philosophy goes hand in hand. It thus implies, if a planet moves in an elliptical path, *the area velocity* equals to the area swept by radius vector of the planet divided by the time it takes the radius vector to sweep this path, which is exactly what the second law states. However it also implies that when a body takes an elliptical path its speed around its parent body is not the same. Some times the body goes faster at other times it goes slower. This is what was causing discrepancies for astronomers before Kepler's time.

THE CONNECTION NEWTON MADE BETWEEN HIS LAWS AND KEPLER'S THIRD LAW

Now to reconcile Newton's law of gravitation with Kepler's third law it takes the following thread of reasoning.

Kepler's third law states, "The ratio of the squares of the revolutionary periods for two planets is equal to the ratio of the cubes of their semi-major axes."

Kepler as we have seen solved the problem empirically from Tycho Brahe's data. His third law worked for the Sun and planets only, because the Sun was so much more massive than the planets. The barycentre of any planet and the Sun is close to the centre of the Sun, so the Sun does not appear to move and the planet thus appears to go around the Sun. This is why Kepler's third law would not work for bodies where the masses were not so markedly different like, say, the Moon-Earth system. Nor would it work for two bodies of equal mass. Here the both the bodies would go around their barycentre, instead of one going around the other This is where Newton stepped in with his concept of gravitation. By modifying Kepler's third law to include gravity, Newton was now able to include all bodies, which went around a common centre of gravity, no matter what their masses relative to each other.

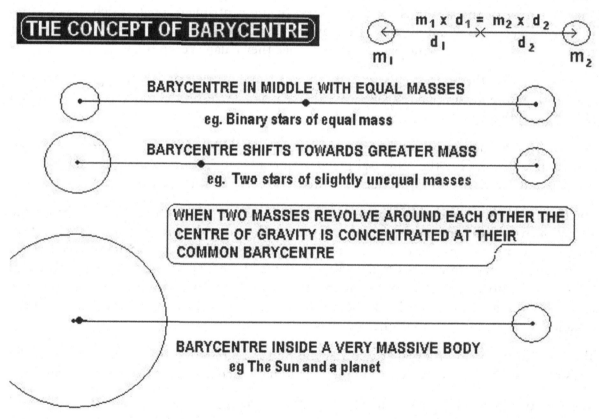

6.16 The concept of barycentre between two masses.

Newton's law of universal gravitation states that all matter, irrespective of its size, attracts every other matter, no matter how big or small, with a force that is proportional to the product of their masses and inversely proportional to the distance between them. He put it like this, $m_1 \times d_1 = m_2 \times d_1$ (where m_1 and m_2 denotes the mass of the first and second bodies respectively and d_1 and d_1 are the distances of the centre of these bodies from their barycentre.)

Now we come to the essential part, which reveals Newton's connection with Kepler's third law. Newton had concluded that the attraction of two bodies would not only depend upon their respective masses, but also on the distance between them. To understand the connection of Newton's law of gravitation and Kepler's third law we must once more look back to the Kepler's first law, but this time as explained by Newton. The planets, as they move in an elliptical path around the Sun, experience acceleration and deceleration respectively as they come closer to the Sun and move again away from the Sun. This revolution results in what is known as centrifugal force. It is suffice to know here that the centrifugal force tends to "push" a body away from the central point, around which it is moving. Thus, the centrifugal force is opposed by the force of gravitation to keep the planet orbiting the Sun. In other words, the gravitational force of the Sun works by pulling on the planet and draws the planet towards it, whereas the centrifugal force acts in exactly the opposite way. This conforms to Newton's third law of motion, which says, "Every action has an equal and opposite reaction." The concept becomes clearer if we consider what happens if we swing a stone tied to a string, with a spring interposed in the middle. The harder we swing, the more will the spring stretch. The stretching of the spring is a manifestation of the centrifugal force, which opposes the attractive force of gravitation. Thus, we may say that centrifugal force must be equal to the gravitational force to keep the planet in orbit since they balance each other.

Though we know the path of a planet to be elliptical, in most cases it a close approximation of a circle, therefore for simplicity's sake we will take it as a circle. So the solution can be worked by anyone with a little mathematical skill.

The force of attraction of gravity is represented by $F=GMm/r^2$, where F represents the gravitational force, G is the universal gravitation constant, M and m represent the mass of the Sun and the Earth respectively, r is the distance between their centres.

For those who are wondering where the r came from, we will elaborate a little. Since the attraction would depend upon $m_1 \times d_1 = m_2 \times d_2$. Now d_1 and d_2 represents the separation between the two centres. If we add d_1 and d_2 it still represents the same separation, but makes the mathematical picture look simpler. We can then add these two values, which now become transformed into a new value, but represents the same thing. By convention we call this new value r meaning radius; then it can be written as $d_1 + d_2 = r$, where 'r' now stands for the distance between the centres of the two masses.

The centrifugal force for a planet orbiting the Sun is represented by mv^2/r, where m is the mass of the Earth, v is its velocity and r is the radius, which is its distance from the Sun.

Since the two forces, the gravitational force (GMm/r^2) acting on the planet and the centrifugal force (mv^2/r) are equal. Then these values will take the form of the following equation:

$$GMm/r^2 = mv^2/r.$$
$$\text{or}$$
$$GM/r = v^2$$

Remembering that $v = 2\pi r/p$ for a circle, where $2\pi r$ is the distance of the circumference of the orbit of the planet and p is its period.

Now the equation will look like this:

$$GM/r = (2\pi r/p)^2$$

which is the same as saying

$$GM/r = 4\pi^2 r^2/p^2$$

or
$$p^2 = (4\pi^2/GM)r^3$$

Note that all values inside the bracket are constants and thus may be ignored, since their values would not contribute to changes in values of p or r in any way.

So now the equation looks like this:

$$p^2 = r^3$$

This brings us back to Kepler's third law.

Once the Earth's period of rotation and this distance from the Sun has been calculated in this way, we can then find out the values for the other planets from $p_1^2/p_2^2 = r_1^3/r_2^3$. You will remember that Kepler's third law on planetary motion deduces the ratio as being $p_1^2/p_2^2 = r_1^3/r_2^3$.

Though the mathematics involved for the ellipse instead of a circle is a little too complex to detail here, the answer will be the same. Such detailed calculations were used by John Couch Adams to predict the path of the new planet that lay beyond Uranus. Using the same method, the revolution of binary stars around each other and bodies with highly elliptical paths like comets may be calculated with great accuracy also.

NEWTON'S OTHER ACHIEVEMENTS

Though Newton's most fruitful period would be the 18 months that he spent at Woolsthorpe during the Great Plague of 1666, which was reminiscent of Einstein's golden year of 1905, Newton had other achievements to his credit.

In 1686, Newton started writing his book *Philosophiae Naturalis Principia Mathematica* or *Principia* as it is known for short. It was published in the summer of the following year.

In 1687, King James II came to the throne of England and Scotland. He was a Catholic monarch ruling a predominantly Protestant kingdom, because of the religious divide his policies came into conflict with his subjects. This resulted in Newton entering politics. He was elected to parliament for the university. By 1688, James II had fled the country. Upon the dissolution of the parliament in 1690, Newton returned to the university to pursue his work. Two years later, he was incapacitated by a serious nervous breakdown, which lasted a while. It is said that he was suffering from nervous trouble and insomnia. Some attribute it to the dissolution of a friendship with a Swiss mathematician with whom he was in correspondence; be that as it may, by 1694 he had recovered.

He was offered the post of warden of the mint 1696 by the then Chancellor of the Exchequer, Charles Montague the 1st Lord of Halifax, who had also been a fellow of Trinity college. Newton accepted the post and in 1699 became the Master of the Mint on the death of the then incumbent Master Lucas. The post at the mint was of a sinecure nature. This meant that though Newton was paid for the post he now held, but he had no serious responsibilities. However, Newton took his new position in earnest.

In 1701, he resigned his position at Trinity College having moved to London to take up his duties as Master at the mint. He was re-elected to parliament in 1701, but never took special interest in politics. In the mean time, he continued to pursue his duties at the mint very efficiently and brought to book many counterfeiters.

At the time Newton was appointed warden in 1696, it is estimated that 20% of the coinage was counterfeit. Newton set about on the task of remedying this situation. He gathered evidence himself by frequenting lowly taverns and alehouses. Newton was made justice of peace, which gave him the authority to cross-examine witness, informers and suspects, also bring to judgment such offenders. In the next eighteen months, he went through about two hundred such cases and brought about many convictions. The penalty for counterfeiting was death by hanging in those days.

One of the important and difficult cases where Newton gained success was against one William Chaloner. Chaloner was a counterfeiter. He was quite wealthy and passed himself off as a gentleman. He also had influence through friends in high society. Reminiscent of the strategy of Titus Oates at the time of Charles II, he engineered cases of Catholic conspiracies and brought these "conspirators" to justice. On the side, he himself pursued his business as a counterfeiter. Chaloner was obviously not a very nice person. Once he had the audacity to accuse the Royal Mint of providing the means of supplying the necessary tools to the counterfeiters. He went on to boldly propose, the government should adopt his plans of minting coins, which would be proof against any form of counterfeiting. No doubt, a person of his character would have plans to turn it to his advantage. In due course Newton brought him to court, but he went scot-free through his connections in high places. This made Newton furious. The next time around he was able to bring Chaloner to justice, on this occasion with irrefutable proof. He was tried and finally hanged.

In 1717 as Master of the Mint, Newton shifted the currency of Great Britain to the gold standard. The suppression of counterfeiting and reforms contributed greatly to the stability of the country's economy. When Queen Anne visited Cambridge in 1705, she knighted him. This signal honour was for his work as Master of the Royal Mint, rather than for his contribution to science.

In 1703, he became the President of the Royal Society. Newton continued to be re-elected to the post each year until his death. He also continued to hold the position of the Governor of the Royal Mint. His *Opticks* went through three editions during Newton's lifetime. His Principia went through further editions during his lifetime. In 1713, a third edition came out and one in 1726, the year before his death. After a sudden illness, Newton died at the age of 84 on March 1727. He was buried in Westminster Abbey.

NEWTON THE MAN

Sir Isaac Newton was a key figure in the development of modern science, but for all his greatness, Newton was not a pleasant man. He did not appear to have any relationships with women. This is not necessarily due to his intellectual abilities, because intellect has nothing to do with the ability or inability to develop relationships, as some people are prone to think. This aspect of his nature was probably due to other causes in his life. For one, he never knew his father. His resented his stepfather and they did not get on well with each other. All this indicates that he lacked a father figure. Also, his mother probably did not give him the love that a mother should, because she did not understand the genius in him. Newton therefore grew up as a lonely child and thus took to living in a world of his own, especially as he did not find anyone to share his intellectual abilities. He therefore learned to keep to himself. That is why people think he was secretive. Many people have said that Newton did not take criticism well. They give the instance of the time when he came to conflict with Robert Hooke. This accusation is wrong. We should remember, people during those times did not always understand what Newton was saying. So some times when people like Robert Hooke, who considered himself an expert on light, criticized him, it irritated Newton. This was quite natural, because he felt he was being criticized by some one, who himself was out of his depth. If Newton had ignored them, people would have said he was a gentleman. Here Newton expressed his annoyance in what appeared to be an unseemly flare up of pent up emotion, which in his case often crossed the boundaries of social etiquette. Thus people tend to judge Newton harshly in this matter, but we should remember Newton was not an average man. It is because of this, it is difficult for an average person to understand his reactions. Newton was submerged in his work. In all things, he felt very deeply. He did his work thoroughly. It was only when he was sure of what he was about to say, only then he would express his findings. Thus when faced with what appeared to be facile criticism, it is not unnatural for a man like Newton to express himself as he did. In fact, he found all this unsettling to him, so he withdrew into virtual isolation. All this turned Newton into a hard man.

For all his hardness, we get a glimpse of Newton's softer side. Once when Newton was 19 he lodged with a local apothecary and became engaged to his stepdaughter, Anne Storer. However, when his destiny beckoned, Newton went off to Cambridge University. There he became engrossed in his work. Anne married another man. We would like to think Newton carried with him those memories of his youth and thought of her at some quite moment in later life.

Newton had a darker side to his character. He could go a long way to get what he thought was rightly his due. He did not think twice about using devious means. When Newton came into conflict with Leibniz as to who was the first to develop what we call calculus. Newton claimed that he had developed calculus long before Leibniz in 1666, which he probably did. But he had nothing to show for this except for some bit of calculation on a tangent. He described this to a student of his some twenty years later. By then Leibniz had already published his work. Newton called his new discovery fluxions. We do not hear anything from him till 1693, but later he was to give a full account in 1704. He had kept it as a secret mathematical tool to arrive at some of his answers without letting anyone know how he got them.

When in 1684 Leibniz forestalled Newton by publishing his new discovery Newton was beside himself. He drummed up support in his favour. Incidentally, many of the letters purported to be written by his friends in his support appears to be written by his own hand! On Leibniz's part, it seems that he had seen and made copies of parts of Newton's work as early as 1674. In fact, some of these copies have been recovered from his papers. However apparently they did not give enough details as to how one should use this new mathematical tool, but then Leibniz was a mathematical genius of that age. The controversy continues to drag on even to this day. Unless new facts emerge, no one will know the true answer. The controversy however was to take a greater toll on the mind of Leibniz. He had made the cardinal error of applying to the Royal Society for resolving the dispute. Poor Leibniz did not stand a chance. Newton was the president of the Royal Society. He made sure the members who were appointed to arbitrate the issue were those who would decide in his favor. The situation was somewhat like that between Reverend Nevil Maskelyne and John Harrison, which we have described earlier.

To this day, it is not possible to tell who was right and who was wrong. The fact remains that Newton was something of a Dr. Jekyll and Mr. Hyde when it came to dealing with conflict of interests. This becomes obvious in his dealings with John Flamsteed, the Astronomer Royal and one Ralph Cudworth, a Platonist at Cambridge. In the instance of Cudworth, it is claimed by some, Newton stole a theory of atomism and not as Newton claimed by "turning to nature for truth".

In case of John Flamsteed, when Newton was preparing the *Principia*, he had helped Newton by supplying certain facts and figures. But later for some reason, not surprisingly knowing Newton, Flamsteed became reluctant to part with further information. Newton was so incensed that he used his influence and got himself appointed to the governing body of the Royal Observatory and tried to force the issue by trying to publish Flamsteed's data. When he did not succeed, he had the papers seized and arranged them to be published by Edmond Halley, who was Flamsteed's enemy. Flamsteed took the case to court, obtained judgment in his own favour. He got back his papers before Halley could publish them. Newton retaliated by deleting Flamsteed's name from all references in his *Principia*.

There was again another example of his allowing Robert Hooke's papers to be destroyed by default. Robert Hooke was a prominent scientist in his own right, but with whom Newton did not see eye to eye. After Hooke's death, his papers and writings lay with the Royal Society. Rightfully Newton as the President should have seen to it that the works of another great scientist of the day were preserved, but instead he saw to it that they were lost through neglect. This was most unfortunate for us.

When we think of Newton, a picture of a great man of science is conjured up in our minds. However, if we think of Newton in the perspective of his life, we will see a different man.

On one hand, he was a man of great intelligence and a great scientist; efficient and thorough in whatever he did; but at the same time lonely and loveless. On the other hand, he was a ruthless and ambitious person, who would brook no interference. On reflection, if one can imagine him under another setting, at a different time and in some other country, he might have been a feared dictator or even a successful leader of a mafia. Nevertheless, we must try to temper our understand of his controversial character, against the broader background of his life and in the light of those historical times.

Newton dabbled in alchemy. It appears that chemistry was not his forte. Boyle's *Sceptical Chymist* had already been published in 1661. He may have read it, but he apparently did not avail of these ideas. The following quote by the John Maynard Keynes (1883-1946), the British economist, even though made in a different context, would also sum up Newton's role as a practicing alchemist very well when he said, "Newton was not the first of the age of reason: but the last of the magicians".[2]

In spite of his scientific interests, it did not however stop Newton from having a deeper religious side to his nature. Some time in 1670's he started taking a deeper interest in religion. He had profound believe in God and the Bible, but rejected the Church's concept of the Trinity, which is God the Father, God the Son and God the Holy Ghost. He thought that Christianity had wandered off course. According to him, it happened when the Council of Nicaea in the 4th century put foreword these erroneous doctrines on the nature of Christ. That is why he excused himself from taking the cloth, which was one of the criteria for becoming a professor at Cambridge.

He argued his point to King Charles II, who allowed this great man to take up this post without becoming a pastor of the Anglican Church by royal decree - the Lucasian Professor need not be ordained. It is said he spent much time on thinking and writing on religious topics. He wrote *Observations upon the Prophecies of Daniel* and the *Apocalypse*, which was published posthumously. Newton also believed God created the universe after which it ran on its own. However, he rejected the idea of Leibniz and Spinoza that God and matter were identical, which is otherwise known as hylotheism. In the end, we get the feeling that he believed deeply in God and tried to understand His work through science.

---

2   This is the full quote, "Newton was not the first of the age of reason. He was the last of the magicians, the last of the Babylonians and Sumerians, the last great mind that looked out on the visible and intellectual world with the same eyes as those who began to build our intellectual inheritance rather less than 10,000 years ago."

7 UNION OF SCIENCE WITH PHILOSOPHY - A BIOGRAPHY OF EINSTEIN

"You see, son, when a blind bug crawls along the surface of a sphere it doesn't notice that its path is curved. I was fortunate enough to notice this."
Albert Einstein to his son.

INTRODUCTION

In antiquity, the responsibility of interpreting natural phenomena in the physical world rested on philosophers. In doing so, they depended solely upon their senses and their thinking powers in order to explain the world around them. It was only rarely, someone like Eratosthenes tried to test his conjectures by measurement. Such instances were rare exceptions. It was not until the sixteenth century these natural philosophers came to understand the importance of measurement. They realized measurement formed a sound basis on which information could be transferred, from one person to another, without the risk of distortion. This lent objectivity to a degree hitherto unknown. Measurement now became the new corner stone of science. At first measurements were restricted to only those attributes for which there were measuring devices. Like rulers and tapes for length, breadth, height; scales for weight; and clocks for time. Attributes like the degree of temperature, intensity of light, loudness of sound, hardness of materials etc. could not be measured. People took such attributes for granted. It never crossed their minds that such things needed to be measured at all. Gradually as people started to explore nature and came to understand, more and more, they now began to appreciate that these latter attributes were also properties in their own right. They now looked around and devised ways to measure such attributes. As the ability to measure different attributes widened its scope, people equated measurement with science. They thought it was the path to knowing reality, but in doing so little did they realize, that they would lose their bearings. Now people forgot that science was a discipline, an objective approach along a path of reason, designed to undercover the meaning behind all things. Its ultimate aim was to reconstruct the order of things, so that we may understand the sense behind it and not just measurements and facts. This lapse led to the gradual separation of science from philosophy. Philosophy was now viewed askance by the scientist. So much so that by the later part of the 19th century science had experienced a shift and scientists no longer considered themselves natural philosophers. It looked like philosophy had lost its place in science and materialism ruled.

When Michelson and Morley showed one could not add to the speed of light, it was a discovery that would shake the very foundations of classical physics. The men of science, however, did not react, as scientists should have done. As one would expect at first, they did not believe the results. It was left to Albert Einstein to reconcile the so-called "practical hard headed" objective science with philosophy once more. This had indeed been long overdue, since it is not just mathematics and physics, but philosophy as well that lies at the root of science. Without the philosophy, there can be no understanding of the meaning that lies at the heart of science.

EINSTEIN

Albert Einstein was born on 14th March, 1879 at Ulm, Württemberg in Germany. He was the son of Hermann and Pauline Einstein. His family moved to Munich the following year. Here his father started a company called Elektrotechnische Fabrik J. Einstein & Cie. They manufactured electrical equipment.

EARLY LIFE

When Einstein was five years old, his father showed him a compass. He was later to recall, it had left "a deep and lasting impression" on his mind. Even at that tender age, Einstein realized there was something in emptiness of space around the magnetic needle that was controlling its movements. Geniuses are born and not made. A few years later, he was making mechanical devices and models as a hobby.

In 1889 Max Talmud a family friend was to introduce the boy Einstein to science through works like Euclid's *Elements* and Kant's *Critique of Pure Reason*. These works were to arouse this 10-year-old boy's interest in philosophy, science and mathematics.

Website: http://en.wikipedia.org/wiki/File:Einstein1921_by_F_Schmutzer_4.jpg

7.1 Albert Einstein during a lecture in Vienna in 1921 (age 42). Photo by Ferdinand Schmutzer. [Public Domain]

HIS EDUCATION

While in Munich Einstein started to attend the Luitpold Gymnasium, which was a secondary school. At the period, which became known as "current wars", his father's business venture had failed in 1884. The world was moving onto alternating current, while they were still manufacturing direct current equipment. They moved to Italy in the hope of gaining buisness. His family first went to Milan and then to Pavia, where they finally settled. Einstein had been left behind to complete his schooling in Munich. In response to his profound childhood impression Einstein was to write his first paper *Investigation of the State of Aether in Magnetic Fields*.

The Germans had a penchant for military like discipline in everything they did, even for their children's schooling. Einstein found the atmosphere at the Luitpold Gymnasium too restrictive for his liking. Their mode of teaching also did not appeal to this budding genius.

The learning there was more by rote rather than through understanding. Using a doctor's certificate as an excuse for his health, he left before finishing school in the spring of 1895 and rejoined his family at Pavia. He was sixteen now. Later in life, he was to mention, it was here that he did his first thought experiment – "how it would be like if he travelled on a beam of light". These thought experiments were to become a hallmark with him and would help him to construct his theories later in life. It must be understood that thought experiments were not some new discovery by Einstein, but often used by scientists, especially theoreticians, to develop theories. Einstein attempted to enter the Swiss Federal Institute of Technology in Zurich from Pavia. Without a school completion certificate, he was required to give an entrance examination. He failed to qualify, even though he did remarkably well in physics and mathematics. Now he had with no other alternative, but to complete his schooling. This he did at Aaran, Switzerland. While there, he lodged with Professor Jost Winteler. He became enamored with the professor's daughter, Sofia Marie-Jeanne Amanda. However, nothing came of it as she moved on to a teaching job in Olsberg.

During this period of his life, he was quite active intellectually. He was already studying James Clarke Maxwell's electromagnetic theory. He qualified in 1896. By nature, Einstein was a peace-loving person. To avoid the compulsory military service as a German citizen he applied for Swiss citizenship, which he would eventually receive four years later. In the mean time, he enrolled himself at the Swiss Federal Institute of Technology in Zurich to train as a teacher in physics and mathematics. Here again he was not happy with the teaching, which his great mind found uninspiring to say the least. To satisfy his intellectual craving, he turned to study the works of Ludwig Boltzmann, Hermann Heimholtz, Heinrich Hertz and Gustav Kirchhoff. It was also here that he met his first wife, Mileva Marić (Marity), a Serbian Cyrillic girl from a wealthy family from Austro-Hungary, who had enrolled there. She was the only woman enrolled in the mathematics course, something very rare at the time. Einstein received his diploma in 1900. The following year on 21stFebruary 1901 he received his Swiss citizenship.

THE YOUNG SCIENTIST
Later that year, Michele Besso, a friend introduced him to Ernst Mach(1838-1916). Mach, who was both a philosopher and a physicist, had enunciated a Principle, which may appear too abstruse for the average person's understanding. What Mach was saying will become a little clearer in the following example. Say you are standing still with your arms by the side in a field on a clear night. If you happen to look up you will see the stars above you unmoving. If, now, you start to turn around fast and look up, you will see the stars spinning and at the same time, your arms, which were hanging by your sides, will move away from your body. This seems only natural for us, but Mach asked, "Why would the arms be pulled away from the body when the stars were spinning and why was it the stars stood still when the arms were by the side?" Now the reader will understand why I have used the word "abstruse". What he wanted to imply was, "mass out there (the stars) influences inertia here (on Earth)." Some years later Einstein would turn this vague idea into something concrete, when he used it as a stepping-stone to formulate his theory of general relativity.

In 1901, Einstein sent a paper on *capillary forces* to the prestigious journal Annalen der Physik. It was accepted. During this time Einstein was also on the lookout for a teaching post, but unable to obtain one. Through the influence of one of his father's friend, he obtained a temporary job as a clerk in the Swiss patent office, appropriately known as the Federal Office for Intellectual Property. His work here involved preliminary screening of patent applications. By 1903, he was made permanent; however, he had been passed on for promotion until he "fully mastered machine technology". This was not surprising, because Einstein was a theoretician and not a technician.

The nature of his duties there left him ample time to pursue his intellectual interests. Einstein and some of his friends at the patent office, including Michele Besso, met from time to time to discuss various topics on science and philosophy. They discussed works of Hume, the Scottish philosopher whom we have already met before, and the works of other scientists also. Amongst them was Poincaré, whose work on relativity had foreshadowed that of Einstein. It was from these discussions that Einstein was able to draw conclusions on space and time and the corpuscular nature of light. It is quite probable, while he was dealing with patents on transmission of signals and synchronization of time, his boyhood questions on light came to the forefront of his thoughts. These would eventually form the basis for his special theory of relativity.

EINSTEIN'S FIRST MARRIAGE

At this time, Einstein wanted to marry Mileva. Despite his mother's earlier objections to Mileva being non-Jewish, "older" and "defective" Einstein and Mileva were married on 6thJanuary, 1903. Initially their relation was not only on a personal level, but on an intellectual one as well. He described her as a woman "who is my equal and who is strong and independent as I am". Before their marriage, they had a daughter, Lieserl, in 1902. We only know about her existence through Einstein's correspondence with his wife at the time. Her fate however is unknown. It appears that she had contacted scarlet fever in 1902 and probably died as a result. Their first son after marriage, Hans Albert, was born on 14thMay, 1904. Their second son, Eduard, was to be born on 28thJuly, 1910.

EINSTEIN'S GOLDEN YEAR

In 1905, Einstein published four papers. Each of them was to proclaim a great discovery. The first, established of the equivalence of mass and energy, embodied in the famous equation $E=Mc^2$. The second, the theory of special relativity, which showed the speed of light was the same for all observers from any inertial frame. Third was his theory for the occurrence of the phenomenon of Brownian motion, which added support to the atomic theory. Last the foundation of the photon theory of light.

BROWNIAN MOTION

The English botanist Robert Brown (1773-1858) in 1827 had noted that pollen of the herb *Clarkia pulchella* suspended in water, when observed under the microscope, were seen to execute a continuous random motion in a zigzag fashion. Brown was not the first to have observed this phenomenon, as he himself conceded. F W von Gleichen had observed it some 60 years before. However, it was Brown, who carried out a detailed study of this phenomenon. He concluded that the zigzag motion could not be attributed to the fact that the pollen were living. Pollen that were dead, also exhibited this phenomenon. He at first thought, the dead pollen retained their vitality in some way even after death. Later, when he noted this phenomenon in other minute inanimate particles, like smoke particles and minerals in suspension, he changed his idea. Brown was able to confirm this phenomenon had nothing to do with life, but was applicable to all minute particles in suspension. He, however, could not give a reason.

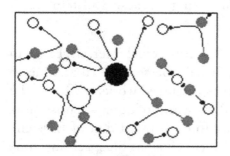
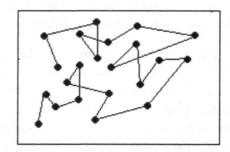

BROWNIAN MOVEMENT

HISTORY OF A PARTICLE SUSPENDED IN A FLUID MEDIUM OVER TIME

● SUSPENDED PARTICLE
● MOLECULES OF THE FLUID MEDIUM
○ RESULTANT POSITION AFTER COLLISION
●—— DIRECTION OF THE FORCES AT PLAY

7.2 Brownian motion.

This phenomenon was to intrigue many scientists of that period and many were to investigate it, but they also could not find any explanation. Many ideas were put forward, such as capillary forces or thermal convection, but none turned out to be satisfactory. In 1877, Delsaulx pointed in the right direction by saying, "In my way of thinking, the phenomenon is a result of thermal molecular motions in the liquid environment of the particles ..." This concept would later provide the basis of the theory for Brownian motion. The idea was drawn from the kinetic theory of gasses, which was developed in the later part of the nineteenth century and elaborated by J. C. Maxwell (1831-1879), L. Boltzmann (1844-1906) and R. J. E. Clausius (1822-1888). They showed the molecules of a gas carry out ceaseless translation motion at all temperatures above absolute zero. However, the direction of the movement of individual molecules in a gas could not be predicted. This concept was now translated to the molecules of a liquid. Here, the only difference being, the individual movements of the molecules are restricted due to their greater density in a liquid in comparison to a gas. Thus, any suspended particle at any given moment in time will be subject to "bombardment" by the molecules of the liquid from various directions. This is a continuous process, which results in the zigzag motion of particles suspended in it.

This implies Brownian motion will continue as long as the system exists. This had been already demonstrated by Cantoni and Oehl in 1865. They showed Brownian motion was still present, even after a year, if pollen grains were suspended between two microscope cover slips, which had been sealed on the sides. In this way the water, in which the pollen were suspended, would not evaporate.

What was happening here is the random motions of millions upon millions of molecules of a liquid continually bombard the suspended particles. In such a situation, it is not possible to know either the direction or amount of displacement of any individual suspended particle at any moment in time. Thus, any quantitative analysis was out of question. Einstein, Marian Smoluchowski (1872-1917) and Paul Langevin (1872-1946) addressed the problem independently. It was Einstein who was the first to publish his theory on the quantitative translation of Brownian motion in a paper *"On the Motion—Required by the Molecular Kinetic Theory of Heat—of Small Particles Suspended in a Stationary Liquid"*, Annalen der Physik **17**: 549–560. Here Einstein's theory did not limit the occurrence of Brownian motion to liquids only; it was also applied to minute particles suspended in gases as well

THEORY OF RELATIVITY

Many who are not conversant with science find the very word "relativity" daunting. To them it is something that lies in the rarified realms of scientific theories, which are not meant to be understood by all. This need not be so, at least as far as the gist is concerned. This part is only for those who are unacquainted with physics. I will try to explain as best as I can even though I am not a physicist.

In essence, relativity gives us insight into the nature of space and time, not how we perceive these entities in our everyday experience, but as they really are. Our understanding of space and time is linked to motion. If we pause to think about it, we will come to realize that without motion, everything would appear to stand still and we would not be able to perceive either space or time. Events would stop. Therefore, our understanding of relativity comes from how we interpret events in the background of space and time.

If we look at it from the human point of view, an event would be meaningless unless it can be sensed. One of the ways we can sense an event is by viewing it. This is our concern here. Whenever an event occurs, it has to happen within the ambit of space and time. Thus, any event can be "fixed" by three coordinates of space and by the time coordinate.

In our present discussion, we will be using the word *reference frame*. In our present context, all it means is the position and time of the observer and the same for the event concerned in relation to the coordinates of space and time.

When we as observers view an event from another place, we receive information through the medium of light. This implies the information about the event has been transcribed or transferred to us from one reference frame to another. This, through traditional usage in the language of physics, is known as *transformation*. In everyday life, we experience such transfer of information of events, around us, all the time. It is not only just a visual experience for the observer; but beneath it all there lies much more than what our senses tell us – the reality of space and time. This is what relativity explains to us.

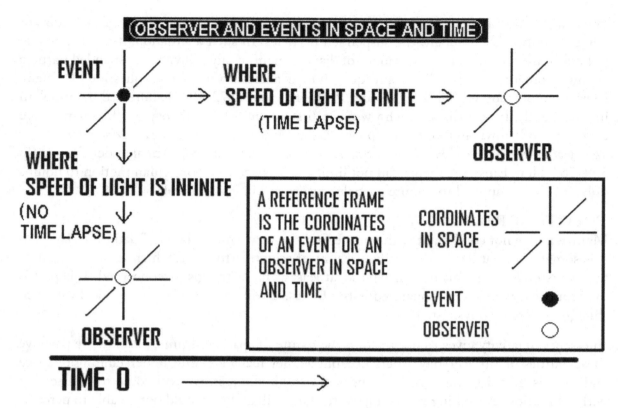

7.3 Transformation: Without and with taking into account the speed of light.

What the observer experiences depends upon his or her relative speed with respect to the event. Two observers, each from a different reference frame, travelling at different speeds will view the values of space and time differently with respect to any particular event, which they happen to observe simultaneously. This apparent distortion of space and time has a sound basis in the nature of time and space itself. We, however, are unable to appreciate all these subtleties of nature. These effects can be appreciated, if the observer happens to be travelling near enough the speed of light. We call such speeds, *relativistic*. When we travel at the ordinary speeds, which we experience on Earth in our day-to-day lives, such distortions are so minute that our senses are unable to appreciate them. On Earth, it is only under laboratory conditions that we can demonstrate such effects through speeds, which approach the speed of light.

Looking at it from the evolutionary aspect all this knowledge is not necessary for our biological survival, because the living do not experience relativistic speeds. Thus, nature had not intended us to understand how relativity works. This is why the explanation of how this transformation works by Hendrik Lorentz (1853-1928) and George Fitzgerald (1851-1901), which culminated in Einstein's theory of special relativity and later general relativity, was such a great achievement in human understanding.

To appreciate how relativity works we have to know the part played by each of these individual components, which go to make up this transformation and touch a little on the historical background. In this way, we may have some idea of how relativity was viewed in the past and the events that led us to our present understanding.

VIEWS OF THE ANCIENTS

In ancient times and for a long time afterwards, physical phenomenon such as light, sound, magnetism and bodies falling to Earth were taken for granted. They were treated on the same footing as the ground people stood on or rain that fell from the sky or the air they breathed. No one questioned the "why?" of it all. At that time other phenomenon like electricity, gravity and radiation were not even known. Motion, on the other hand, had always been an integral part of man's world and that is why people tried to understand it. Thus, not unnaturally, the motion of bodies took a prominent place in the investigation by the ancients; especially as we have seen, the motion of celestial bodies.

I will begin with Aristotle's views of the world. It formed the mainstay of ideas and philosophy for a few centuries before and many centuries after Christ. Moreover, it is to Aristotle that our present ideas can be traced.

Aristotle said that the preferred state of a body was the state of rest. This was believed to be due to the property called *inertia*, which resists change to this preferred state of rest. His view seemed logical to the people as they found that the bigger a body, the more inertia it possessed. Thus, more the force would be required to move it. The Earth was so big that it was assumed that it could not be moved. They believed it to be fixed in a state of rest lying at the centre of their universe. It is because of this idea they thought, the state of rest was in some way a preferred state in nature. Thus, they never arrived at the concept that all bodies in this universe were coasting through space, where there was no friction and they would continue in that state, unless acted upon by a force. Therefore, their ideas stopped with the belief of a motionless Earth at the centre of the universe.

Their ideas of space did not go very far. To them the heavens were made of crystalline spheres, so space was limited. On the other hand, time according to their view had no beginning and no end, therefore infinite.

People understood that to see they needed light. Since their ideas about light stopped here, they did not have any concept of light being a form of energy and as such, it possessed properties of its own, including speed - let alone knowing whether this speed was finite or infinite. To them all events, no matter how far or how near, were thought of as an instantaneous experience for the observer. Thus, the problem of time lapse during transformation of an event from one spatial and temporal coordinate to another never crossed their minds, which is to say from the event to the viewer. All this implied that all events were to be interpreted in reference to a fixed Earth. When Aristotle's civilization faded from history, their ideas were salvaged by the Christian Church to be passed on over the generations, until people started to question them.

VIEWS DURING THE 17^(TH) CENTURY

Galileo was the first person to take up the study of motion on a serious footing. He proved that bodies fall at a certain given rate irrespective of their mass. His student Torricelli went on to show that a feather and a coin did indeed take the same time to fall, if air resistance was excluded. Ultimately, this was to lead to our present understanding of behavior of bodies in uniform motion. What Galileo understood and Newton underscored in his three laws of motion gave rise to the important concept that *"all bodies continued in a state of rest or uniform motion unless otherwise impressed upon by an external force"*. This is what Aristotle had failed to appreciate. The term inertia was however retained, even though it did not have the same meaning the ancients had given it. The term inertia was now used to describe the property of bodies, which not only resisted any change in the state of rest (in a frictionless scenario), but also that of uniform motion. Thus bodies in such reference frames, like that of rest or of uniform motion, were described as *inertial frames* or simply as *inertial*. It also implied that if a body was in an inertial frame with respect to another, then the latter body was also inertial.

Now an inertial frame has certain properties from which they may be distinguished as inertial. Galileo recognized this fact. In an internal frame, the internal features do not suffer any change with respect to time. Thus if one happened to be a passenger in a train with no windows and travelling at uniform speed on ultra-smooth rails. In such an inertial system, it would *not* be possible for the passenger to know whether the train was at rest or travelling with uniform speed from any phenomenon that occurred inside the carriage. As for example, the water in the glass in front of a passenger would not move. A ball dropped would drop straight towards the foot and so on.

Counterpoised with the idea of uniform motion there is motion, which is non-uniform. I will discuss this point here, even though it does not come up in our immediate discussion. We will however come across it later. So as we are on the subject of motion it is best to get over with it now. Here by non-uniform motion, we do not mean erratic movement that results in changes in a particular direction or which changes against time in an irregular fashion. We are speaking of motion in a particular direction, which changes uniformly with time. This type of motion can be either acceleration or deceleration.

Now coming back to the example of the train, if the train happened to speed up (accelerate) or slow down (decelerate), it is no longer in an inertial frame. It now becomes an *accelerating frame* or *non-inertial frame*. Unlike in an inertial frame, the passengers, in a non-inertial frame travelling in a closed carriage would realize that the train was moving. On accelerating they would feel the jerk backwards or if decelerating they would jerk forwards as the uniform motion changes to acceleration or deceleration. Again, he or she would also see the water in the glass spill or even the glass fall on the floor. The ball would no longer drop straight. As long as the acceleration or deceleration lasts, the passenger who is sitting can feel the effect of some force at work. A ball on the floor would tend to roll back when the carriage accelerates or roll forwards if it decelerates.

The forces that produce acceleration or deceleration was able to break the state of inertia of a body, therefore they came to be known as *inertia forces*. Though the term may sound contradictory, its usage has been retained, because of tradition. The next question is from where or how do these forces arise.

Newton was well aware of these forces. He considered inertia forces as absolute. Absolute, in the sense that such motion appears to stand alone, independent of all other factors. Even without a reference frame to compare, such form of accelerated motion could be understood by an observer. Here Newton was not talking of carriages. We all know horses or the engine supplies the extra energy to break the inertia. He was talking of celestial bodies. In his Principia, Newton while explaining inertia force gave the example of a bucket full of water, suspended by a rope. When the bucket was twisted a number of times, the rope underwent torsion. The bucket was then released. As the bucket started spinning, the level of water was seen to rise further and further towards the edges. This was the result of centrifugal force, which manifests it self by the water climbing the edge of the bucket. Another example of such a force is the Coriolis force, which is evident by movement of the atmosphere due to the Earth's spin. These forces can, however, be "wished" away if the spinning of the body stops, so inertia forces came to be known as *fictitious forces*. Although today, we know better. There is nothing fictitious about these forces, because they all have their origins in energy. In the case of the bucket, the energy in twisting the rope was transferred to the bucket, which was then reflected in the water's behaviour. However, in Newton's time, man had not yet come upon the laws of conservation.

Again, like the ancients, Newton did not take into account that the speed of light was finite. Thus, if an event occurred and an observer viewed this event, he presumed the information had been immediately transferred to the observer. In doing so, it was assumed at the time the speed of light was infinite. This was in spite of the fact that it was known to Newton, the speed of light was finite. Ole Roemer (1644-1710), a Danish astronomer, had already discovered this in 1672 and his work published in 1676.

A few words about Roemer would not be out of place here. It would be laconic to call him an astronomer only. As a schoolboy I only knew of Roemer as the man who had calibrated the thermometer with 0° being the freezing point of water and 80° as the boiling point. Later I came to know he discovered that light had speed, but Roemer was much more. Roemer was appointed as teacher to Louis XIV's eldest son, a post he held for a time. He later introduced a uniform system of weights and measures in Denmark and convinced the king to introduce the Gregorian calendar. Roemer also invented a scale for the thermometer, which Fahrenheit improved after he met him. He was a capable administrator, who had the vision to introduce the first street lighting in Copenhagen. Albeit the fact that they were oil lamps. He overhauled the city's sewer system and water supply. He introduced rules for building houses. Roemer also saw to it that the city had an efficient fire fighting depart. He was appointed to a high post in the police department of Copenhagen. He sacked the entire police force, because of their low moral and inducted new blood. As a social reformer, he tried to control vagrants, poverty and prostitution in his city. This goes to show how little textbooks tell us about a man.

Returning to Newton; he gave us gravity. This implies that bodies would be attracted to each other. So as all celestial bodies would attract each other, eventually with time all bodies in the universe would come together as one entity in a big crunch. This would bring about the demise of the universe, as we know it. Such a scenario would inevitably happen in a finite universe. Newton understood this predicament. To get out of this problem Newton assumed that the universe consisted of infinite number of bodies in infinite space. Thus if space was infinite, then there would be no centre and consequently there would be no question of all celestial bodies ending up as one, as would happen in Aristotle's finite space. In the new picture that Newton drew, space that had been finite before, now emerged as infinite. Time, as we have already seen, was also treated as infinite, just as it was in the past. Nevertheless, they remained independent of each other, according to Newton's interpretation.

We have already said if a body is moving in uniform motion with respect to an inertial frame, then the first body is also inertial. We have also seen that an observer experiences no changes inside an inertial system. But, what changes, if any, would an observer experience if he or she were to observe one inertial system from another inertial system, say another ship, in relative motion to another ship.

If two ships are travelling on the open sea at the same speed, within hailing distance from each other, without any other reference to compare. An observer from one ship could not tell if the second ship was moving or not. The same would be true for an observer on the second ship. To the respective observers, even time would appear to be the same, since there is no apparent motion between the two.

7.4 Inertial frame – Uniform relative motion.

This form of transformation does *not* take into account the time lapse between the event and the observer, due to the speed of light being finite. This form of transformation is known as *Galilean transformation*. The crucial point to appreciate here is that it gives rise to an ambiguity, because it does not recognize that the value of space in the form of distances covered by light, which results in a time lapse between the event and the observer. This introduces a distortion. For short distances and the low speeds that we have been discussing until now, it may well appear near the truth. The fallacy of such a perception becomes apparent only at speeds approaching that of light.

EVENTS LATER

Since Galileo's time, a lot of water had flown in the world of science. Now the world was aware of other factors outside mechanical laws, such as the existence of the laws of light, electricity and magnetism. The work on these fields had opened new vistas in science. With such new knowledge there would soon come the question of whether what applies for mechanical laws can also apply for the laws of electrodynamics as well, where Galilean transformation is concerned.

We have seen people already knew that sound needs air to propagate and how Christiaan Huygens introduced the idea of ether and assumed that light also needed a medium to propagate. By the end of the 19th century many scientists like Lord Kelvin and Planck and others even thought that there were many types of ether, each different for light, heat, electricity, magnetism etc.

James Clerk Maxwell (1831-1879), a brilliant young Scottish physicist, working on the discoveries of Faraday and others, worked out a set of mathematical equations, which united electricity and magnetism. He also showed that the electromagnetic field had the same velocity as light. This led him to conclude that ether was the same for all. He also had tried to find the relation of the Earth's motion with respect to ether in 1878, but failed.

With this in mind three years later in 1881, Albert Abraham Michelson (1852-1931) designed an experiment. Albert Abraham Michelson, who was a physicist, knew that it was only possible to know the relative velocity of a body. So he set out to find a way to determine absolute velocity of a body. He hoped to do this by establishing the speed of the Earth in ether during its orbit and through the expected effect it might have on the velocity of light. The idea behind it was that this would establish fixed coordinates in ether. This could then be taken as standards for fixed positions in ether, which then could be referred to all other velocities, thereby obtaining the absolute velocity of a body. So that scientists could talk of absolute velocities instead of relative velocities. In doing this he had to take the help of the *addition law*.

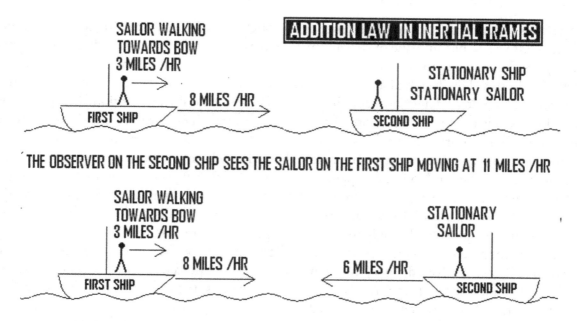

7.5 Addition law in inertial frames.

This is a throw back on Galilean transformation. The law of addition means that in our previous example of a ship, if a sailor on the deck was walking towards the bow. It is obvious his speed would be added to his ship's speed. If, say, the ship's speed was 8 miles per hour and the man's speed was 3 miles per hour. Then to an observer on a stationary ship, his speed would appear be 11 miles per hour. If, now, the stationary ship started to move away from the other, say, at a constant velocity of 6 miles per hour. Then the man on the first ship would appear to move forward at 5 miles per hour. On the other hand, if the second ship were coming straight head towards the first, then the man would appear to be moving at 17 miles per hour. But the question is whether this law of addition (or subtraction), which applies to matter, applies to light? Apparently not, as we shall soon see!

Michelson's experiment involved his trying to measure the speed of light in the direction of the Earth's motion and then perpendicular to the Earth's motion. He also tried in other directions as well. To his surprise, he found the speed of light was independent of the direction of the Earth's motion! He was surprised, because he had expected that the speed of light would follow the additional law. Therefore it would be greater when the beam was sent in the direction of the Earth's motion and would appear less when fired in the opposite direction; as implied by the addition law. His findings ran contrary to the laws of mechanics. The reason for this behaviour may lie in the fact that light is not matter. It is energy.

The Dutch physicist, Hendrik Antoon Lorentz (1853-1928), who would contribute to the explanation of this paradox later, was at first to express his doubts about the accuracy of this experiment in 1886. At the time however others, amongst them Lord William Thompson Kelvin (1824-1907), Michelson's teacher, encouraged him to repeat the experiment. This time Michelson took the help of Edward Williams Morley (1838-1923), who was a chemist. Now with Morley's help he was able to refine the experiment further.

Today this experiment is known to science as the Michelson-Morley experiment. This famous experiment was to become one of the landmark experiments, which would redirect the course of physics. To understand what they did, it is important to understand the actual experiment itself. To do this they designed a most sensitive experiment to find the speed of light, so that they would be able to detect how it behaved in relation to the Earth's motion. The apparatus that they used is called an interferometer. It was imperative that the apparatus should be free from any vibration or subject to any form of distortion during the experiment. To this end, a set of ridged steel pipes were set on a stone base. The whole set up, in turn was floated on mercury in a container. Now the apparatus would not suffer any distortion due to vibration or other forces when it was turned around. A light source was mounted on the platform and placed so that it lay on the focal point of the biconvex lens. In this way, the light rays came out parallel after passing through the lens. This parallel beam was made to fall on a partly silvered mirror placed at an angle, so that it is split by the mirror. This splitting occurs, because the mirror is lightly silvered. Thus, the beam is only partly reflected, while the rest of the light is allowed to pass through the mirror. Of these two beams, the one reflected and the other, which emerges from the partially silvered mirror, each are now allowed to strike another mirror and returned along the same path. The returning beams now become combined and can now be observed through an observational port. Any difference in the speed of light will be detected from the pattern of interference of light waves seen in the form of rings and can then be measured. This is why it is known as an interferometer. If there is no difference in the speed of light, there will be no inference pattern. This latter is what Michelson-Morley experiment confirmed. No matter in which direction the light was fired the result was the same. Though later many others have performed the experiment, improving its accuracy and under stringent conditions, but all their results have been the same.

It was seven years later that George Francis Fitzgerald (1851-1901), an Irish physicist, came up with an explanation for this paradox in 1889. He wrote, "….*the length of material bodies changes, according as they are moving through the ether or across it, by an amount depending on the square of the ratio of their velocities to that of light*".

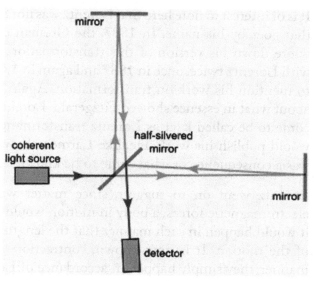

Credit: Courtesy Wikipedia User: Stannered
Website: http://en.wikipedia.org/wiki/File:Interferometer.svg

7.6 A simple Michelson interferometer diagram. [CC-BY-SA-3.0]

Unaware of this Lorentz proposed an identical idea in 1892. In doing this, Lorentz now accepted the authenticity of Michelson-Morley's experiment. Later when Lorentz became aware that his idea had already been published by Fitzgerald, he wrote to him acknowledging the fact. Amusingly, though Fitzgerald had sent his paper for publication, he was not aware it had been published. He said to Lorentz, he was glad to know someone had agreed with him and people would not laugh at his idea.

So why was it, light did not follow the laws of mechanics as far as the addition rule was concerned? Lorentz came forward in 1899 and tried to answer this baffling question. He proposed - suppose the speed of light did vary with the motion of the observer and again suppose that the linear dimension of a body also contracts in the direction of this motion at the same time, it would then explain Michelson and Morley's experiment. The idea that bodies would contract was an assumption by Lorentz. He thought if this assumption happened to be correct, then only it would explain the velocity of light as being constant. As because, in such a scenario, the length of the distance covered by light and the length of the material object would both be reduced proportionately. Thus under the circumstances any scale used to measure them would also contract proportionately. Therefore, the speed of light would appear constant. The situation would be somewhat like the following instance. We all know if a fresh piece of cloth, made from cotton, was soaked in water it shrinks. Now if we try to measure it with a tape made of the same material, before putting them both in water and then after putting both in water. We then would not be able to detect any difference in the length of the cloth, when dry and when soaked. So what Lorentz was saying could not be proved. Thus until further proof was available, the assumption of his weakened the idea from the scientific point of view. Nevertheless, it lent scope to develop the idea of relativity.

It is of interest to note here that Lorentz was not the first to have thought of the transformation that goes by his name. In 1887, the German physicist Woldemar Voigt (1850-1919) first wrote down his version of this transformation. Subsequently though Voigt communicated with Lorentz twice, once in 1887 and again in 1888, but apparently did not think it important to mention his work on transformation. Again, Joseph Larmor (1857-1942) in 1898 wrote about what in essence showed Fitzgerald-Lorentz contraction was a consequence of what has come to be called later as Lorentz transformation. It would be another year before Lorentz would publish his work. He, like Larmor, showed that the FitzGerald-Lorentz contraction was a consequence of what came to be known as Lorentz transformation.

Lorentz went on to suggest, since matter was composed of charged particles held by electromagnetic forces, a body in motion would rearrange its internal structure. He thought, it would happen in such manner that the length of the body would diminish in the direction of the motion. It is now known, contraction of bodies in motion *does not happen* in this manner, they simply happen in accordance of Lorentz transformation as described earlier.

It was Einstein, who through his brilliant insight into the nature of motion in relation to space and time was able to reconcile the paradox in an elegant manner. He showed that two observers, each in a different inertial frame and travelling at speeds near to the speed of light would not agree on either the spatial or the time coordinates that separated them. However, they would both agree on the speed of light.

This apparent inconsistency between space and time for the two observers can be reconciled, only if the values of space and time were mingled in a complex manner into what we would today call as *space-time continuum*. This could be shown not only mathematically, but also through geometry. This geometry, however, is not Euclid's geometry we learn at school. This is

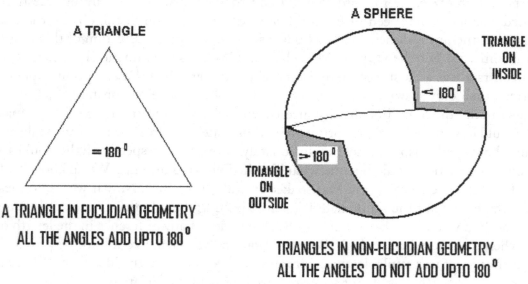

7.7 Non-Euclidian geometry.

geometry on a curved surface, where the angles of a triangle do not add up to 180°. On studying the diagram one will see that the angles of a triangle on a convex surface, adds up to more than 180°, whereas the same on a concave surface adds up to less than 180°. To this representation of three dimension of space we must add the dimension of time. Thus completing the four dimensions in which we exist.

David Hilbert (1862-1943), a mathematician from Göttingen, once remarked, "Every boy in the streets of our mathematical Göttingen understands more about four dimensional geometry than Einstein. Yet, despite that, Einstein did the work and not the mathematicians".

Göttingen is a place most of us may not have heard about. It is a name, which comes up quite often in association with great names in science and mathematics. So I have taken a little of the reader's time to describe its history. Göttingen was a sleepy town in the valley of Liere in Saxony. It had a university, which was founded in 1734 by George II of Great Britain and Elector of Hanover. His grandmother was the daughter of James II, who had married a German prince who had the right to elect the Holy Roman Emperor, therefore the title "Elector". Göttingen was eventually to become a renowned place of learning. It also had a legacy of great mathematicians, starting with the famous Karl Friedrich Gauss (1777-1855) who had been accorded a position with Archimedes and Newton as one of the three of the greatest mathematicians of all time. There were others, who followed. Amongst them was Riemann. It was he who expanded upon the ideas of the Russian mathematician, Nikolai Ivanovich Lobachevski (1792-1856), regarding non-Euclidean geometry. We will see a little later that Hermann Minkowski (1864-1909), a German mathematician from Göttingen, would also contribute to the mathematical interpretation of Einstein's theory.

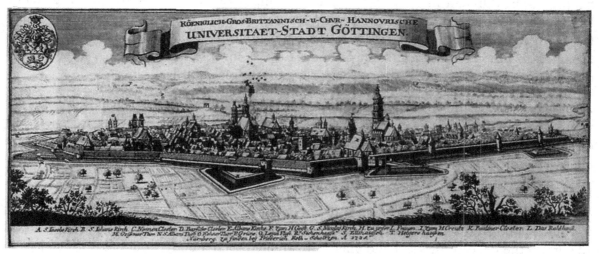

Website: http://en.wikipedia.org/wiki/File:Goettingen_-_Ansicht_von_Suedosten_(1735).png

7.8 Göttingen University. [Public Domain]

Though a similar idea to that of Einstein had been proposed earlier by Jules Henri Poincaré (1854-1912), but Einstein's paper explained his theory from a more rigorous angle. It gave a more precise analysis of the concepts of space and time. This enabled Einstein to give a proper explanation of Michelson-Morley's experiment. Einstein published his paper, "*On the Electrodynamics of Moving Bodies*" in the Annalen der Physik **17**: 891–921.

This is best summed up in Einstein's own words nearly fifty years later. "Looking back at the evolution of the special theory of relativity, it seems obvious that by 1905 it was ripe for discovery. Lorentz already knew the value of the transformations, which were later named after him, for his analysis of Maxwell's equations, and Poincaré further elaborated this idea. As for myself, I was familiar only with Lorentz's fundamental work, written in 1895, but not with his later work, nor with Poincaré's related investigations. In this sense my work was independent. The new idea in it was that the Lorentz transformation goes beyond Maxwell's equations and apply it to the fundaments of space and time". Einstein's paper would have gone unnoticed at the time, but for the fact that Max Plank took interest and wrote a paper on the subject in 1908. This was the second paper on special relativity after Einstein.

There was another paper in the same year by Hermann Minkowski. He contributed greatly to Einstein's theory by supplying the elegance it needed through geometric interpretation, which would have far reaching effects eventually enabling Einstein to formulate his Theory of general relativity. The essence of Minkowski's contribution was that he said if a history of a particle is constructed in space-time, which he termed a *world line*. Then according to the special theory of relativity, if different observers view the history of this particle from different inertial frames travelling at different velocities, they will see the history of this particle in space-time exactly as the same. This is as it should be from the philosophical point as well, because otherwise the universe would not make sense. However, from their different inertial frames they will find that their coordinates of the particle's history in space and time will be different for each observer. Here the difference in the coordinates arises due to the distortion of space and time as they experience the events from their individual reference frames, which happens to be different in each case. The relationship between these coordinates is given by Lorentz transformation. Some have rightly pointed out that this was a crowning moment for geometry. Though Lorentz transcription only applies to points only, however it can be extended to other geometrical configurations such as lines (one dimension), surfaces (two dimension) and bodies (three dimensions); therefore to reality as well.

As noted earlier, it must be reiterated that the difference between classical and this new relativistic interpretation is not so obvious where speeds in our everyday experience is concerned. However, when we come to high speeds involving fast-moving particles, the relativistic interpretation holds true. There are many such examples in physics.

EQUIVALENCE OF MASS AND ENERGY
We know if a body was subject to a continuous force, it would go on accelerating and in time, it would approach the velocity of light. Einstein's theory of special relativity brought a radical change to the concept of mass and energy. It implied such a body would also be subject to increasing resistance, which would counter the external force or forces that tended to increase its acceleration, resulting in slower and slower acceleration of the body, until its mass would appear to become infinite when it reached the speed of light. In which event the acceleration would stop. This apparent increase in mass is not real, so it known as *virtual mass*. It only appears that the mass increases. In practice, a body with ever-increasing speed would never touch the speed of light.

Now something else became apparent. The laws of conservation of energy, conservation of momentum and conservation of mass could no longer be considered as being independent of each other. Here they become one; any violation of one of these laws entails the violation of all the laws at the same time. Since the conservation of energy now becomes linked to the conservation of mass, their relation could now be given by a formula.

A body at rest has a definite mass, which is known as its *rest mass*. It was Einstein's postulate that the rest mass was the measure of the energy residing in the body or its *internal energy*. This resulted in the most widely known equation $E = mc^2$, where E is the energy in ergs, m is the rest mass in grams and c is the speed of light in centimeters per second. The actual value of the speed of light is 300,000,000,000 cms./sec. Therefore a mass of 1gram comes to the mind-boggling figure of 90,000,000,000,000,000,000,000 ergs in terms of energy! This gives us an idea of the vast amount of energy contained in one gram of matter. To those who till now have been thinking that Einstein's theories are far removed from our everyday world, they will find this simple and yet profound implication of this formula can offer a solution for our energy starved world.

THE PHOTON

In the same volume that published Einstein's special theory of relativity there was another paper, which preceded it. It was an idea that was to have far-reaching consequences. The implication of which, even the great man would not be able to accept many years later – the quantum theory. The title of this work was *"On a Heuristic Viewpoint Concerning the Production and Transformation of Light"*, Annalen der Physik 17: 132–148. Newton had suspected long ago that light had a corpuscular nature and Max Planck (1858-1947) had assumed, in 1900, light was emitted in quanta, Einstein proved in this paper light actually *does* consist of quanta of energy, which are emitted as discrete indivisible portions of energy. Einstein explained this through the photoelectric effect. Briefly, this is an effect that produces an electric current when a beam of light is shone on certain substances. When light strikes the material, if the energy is adequate, then an electron is knocked out from an atom on the surface of the material. This sets off a current. If light energy were waves as was thought, then as the light source receded, the photoelectric effect would wane gradually in accordance with the inverse square law. Eventually a point in time would come, when its effects would no longer be apparent. Experience shows that this is not what happens. Even if the source of light is taken to a great distance, where the inverse square is no longer effective for all practical purposes, the photoelectric effect is still seen to occur. So the photoelectric effect cannot be explained by the wave theory of light. Instead, this effect appears to depend upon the frequency of the light wave used, which is the measure of its energy. According to Einstein the farther the distance the fewer the number of particles travelling in any given front of light energy in space and so the less the chances of the photon encountering a electron, but when it did it would knock it out an electron from the atom of the substance. He thus concluded light was particulate in nature.

Yet we also have proof of the wave nature of light. This paradoxical nature of light had already been suggested, but not proven by Newton. It remained ignored for many years and Huygens proof of the wave nature of light had held sway. With Einstein's proof, Newton's idea was finally vindicated. So eventually, the final victory went to Newton.

Einstein's papers created apparent paradoxes, but he was quite comfortable with what appeared paradoxes to others. Both Lorentz and Planck had tried to eliminate what they thought to be paradoxes. Lorentz tried to explain it by saying it was rearrangement of charges. While Planck assumed light was emitted in quanta, even though he thought light was purely an undulatory process and there was nothing corpuscular in its nature. This is where Einstein profound insight into the matter comes in. Unlike Lorentz and Planck, he accepted his conclusions as being a part of scheme of natural laws. To him, where the knowledge of reality is concerned, not only should there be confirmation of observed phenomenon through experimental evidence, but they must also be open to mathematical interpretation, which is the language of nature. It is only then that reality can be confirmed. At the same time Einstein appreciated, the inner beauty of any theory was of equal importance. Thus for Einstein, the external confirmation and the internal perfection lead to the naturalness of any theory, which reveals the harmony in nature. Both to him merited equal importance. In this way, he was able to reconcile philosophy with the practical aspect of science.

The various aspects of nature that we see are not separate entities of nature. All nature has evolved from one source - the big bang. It therefore stands to reason that all things, which have emanated from this single event, must therefore necessarily be interconnected and thus explainable on a common basis. I will describe a simple example from geometry. Let us imagine a finite straight line. The line we know is made up of infinite points. Now let us take the virtual counterpart of any of these points from somewhere from its middle and raise it to a certain finite height. Then connect this point to the edges of the straight line. What we have done is that we have created a triangle from a straight line. Thus it becomes obvious

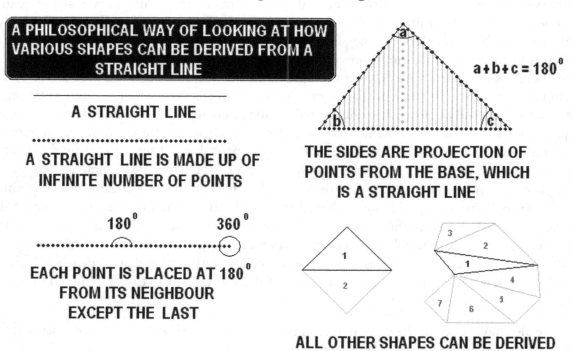

7.9 A philosophical way of looking at how the properties of a triangle are derived from a straight line.

that the triangle is derived from the straight line. Now we know the straight line subtends an angle of 180° from the virtual point before it was raised from the straight line. The triangle thus formed from the straight line also "inherits" the 180°, because of the fact that its source of origin suspends this angle. If we happen to make a similar triangle on the other side on the same base then it becomes a quadrangle and it "inherits" 180° from the two triangles on each side of the original straight line. So the quadrangle now has 360°. Many of the other properties of the angles of a triangle and other shapes can be derived from this concept. This concept can be considered as elegant, because it holds an inner thought, which explains something that crosses the boundaries of the idea of a line as being distinct from a triangle.

It has been suggested that Mileva had a considerable influence on Einstein's work, but evidence for this is hard to come by. Although it maybe possible that she had been of some help to him in view of his remark mentioned earlier. But what goes on in a woman's heart or a man's mind, especially where there was once love, can never be known by others, unless one of them tells. As neither chose to do so, we will never know what her contribution was, if any. Be that as it may, there can be no doubt as to Einstein's genius. It was generally believed he gave her half his share of the Nobel Prize money the rest he gave to charity. Some sources say recent evidence suggests that he invested the bulk of the money he received as prize in USA, but he lost most of it during the depression. While others have confirmed that in the end Einstein gave Mileva much more than what he got as the prize money. So today we do not know the whole story.[24]

THE YEARS THAT FOLLOWED

Though by now Einstein was promoted to Technical Examiner Second Class at the Patent Office, he still had his heart set on a teaching post. Eventually in 1908, he became a privatdozent at the University of Bern. A privatdozent was a teacher who for a token remuneration lectured on subjects not covered in the curriculum; it was a prerequisite to becoming a full-fledged professor. As can be easily understood Einstein was not entirely happy, but he accepted the offer, as he was able to continue with his position at the patent office. This left with ample time to continue with his researches. In 1909, he published a paper on quantization of light, this along with his earlier paper indicated, the energy quanta as proposed by Max Plank must have well-defined momentum and in certain respects act like point-like particles. Thus the concept of photon for light was introduced, however it was only in 1926 that Gilbert Newton Lewis (1875-1946), an American chemist famous for his discovery of the covalent bond, coined the term *photon* for light particles. With the concept of the photon, the idea of wave-particle duality of energy and later the idea of particle-wave duality of matter as conceived by Louis Victor de Broglie (1892-1987), which laid the foundation for the quantum mechanics. The same year he was to become a professor "extraordinary" at the University of Zurich. This position was still inferior to a full professorship. Here we get a picture of Einstein as a teacher from one of his students, Hans Tanner, who wrote, "When Einstein first ascended the rostrum in his shabby suit, the trousers of which were too short, with an iron watch-chain, we were quite skeptical of our new professor. However, he soon captured our callous hearts with his unique way of reading his lectures. His notes filled a slip of paper the size of a visiting card and simply listed the points to be taken up at the lecture. Einstein's lectures thus came straight from his head and we were able to witness the workings of his brain. This was much more interesting than some of the stylistically faultless, sober lectures which might even excite us but which also gave rise to a bitter realization of the distance between teacher and student. …….. After every lecture we felt that we could even have delivered it ourselves." It is said that his students were free to interrupt him during his lectures to clear up a point. He would often take a student by his arm in a familiar fashion to ask him some question or to discuss a point.

In Zurich Einstein met his friend, Marcel Grossmann, again. This would prove to have significant consequences later. At Zurich, as always, Einstein did not confine his company to physicists only. He counted intellectuals from various other disciplines amongst his friends. There was Alfred Stern, a historian, Laurel Stodola, a steam-turbine specialist and even an expert in criminal law, Emil Zürcher.

While still in Zurich, he was offered a post of full professor at the university at Prague. Mileva was reluctant to move since she felt well settled in Zurich and Einstein found the atmosphere there much to his taste. In spite of all, he did not refuse this post as professor of theoretical physics. Here it was expected that the new professor should call on his colleagues and their families. There were about forty such families in all. As can be easily understood Einstein found these formal visits rather tedious. They did not have the freeness, he so much liked when visiting his friends. Nevertheless he tried to take it in good grace and while visiting he took the opportunity to visit the nearby places of interest as well. Soon he lapsed to visiting the places of interest only and missed out on the social duties expected of him. One can well imagine where age-old rules of etiquette took precedence over freedom; the authorities looked askance at the newly arrived professor. To them his ways not only appeared to disrespect the tradition of the place, where he had so newly joined, but apparently also his seniors as well. Einstein's affability, which was extended equally from the highest official to the janitor, was also a source of irritation to the authority. They saw themselves as guardians of etiquette rather than see the man himself for what he was. It was while he was here he wrote a paper, which explained why the sky was blue.

He returned to Switzerland to take the post of associate professor at the Swiss Federal Polytechnic School at Zurich in 1911. He was then offered a post of professor at Charles University at Prague, which he was to accept. Here he wrote a paper on the effect of gravity on light. He predicted that the path of light would be bent by a gravitational field. In 1912, he returned to Switzerland and became a professor at his old institution of Swiss Federal Institute of Technology in Zurich. After shifting house many times, the family was finally able to establish a permanent home at Zurich just before the World War I started.

It was at the Swiss Federal Institute of Technology he came across Marcel Grossman once more. It was he, who introduced him to Riemannian geometry. From this and also through the suggestion of Tullio Levi-Civita (1873-1941), an Italian mathematician, Einstein came to the realization that the correct generalization of his special theory of relativity would have to include gravitation. In the autumn of 1913, he attended a conference at Vienna, where he first mentioned his new theory of relativity, which now included the effects of gravitation.

INVITATION TO BERLIN
While Einstein was at Zurich, Max Plank and Walter Nernst, the two giants of world of science in Germany, came to him with a magnificent offer. The proposal was that he would be

given of a professorship at the University of Berlin, the post of director at the Kaiser Wilhelm Institute and be made a member of the Prussian Academy of Sciences. They also guaranteed his freedom to pursue whatever field of work he chose, with only token administrative responsibilities. This meant it would enable him to pursue his quest for the generalization of his theory of relativity, a thing that was so close to his heart. Einstein knew such a grand offer was not made to one everyday. Nevertheless, though he knew this offer was magnificent, he hesitated all the same. He was well aware from past experience he would be entering a world, not the one he experienced at Zurich, but a world more suited to a military academy rather than a place of learning. A world where dress regulation, precedence and protocol are all expected to be followed, like a court of some medieval king or emperor. Along with all this, there would always be undercurrents of petty politics amongst those who wanted power, rather than knowledge. Above all, there would also be German nationalism that would pervade all. Einstein knew that all this would disturb the peace of his mind and his easygoing life, which he experienced at Zurich. He therefore hesitated and asked them for time to think it over. They agreed to return later. The child in Einstein revealed itself when he told them, when they came back next, if he carried red flowers it would be a sign that he had agreed. When they returned next, Einstein was carrying red flowers.

By this time, Einstein had become estranged with Mileva and he proceeded to Berlin alone to take up his post.

BERLIN
In Berlin, Einstein came to know his colleagues through the weekly seminars, which were a regular feature there. They were attended by many luminaries like Max Planck, considered as the founder of quantum theory and Walther Nernst, who obtained the Nobel Prize in recognition for his work in thermochemistry. Max von Laue, who along with his collaborators discovered diffraction of x-rays in crystals, which was an important contribution to the structure of matter and Gustav Hertz, another Nobel Laureate, who got the prize "for his contribution to the discovery of the laws governing the impact of an electron upon an atom". There were Erwin Schrödinger, whose "wave mechanics" would help to contribute to the quantum theory and Lise Meitner who discovered uranium fission. His colleagues remember Einstein as a person who was able to grasp the ideas posed by others, which he discussed in profundity and at the same time with equal lucidity.

In spite of his simplicity, the great man had a mischievous boyish streak. He had his own amusing way of taking it out on others. He did not have to shout, get angry or be even annoyed. This is shown up in the instance of his visit to Professor Stumpf, who happened to be a professor of physiology. Einstein came to know that he was interested in the perception of space. Einstein was keen to visit him to hear his views. He decided to visit him at his home. On his arrival there one morning, the housekeeper told him, the Herr Professor was not at home and whether he would like to leave a message. Einstein left saying that he would drop in later in the day. When he returned at 2 o'clock in the afternoon, he was again met by the housekeeper. This time she told him the Herr Professor was taking his afternoon rest. Unruffled Einstein reassured her that he would return later. Finally when he returned the professor was available. Einstein told her politely, "You see in the end patience and perseverance are always rewarded." Stumpf was now ready to receive his famous visitor. No doubt, he was pleased and expected that they would exchange the courtesies of a formal introductory visit. Professor Stumpf however must have been greatly disappointed, because Einstein without much ado launched on a discussion on space, which left the professor completely out of his depth. The professor of physiology listened to his exposition for forty minutes, without comprehending a thing! Einstein must have known very well that a professor of physiology had no knowledge of mathematics and unlikely to be aware of advances in physics of the time. In those days, physics and physiology did not mix. He must have known equally well that poor Stumpf would be totally out of his depth. On the other hand, the delay that kept Einstein waiting did not disturb him at all. Once, Einstein had an appointment with a friend. It was decided that Einstein would wait for him at Potsdam Bridge. When his friend told him, he was afraid that he might be delayed and it might hinder his work. Einstein reassured him that it would be no problem and went on to say, "… the kind of work I do can be done anywhere. Why should I be less capable of reflecting about my problems on Potsdam Bridge than at home?"

Within a year of Einstein's going to Berlin the Great War broke out. It was a cause for great disappointment for Einstein. War always brings out extremes in people, whether it be the good or the bad in them. Now, the hitherto peace-loving men of science, whose only battles took the form of fighting out their ideas in discussions and conferences took an ugly turn altogether. They started speaking of Germany's subjugation of Europe. They started taking pride in German might and dreamt of victories over France, Russia and other countries. These erstwhile men of peace were now proudly discussing the results of battles in which people lost their lives, not in hundreds or thousands, but in tens of thousands and hundreds of thousands. It was a sad reflection of human nature. At this period, Einstein got in touch with people who had similar ideas with him, amongst whom was Romain Rolland (1866-1944), who was a Nobel Laureate in literature and had worldwide influence as a great humanitarian and a writer of profound depth. They and other like-minded people represented the intellectual anti-war face of humanity.

While he was still at Berlin, it was after many revisions of his idea that he finally published his paper on general relativity in the year 1915. His theory explains gravitation of a mass, which causes distortion of space and time and this in turn affects inertial motion of bodies.

GENERAL RELATIVITY
Understanding general relativity is a difficult subject for any layman, who is like me. We can only get a glimpse of the matter. To make it easier to understand, I have found it is best to treat it from three different aspects. One is the concept; second the mathematics, but because the mathematics involved is beyond my ability to understand, I have discussed the idea behind the mathematical approach only; and lastly, the description of experimental proofs of the theory.

THE DESCRIPTION OF THE CONCEPT
The theory of special relativity was only special in the sense that it dealt specifically with inertial frames only and not with other forms of motion viz. accelerating or decelerating frames. Though special relativity may be extended to accelerated frames as well, however to do so, it would take complex mathematical methods. Einstein realized this and felt the need to extend his ideas to non-inertial frames in a simpler way.

We have already seen earlier, in case of inertial frames the motion of a body can only be considered as arbitrary in the sense they are relative in nature. What about non-inertial reference frames? This thought came to Einstein. We do not know exactly when this idea occurred to him, but it must have crossed his mind when he wrote his paper on special theory of relativity in 1905. It is known Einstein was certainly mulling over these questions by the year 1907. He felt the difference between two non-inertial reference frames also could be considered as arbitrary. Arbitrary in the sense, contrary to what Newton had envisaged earlier, it could be shown there were no absolute criteria of motion in non-inertial reference frames also. As for example, what would an observer experience if one considered the case of some bodies, which were falling under any uniform field of force? They would all be accelerating at the same rate. Now imagine if an intelligent being was living in one of them. It would be in no way possible to tell whether he or she was in an accelerating frame just by observing the other bodies. That these bodies were accelerating would only be perceptible if we as observers happened to be looking at them from a different frame of reference, such as an inertial frame.

This falling of bodies described is an observation of relative motion only. However, it leaves us without the experience of a person travelling within such a body, which is accelerating. We have already discussed earlier what happens in a non-inertial reference frame. So if we can establish that experience inside a non-inertial frame is also arbitrary, which is to say like that of the experience of an observer in a carriage in uniform motion is no different for the observer than inside a stationary carriage. It is then only the ideas of inertial frames can be extended to non-inertial frames.

Let us take the example of a railway carriage travelling at a uniform velocity on a strait track. Here we know everything remains as it is and there is no distinction between when the train is at rest or if travelling at uniform velocity. If the train now happens to negotiate a curve, then things inside the train would become displaced. Water from a glass would spill, objects would slide, people would feel a force that will make them lean to one side etc. this implies acceleration, but mark that no additional force is being applied here for such movements to happen. Now we have already shown earlier that without a force things cannot move! Since no force is being applied from outside, the question arises where is this force coming from? The answer is that such a force is derived from inertia. Hence, one might say that the bodies in the train were travelling in a straight line and it is only when the train was negotiating the curve their displacement became evident. Since these forces do not arise from any external agent, they came to be known as fictitious forces and as we have seen earlier, they can be abolished if the non-inertial frame changes to an inertial frame again. Thus when the train after negotiating the bend, again continues in a straight line, the inertia forces disappear. Thus, these fictitious forces only arise when acceleration happens due to whatever cause.

When a body is subject to any field of force, the body reacts by acceleration. Take for example a charged body; it will respond to any source that has a charge. A magnetic body will respond to any magnetic field and so on. However, such fields are selective and do not influence all bodies. A magnetic field will not influence a charged body or vice versa. Consequently, we must look for a field of force that extends to all bodies. Only a gravitational field satisfies such a criterion, because it acts on all bodies.

Thus in summing up we may note that bodies inside an inertial frame, which changes to a non-inertial frame, are affected in the same way as gravity. Namely, whether a body in the train has greater mass or lesser mass they undergo the same acceleration when the train starts to negotiate the curve, just the same as it would happen in a gravitational field. Secondly, when any non-inertial frame is converted to an inertial frame fictitious forces vanish. The same will be true if the gravitational field vanishes, because the body will return to an inertial frame.

Any effect of a gravitational field on a body is opposed by its inertia. Inertia as we have already seen is the property possessed by all bodies and it resists change in motion. So here, it resists changes in acceleration due to the force of a gravitational field. Again we already know greater the mass, greater is its inertia, so less the change in motion that will result from any particular force acting on it. However this is counterpoised with the fact that in spite of this greater inertia, the greater the mass the more its gravitational effect on other bodies. These two opposite forces therefore cancel out each other. Thus, it happens that regardless of the mass of a body the acceleration is the same for all bodies in any given field. Therefore, on Earth the acceleration experienced by all bodies is equal to 32 feet/second/second.

Since the inertia, in case of acceleration, manifests itself by resistance in the opposite direction to which a force acts or in other words opposite to the direction of motion. It is this counter force created by inertia, which is felt as inertia forces. Thus, there is no distinction between acceleration due to a gravitational field and acceleration caused by any other force. Einstein gave an example. Imagine a lift hanging motionless in a gravitational field. A man standing inside the lift realizes the force of gravity is there when he experiences his feet pressing on the floor of the lift. Now in Einstein's thought experiment, if we make the gravitation field disappear and instantaneously replace it by a force equal to it. This would make no difference. Here we must remember this is a thought experiment, so we must not be too particular and think about how a field other than the gravitational field would act on the lift or on our bodies. We must keep to the point that Einstein was trying to make. The new inertial forces produced by acceleration will be equivalent to inertial forces produced by gravitation and thus in no way distinguishable from each other. Therefore, gravitational forces are same as acceleration forces, the fictitious forces in them have similar origins, which is to say they arise due to inertia of bodies. This establishes the principle of equivalence of acceleration and gravitation. The problem was that how could it be shown?

In special relativity, we have seen that a body in uniform motion traces out its history in space-time as a geodesic. This can be taken as the representation of inertia, because only a body in an inertial frame travels in a straight line and remember a geodesic is a straight line.

We can say it is here that Einstein's genius comes into play. It may be summed up in what Einstein said to his son. One day when Einstein's son asked him "Daddy, why are you so famous?" he laughed and replied, "You see, son, when a blind bug crawls along the surface of a sphere it doesn't notice that its path is curved. I was fortunate enough to notice this."

The theory of special relativity is a special case of a wider generalization. In the presence of gravity, the history of a body in the form of a geodesic can no longer be a straight line, but curved. Einstein realized it implies space must necessarily be curved also. Since relativity mixes space and time into one, therefore it follows that gravitation cause's space-time to curve. Since gravity "emanates" from bodies, then bodies must affect space-time in the same way.

If all this is true then we should be able to see the manifestation of general relativity. In fact, such is indeed the case. We have proof of this in various ways such as bending of light rays, explaining perturbations of the planet Mercury's orbit, slowing of clocks under separate conditions of gravity, stretching of light waves. All of them has been shown to be true experimentally, but before the experimental proof comes the mathematics. Though I do not understand the mathematics, I have mentioned it briefly, because it is complementary not only to the theory just described, but also to the experimental verification that comes after it.

THE MATHEMATICS
If the reader is like me, he or she may skip this part.

Apart from wide spread interest in general relativity amongst physicists all over the world; interest amongst mathematicians was also aroused. This happened because of the form of mathematics it took to explain Einstein's theory. While Einstein had taken the help of much simpler mathematical tools to explain his theory of special relativity, he now needed the help of four-dimensional Riemannian manifold of space-time to explain his theory of general relativity. Manifold here means any surface. This explanation required special calculus known as *tensor calculus*. While working on his special theory of relativity Einstein had no need for accessing such intricate mathematics, because special relativity did not include gravity and therefore there was no distortion of space. Thus, it did not need any complex mathematics. However his mathematics now took a form he and other physicists were not aware of yet. Unknown to the physicists this form of mathematics had already been worked out by the mathematicians. Not because they needed it for any purpose, but they reveled in such mathematical gymnastics, which we as ordinary mortals could never appreciate. Einstein was no mathematician, as we have already seen earlier from David Hilbert's remark. Moreover, Einstein by his very nature wanted to keep things as simple as possible. He was "skeptical about the need to introduce complex mathematics as he privately thought they were introduced to mystify the confused the reader".

In 1911, he made an early attempt to extend his special theory of relativity to include gravity. Later in Prague, he met Georg Alexander Pick (1859-1942), an Austrian mathematician, who drew his attention to the mathematical methods of Gregorio Ricci-Curbastro (1853-1925), an Italian mathematician and Tullio Levi-Civita (1873-1941), another Italian mathematician. These methods would prove useful in propounding Einstein's general theory of relativity. It was in Zurich that Einstein met his old friend Marcell Grossmann (1887-1936) and David Hilbert, who helped him to learn this form of mathematics that would make it easier to explain his theory more elegantly.

Technically speaking, general relativity is a metric theory of gravitation whose defining feature is its use of the Einstein field equations. The solutions of the field equations are metric tensors, which define the topology of the space-time and how objects move in it.

A metric specifies the distance between two points on a given surface. Thus, the metric that is specified for a given surface defines the nature of the surface

These ideas have their mathematical basis in tensor calculus. Here the geometric property of four-dimensional space-time is characterized by ten functions that make up the so-called *metric tensor*.

This is what it is about in simple terms. A *tensor* is a set of functions or properties or components fixed relative to a reference frame, which can undergo transformation to another coordinate in another reference frame, according to certain laws. It is to be noted that each component of the tensor is related in a linear and uniform fashion whenever the tensor is transferred to another reference frame. This means if the components of one tensor equal those of another in one coordinate system, then they will be equal in all other coordinate systems to which they can be transferred. Thus if a tensor is transformed to another coordinate system, then the properties are retained. If in case one property vanishes, then this property will vanish in the other coordinate systems concerned as well. This makes the tensor invariant. This is of important significance to general relativity, because each observer has his or her own coordinate system and since the laws of physics are the same for all observers these laws are expressed as tensors, which express the independence of the coordinate systems as well.

From the metric tensor, we can know the separation of any two events. Einstein in his field equation formulated in which these ten functions and their ten other derivatives described the material content of space-time continuum. The last ten functions described by Einstein goes to make up the *energy-momentum tensor*, which gives us the measure of energy, consequently the mass and momentum present in a given part of space at a given time.

All this, however, does not change the case for special relativity. Any small locality of curved space can be described as flat. Just like the surface of the Earth is curved, but when considering small areas on its surface it can be taken to be flat. In reality, "flat" space-time still has a curvature, because of gravitational force acting on it even though it is a very, very small area. Thus, special relativity is a limiting case of general relativity.

In theory, the field equations can be solved and therefore we can obtain the metric tensor. Thus from the latter the geodesic of space-time describing the motion of any matter may be worked out. The history of photon can also be obtained in this way, but from a special class of geodesics known as *null geodesics*.

Summing up the theory of general relativity rests on two postulates, first the metric tensor will satisfy the field equations and secondly, the history of a particle is given by a geodesic in space-time having the previously mentioned metric tensor. Though the second postulate has an important role in supporting the theory, the question came up later in Einstein's mind whether he could do without it. Between the years 1938 and 1949, Einstein and his colleagues were able to do just this very thing. In a set of papers, they showed that if mass points can be represented by singularities of metric tensor, then, if, we want to obtain a solution for field equations, the singularities must satisfy certain conditions. These conditions prove to be the equations of motions of mass points!

EXPERIMENTAL VERIFICATION

One of the earliest verification of the theory of general relativity lay in explaining the perturbations of the planet Mercury's path around the Sun. It had been long known that these perturbations involved the precession of the long axis of the orbit of Mercury. This was an observation that had been made by the French astronomer, U. J. Leverrier, long before the advent of the theory of general relativity. Explanation of this discrepancy in Mercury's movements could not be reconciled through Newtonian mechanics. The discrepancy disappears, only if calculations were performed relativistically.

General relativity also predicted that light rays would bend if they passed near a massive body, say, near a star such as the Sun. The only way this could be tested at the time was to observe stars very close to the Sun's disc to see if the star light bends. Under normal circumstances, stars are not visible in the full glare of the Sun during the day. It is only during a total solar eclipse stars very close to the Sun are visible. So to avoid any parallax errors a picture of the sky must first be taken of the part of the sky at night where the Sun would be present during a total eclipse. Then from exactly the same place on the Earth's surface and pointing at the same direction of the sky during a total eclipse another picture must be taken. The verification in this way came 1919, when the British astronomer and physicist, Sir. Arthur Eddington mounted an expedition, one to the Isle of Principe in the Gulf of Guinea off the coast of Africa and the other at Sobral in Brazil, from where the total solar eclipse was due on 29th May that year. At Principe, the sky was overcast and there was a downpour. No stars were to be seen. Luckily, as the total eclipse neared its end the sky cleared and some photographs could be taken. Whereas, at Sobral the sky was clear and many photographs were thus taken. The expedition was hailed as a great success. The predicted bending of starlight was 1.74 seconds arc, which was in close accordance with the predictions of general relativity.

Nevertheless, there had been errors in these observations. However these were, not fully realized at that time. Partly this happened, because the heat of the weather had caused the apparatus to be heat up, so errors had crept in. This distorted the results. Partly, as people said, it was a case of Eddington's mind being guided by the predicted answers! This is a trick of the mind, which experimental scientists have to face. They have to consciously guard against it. More accurate observations, during the total eclipses of 1922, 1929 and 1951, have confirmed that light from a star does indeed bend as it passes very near a massive body like the Sun.

Another conclusion of general relativity is that a light ray of a given frequency sent perpendicular to a gravitational field of the Earth would lose energy as opposed to it travelling in a non-gravitational field. This means that the frequency of the light wave decreases as it looses energy in a gravitational field; this is exactly what happens!

Then there is another phenomenon, where a clock slows down under a gravitational field. This has been verified experimentally. A very accurate clock on top of a very high tower runs faster than one on the Earth's surface. This is called *gravitational time dilation*. This is also true.

Then there is another phenomenon known as *frame–dragging*. Here a rotating mass drags the space around its immediate vicinity.

General relativity predicts that the universe is expanding. This has been confirmed, as we will discuss in the second part.

IN PERSPECTIVE
In Einstein's own words the "time was ripe" for the discovery of special relativity by 1905.

If Einstein had not come forward someone else would have done it. However, in case of general relativity it was quite different issue all together. By 1916, Einstein had developed the mathematical framework for his idea on general relativity. Minkowski, who as we have seen had been of help to him in giving elegance to his work on special relativity, had died in 1909 as a result of appendicitis. This time, it was David Hilbert, who helped him in the intricacies of the mathematics that would eventually give full expression to his theory.

Einstein's idea on general relativity was far in advance of his time. It may be certain that without Einstein, scientists would not have been able to even think of this idea for at least another half a century. It would then come to the notice of scientists not through a conception of the mind, but through information coming in from various experiments spread over time. Like the effect of gravity on clocks or the bending of light, as deduced from astronomical observations. People must have been puzzled at first just like in the case of Michelson Morley's experiment, which lead to special relativity. By the time, scientists realized what was happening a lot of time would have passed. Science would have taken a different course. It would then have been a collective effort. With Einstein in the picture, it became an example of how the human mind can reach out to beyond the realms of accepted methods. It may be certain that In the case of general relativity, the idea came earlier and confirmation came later. Thus in one stroke Einstein had been able to unite gravity with space and time.

THE AFTERMATH
At about this time, because of the poisoned atmosphere the war had created, Einstein looked for the company of like-minded people. One of the homes he frequented was that of his second uncle, Rudolf Einstein. Elsa, his daughter had recently been divorced and was living with her father with her two daughters. It was here that Einstein met his cousin and childhood friend once again.

In 1917 when he heard about the October Revolution, he hailed it as a success. Thinking it would bring about a new world order based on reason and consequently be good for science as well. It appears that he believed in Lenin as a man who "completely sacrificed himself and devoted all his energies to the realization of social justice". He goes on to say, "Men of his type are the guardians and restorers of the conscience of mankind." How mistaken he was in giving his stamp to a man who took power through force and started his reign with the blood of his Emperor and his family on his hands. In his enthusiasm for the new world order of the followers of Marx, Einstein did not realize that to make it work it needed human beings and human beings are inherently flawed. Thus, it will never be possible for any one to make an ideal world. Einstein should have known better! Perfection is not to be found in books and political manifestos, but in ones heart. It takes great moral courage to put perfection into practice. If Einstein had made any blunder in his life, then this was it!

After receiving his divorce from Mileva in 1919, Einstein married Elsa. She became a devoted wife to him and was very protective of him.

HIS LECTURE TOURS
Einstein was invited by academic bodies of many countries to lecture on his theories during the decade that followed. He toured Holland, Czechoslovakia, Austria, America, England, France, Japan, Palestine and Spain.

In England Einstein got a cool reception. It was the first time this renowned scientist was not given any applause. This phlegmatic race knew how to put this German scientist in his proper place as only they know how. However, Einstein spoke of the role of British scientists' contribution to the mainstream of science and their role in the confirmation of his theory of general relativity. He presented his idea of international cooperation in the field of science. His message seems to have gone down well with scientific community.

In France, although Einstein gave many lectures, he was not allowed the honour of speaking at the French Academy. Members' argument being, Einstein was not a member of their Academy. Thirty members said they would leave if Einstein addressed them. While others demanded that even if he attended, they would not allow him to sit with them and he would have to sit in the audience.

These French academicians were people with a high opinion about themselves and consequently had a low opinion of others. Their treatment of Einstein was not much different from that meted out to the untouchables in the caste system of India. It should be of interest to note that the caste system was not a sanction of the oldest of the Vedic texts, the Rig Veda. It was an offspring of people who were like-minded with these French academics.

These pompous and proud people of the Academy, had not yet had the time to come to grips with the theory of relativity, so they did not understand the worth of this German, who espoused peace, freedom and social progress. They were more concerned with their privileges and rights as members of the Academy. They were also conscious of being Frenchmen and as such, they were flushed by their recent victory over Germany. Little did they know that even then, the wheels of fortune were turning and soon in a few years, the shadow of Germany would lie across Europe. Then they would be humiliated once again, as they had been during Bismarck's time and many amongst them would bow to their country's aggressors and advocate cooperation with them.

In Japan, it was completely different experience. Einstein gave a lecture, which then had to be translated by an interpreter. He felt sympathy for his audience, as they had to sit for over four hours listening to something, which required only half the time. On the next occasion when he shortened his lecture, but he reckoned without the Japanese spirit. In most other countries if this happened, people would be relieved. This was not the case in Japan. Einstein came to know that his second audience took it as an affront, because he had cut his lecture short.

Finally, after a visit to Palestine and Spain in 1923 Einstein returned to Berlin.

RECOGNITION

After many delays and dithering, the Nobel committee finally and it seemed cautiously decided to award Einstein the Nobel Prize in Physics in 1921. It was for the "photoelectric law and his work in the domain of theoretical physics.", but not for relativity. The award was long over due. We cannot blame them. They are people who felicitate the great, but they are in no way great themselves, so they do not recognize greatness, even when they see it right in front of them. They give way if there is scent of any controversy and easily bend to political pressure. Many a great man has failed to measure up to their standard, because of such reasons. Gandhi was one of them. Jagadish Chandra Bose, the discoverer of wireless and Satyandra

Nath Bose of Bose Einstein condensation fame were amongst others. Incidentally, some people are not fated to be recognized. Even the Britannica Encyclopaeda, 1962 edition Vol. 8 p. 114A, confuses Satyandra Nath Bose with Jagadish Chandra Bose. However, Satyandra Nath Bose's name will live on in the boson. No doubt, there are many others like them from different parts of the world, who deserved the Nobel, but were passed over. On the other hand, many who did not deserve it have been awarded one.

It was in 1923, Einstein travelled to Sweden to receive the award.

The city council of Berlin had decided to honour their famous resident by gifting him some land for a country house. The matter however was mismanaged at every step, until it became a political issue. Einstein sensibly declined to be a part of it. If this had happened in Calcutta and the Calcutta Corporation had decided on such a thing, in spite of the fact that it is never able to do anything right, Einstein would have certainly had his garden house.

CHANGING ATMOSPHERE IN GERMANY
In 1930, Einstein went to California Institute of Technology at Pasadena; he was to return there in the winter of 1931 and again in the spring of 1932.

At this time Einstein's fame outside Germany and the honours bestowed upon him by those countries, set up a controversy in Germany. Times were changing; the people of Germany had not got over their humiliation after the defeat of their country in the Great War. Therefore, the showering of honours on Einstein by these very countries, which had defeated them, caused resentment. Also the poor economic condition of the average German following the Great War, compared to their harder working Jewish brethren, was rousing undercurrents of resentments. The political situation in Germany was gradually becoming intolerant. When Adolf Hitler came to power, such sentiments would turn into a flood.

Returning to Europe in the spring 1933 he did not go back to Germany, but went to Belgium. He took up residence in a small villa, which was near the seashore at Le Coq sur Mer. He got a warm welcome since their queen, Queen Elizabeth, took great interest in Einstein and his work. He was given protection against possible threats against his life.

That March in Germany the police visited Einstein's house and confiscated it. His works were publicly burnt in the square in front of the State Opera House in Berlin. Nonetheless, some professors continued to teach the concept of relativity, though without mentioning Einstein's name.

Einstein now joined the Institute of Advanced Study at Princeton in New Jersey in America, where he was to spend rest of his life. Just as James Clerk Maxwell had united electricity with magnetism showing they were two faces of the same phenomenon. Here Einstein pursued his goal of linking electromagnetism with gravity, which is known as the unified field theory. He had been trying to do this ever since after his development of the general theory of relativity. However, he would never realize this goal in his lifetime. After the success of general relativity, he had drifted away from the main stream of discoveries in physics. Physics had turned to the quantum theory with which he found difficult to come to terms.

In 1939, it became known that the German physicists, Otto Hahn (1879-1968), Lise Meitner

(1878-1968) and Fritz Strassmann (1902-1980) had discovered the fission of uranium. Its significance was drawn to the American president's attention through Einstein's letter. The atom bomb was never used against Germany, but resulted in the great human tragedy of Hiroshima and Nagasaki for which there was no need and it certainly could have been avoided.

THE END

Einstein retired from his official position at Princeton in 1945, but continued to work there. In 1948 on 13th April, Einstein had abdominal pain. He had an aortic aneurysm, which had to be reinforced to prevent it from expanding and rupturing. At last, the inevitable happened on the 17th April, 1955, when his abdominal aneurysm started leaking. When he spoke to his stepdaughter, Margot, who was in the same hospital suffering from sciatica, he appeared to be better. An operation was suggested to relive his condition, but Einstein refused saying, "It is tasteless to prolong life artificially. I have done my share, it is time to go." He breathed his last on 18th April, 1955.

BOOKS AND INTERNET SITES CONSULTED

BOOKS

Boris Kuznetsov: *Einstein*, Progress Publishers, Moscow 1965.

J. Bronowski: *The Ascent of Man*, Book Club Associates. (reprint) 1980.

Carl Sagan: *Cosmos*, Publishers Book Club Associates, London 1981.

Colin McEvedy: *The Penguin Atlas of Medieval History*, Penguin Books Published (reprint) 1986.

Callender, Craig and Edney, Ralph: *Introducing Time*, Icon Books,

Edited by Christopher Cook: *Pears Cyclopaedia 1984-85*, Pelham Books Ltd. 1984.

Encyclopaedia Britannica 1962 Edition.

Francis Watson: *A Concise History of India*, Thames and Hudson, 1981.

James Burke: *Connections*, Macmillan 1978.

Morris Kline: *Mathematical Thoughts from Ancient to Modern Times*, 3 vols., Oxford University Press. Paperback edition 1990.

Patrick Moore: *New Concise Atlas of the Universe*, Artists House 1982.

R. F. Tapsell: *Monarchs, Rulers, Dynasties and Kingdoms of the World*, Publishers Thames and Hudson Paperback (reprint) 1987.

INTERNET

Wikipedia.

Other internet sites as acknowledged.

APPENDIX

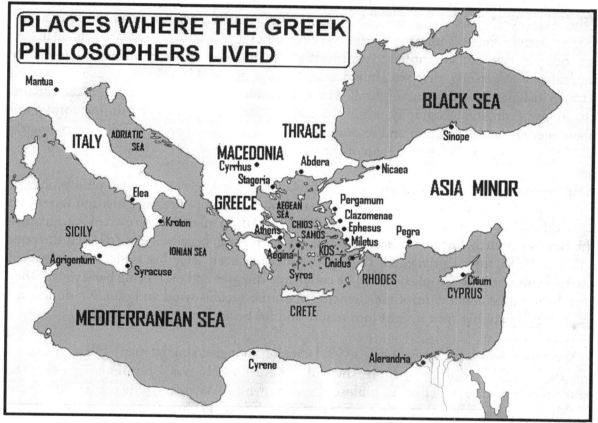

Places where the ancient Greek philosophers lived and worked.

1. **Zeno of Citium**, Greek philosopher (334-262BC). Founder of the Stoic school of philosophy, which stressed on goodness and peace of mind attained through living ones life in accordance with nature.

2. **Epicurus of Samos and Athens**, Greek philosopher (342-270BC) and founder of Epicureanism school of philosophy. He taught pleasure and pain was the measure of good and evil and that the gods were not there to reward or punish. The idea was to attain a tranquil and peaceful life through freedom from fear. He re-found Democritus' idea of the atom and thought the world was made up of atoms moving about in space.

3. Federico II of Mantua. Whom does he represent?

4. Anicius Manlius Severinus Boethius of Rome (c.480-c.525) a Christian philosopher of noble lineage. He was a senator and later consul. His fortune fell when he was accused of conspiracy by Theodoric the Great. His best work is the Consolation of Philosophy. Some of his works were influential and held their place in the Middle ages. They were translated by diverse people, including King Alfred, Chaucer and even Queen Elizabeth I.
or
Anaximander of Miletus (610BC-546BC) was a pre-Socratic Greek philosopher and a student of Thales. He belonged to the group of thinkers, who belonged to what is termed as "The Axial Age". This was a period between 800BC-200BC, when human thinking across Greece, Ionian, Persia, India and China in the modern sense began. He stressed that all nature was ruled by laws and attempted to explain the different aspects of the universe. Remarkably he realized something that we should take note of today. He said what ever disturbed nature's balance would soon succumb. His contribution extended to many fields, such as astronomy, geometry, geography, origins of human kind and tried to explain meteorological phenomenon on observation and reasoning. He gave the gnomon of sundials a geometric basis. Anaximander also drew a map of the then known world.
or
Empedocles of Agrigentum (490BC - 430BC) was another pre-Socratic philosopher of Greece, who is known for introducing the idea of four elements – fire, air, earth and water into Greek thinking. He was born to a good family and had been offered the crown of Agrigentum the city of his birth, but he refused. This was not surprising, because he befriended the poor and was harsh to the privileged. During his life and even later his achievements gained mythical proportions. People believed he could cure diseases and even bring back youth. They even believed he could control the elements of nature such as wind and rain. No doubt this was partly due to his keen insight into nature and his brilliance as an orator.

5. Averroes of Córdoba, Spain (1126-1198) { Córdoba not given on the map}. His actual name was Abū 'i-Walīd Muhammad ibn Ahamad ibn Rushid. He was a prominent Andalusian Muslim polymath. His school of philosophy is known as Averroism, which later influenced Jewish philosophers. He tried to reconcile reason with faith. He believed in the eternity of the world, which to him was a continuous process. This was in contrast to one instant of creation. He also thought there was a universal that was eternal and indivisible, but shared by all.

6. Pythagoras of Samos (570-495BC) was an Ionian Greek philosopher. It is said, Pythagorean philosophy influenced Plato and from him the ideas passed into western philosophy. He was a mystic, who believed numbers were the key to the knowing the universe. To every school child he knows about his famous discovery, the squares of the two sides of any right angled triangle adds up to the square if the hypotenuse.

7. Alcibiades of Athens (450-404BC) was a statesman, orator and general. Alternatively the image may be that of Alexander the Great of Macedon (356BC-323BC), who needs no introduction.

8. Antisthenes of Athens (445-365BC). A Greek philosopher and pupil of Socrates. Alternatively the image may be that of Xenophon of Athens (430-354BC), a soldier of fortune and a historian. He was one of those who led the leaderless Ten Thousand safely back home after the Battle of Cunaxa from the Persian heartland through hostile teritory. He was an ardent admirer of Socrates and known to taken his advice.

9. Hypatia of Alexandria [Image of Francesco Maria della Rovere]. She was the only pagan woman mathematician of antiquity. She also taught astronomy and philosophy. Many of her students came from afar. She was murdered by a mob of cowardly Christian fanatics in 415. With her death ended of what is known as Classical Antiquity.

10. Aeschines of Athens (389-314BC) Greek statesman or the image may be of Xenophon of Athens.

11. Parmenides of Elea. He was an early 5th century Greek philosopher, who believed that knowing through our senses could not give us the read picture of the world. It is only through logos or reason can the truth be attained.

12. Socrates of Athens (c.469-399 BC). The Greek philosopher and soldier, who distinguished himself in both fields. He is credited to be one of the founders of Western philosophy. He had many followers, amongst them Xenophon and Plato.

13. Heraclitus of Ephesus [Image of Michelangelo]. Lived 535-475 BC. He believed that change was central to the universe and all things were made of opposite qualities.

14. Plato of Athens (c.428-c.348) [Image of Leonardo da Vinci]. Greek philosopher and pupil of Socrates, who founded the Academy of Athens. It may be considered the first school of higher learning in the West. His most famous work was his *Dialogues* and his most famous student was Aristotle.

15. Aristotle of Stageria (384-322BC) was a Greek philosopher and a student of Plato. After Plato died he became the tutor to Phillip of Macedon's son Alexander. On his return to Athens and set up school in the garden of Lyceum. His school of philosophy became known as Peripatetic school. His varied interests and ideas influenced Western and Arab thought.

16. Diogenes of Sinope (c. 408-323BC). He lived in a tub and was the founder of Cynic philosophy. It is said he was the only person to have told Alexander the Great to move out of the sun, which he was enjoying and to have lived.

17. Possibly, Plotinus of Lycopolis, Egypt (204-270). He was a Greek philosopher, who founded Neoplatonism. He believed in the undivided "One". His mysticism influenced many religions, including pagan, Christians, Jews and Gnostics mystics.

18. Euclid of Alexandria (fl. 300BC), a mathematician famous for his Elements – book on geometry. Alternatively the image may be of Archimedes of Syracuse (287-212BC) with his students. He was one of the most innovative of ancient natural philosophers, who was also a mathematicians, astronomer and physicist. He is the inventor of the Archimedes screw, reflectors that could burn enemy ships and the principle that goes by his name. Every school child knows his name.

19. Zoroaster of Persia (6th century BC) founder of Zoroastrianism, the religion of today's Parsees', the descendants of Persians, who fled their home to India due to the persecution of Mohammedans. One of his contributions was the idea of Free Will.

20. Ptolemy of Alexandria (98-c. 168), astronomer. The image is also attributed to be that of Apelles – painter of Greece 4th century BC [Image of Raphel].

21. Protogenes – painter 4th century BC and a rival of Apelles [Image of Il Sodoma or Perugino or Timoteo Viti].

The bracketed names [ ] are the names of Raphael's contemporaries from whom the images of the ancients were painted.

REFERENCES

(Endnotes)

1   Wikipedia: Article on *Sun*
Website: http://en.wikipedia.org/wiki/Sun

2   NASA » Home » Multimedia » Worldbook@NASA » Moon
Spudis, Paul D. "Moon." World Book Online Reference Center. 2004. World Book, Inc. (http://www.worldbookonline.com/wb/Article?id=ar370060.)

3   Rev. Theodore Evelyn Rece Phillips and William Wilson Morgan: Article on *Saturn*. Encyclopaedia Britannica. Hazell Watson & Viney Limited (1962). Vol. 20, p 9.

4   Sagan, Carl: *Cosmos*, Book Club Associates, London by arrangement with Macdonald Futura (reprint) 1981. p 78.

5   Sagan, Carl: *Cosmos*, Book Club Associates, London by arrangement with Macdonald Futura (reprint) 1981. p 79.

6   Roemer, Elizabeth: Article on *Comet*, Encyclopaedia Britannica. Hazell Watson & Viney Limited (1962). Vol. 6, p 100A.

7   Yeomans, Donald K. "Comet." World Book Online Reference Center. 2005. World Book, Inc. Website: http://www.worldbookonline.com/wb/Article?id=ar125580.

8   The Oxford Dictionary of Quotations. Book Club Associates by arrangement with Oxford University Press, Fletcher & Sons, Norwich, Third Edition 1981. p. 396:4

9   Watson, Francis: *A Concise History of India*, Thames and Hudson. Camelot Press Ltd., Southampton. Paperback Edition 1981. pp 83 – 84

10   Watson, Francis: *A Concise History of India*, Thames and Hudson. . Camelot Press Ltd., Southampton. Paperback Edition 1981 p. 71.

11   Ashmore, Harry S et al.: Article on *Rudder*, Encyclopaedia Britannica. Hazell Watson & Viney Limited (1962). Vol. 19, pp 615 – 616.

12   Leif k. Karlsen: Navigation Notes, Viking Navigation using the Sunstone, Polarized Light and the Horizon Board. One Earth Press. Issue 93, pp 5-8.
Website: http://www.oneearthpress.com/pdf/nav_notes.pdf

13   Peter Ifland History of the Sextant from a talk given at the amphitheatre of the Physics Museum under the auspices of the Pro-Rector for Culture and the Committee for the Science Museum of the

University of Coimbra, Portugal, the 3rd October, 2000.
Website: http://www.mat.uc.pt/~helios/Mestre/Novemb00/H61iflan.htm

14   Internet: Classic Encyclopedia (Browsable online version of Encyclopaedia Britannica 1911) » Compass.
Website: http://www.1911encyclopedia.org/Compass

15   J. Bronowski. The Ascent of Man. .Book Club Associates by arrangement with the British Broadcasting Corporation, Sir Joseph Causton & Sons Ltd. 1977 reprinted 1980. pp. 65 – 68.

16   Buttery, Jr. Theodore Vern Article on *Atlas*, Encyclopaedia Britannica Hazell Watson & Viney Limited (1962). Vol. 2, p 630.

17   Ashmore, Harry S et al.: Article on *Frederick II* Encyclopaedia Britannica. . Hazell Watson & Viney Limited (1962). Vol. 9, pp. 711 – 713..

18   Wikipedia: Article on "Heliocentrism".
Website: http://en.wikipedia.org/wiki/Heliocentrism

19   Wikipedia: Article on "Tusi-couple".
Website: http://en.wikipedia.org/wiki/Tusi-couple

20   Personal communication: Mr. Arun Mullick FRCS Edin.

21   I regret that I have misplaced the reference of this quotation.

22   Internet: See under web site – http://en.wikipedia.org/wiki/File:Galileo.script.600pix.jpg.jpg prepared by Adrian Pingstone – NASA / JPL.

23   Wikipedia: Article on *Achromatic lens*
http://en.wikipedia.org/wiki/Achromatic_lens

24   Esterson, A. (2006). "Mileva Marić: Einstein's Wife":
Website: http://www.esterson.org/einsteinwife2.htm

INDEX

Symbols

17-year locust 134
95 thesis 98
Æther
  see ether 180
 Einstein, Elsa 220
  Einstein's second wife 220

A

Abbasid Dynasty 50
Abu Said Sinjari 90
Abu Ubayd al-Juzjani 90
Accelerating frame 200
Acceleration 182
Adams, John Couch 15
Addition law 203
Adelantados 79
Admonitio ad astronomos 125
Adonis
  asteroid 34
Africo
  name of wind 69
Aftercastle 74
Agriculture 84
Airy, Biddel George 16
Åkerbald, Johan David 178
al Biruni 90
Alchemy 120, 191
Alcmene 11
Aldabran
  star 106
Alexander the Great 52, 77, 132
Alexandria 48, 50, 87
Alfonso X of Spain 72
Algae 26, 27
Alidade
  parts of an astrolabe 65

al-kitabu-l-mijisti. See Almagest
Allegiance 162
Allegory of Good Government 148
Almagest 92, 95
Al - Ma'mun
  Caliph 50
Al-Ma'mun
  Caliph 95
Alpha Centauri A 18
Alpha Centauri B 18
Alpha Centauri C. Proxima Centauri
Alpha Centauri system 18
Amphitryon 11
Anchor escapement 150
Andromeda Galaxy 5, 102
Andronicus of Cyrrhus 69
Androsthenes 132
Angiosperms 27
Annalen der Physik
  Journal 194
Anne
  Queen of Great Britain 152
Antikythera mechanism 64
Antila (Air Pump)
  constellation 13
Aphelion 22
Apogee 30
Apollo 23
  asteroid 34
  of Greek mythology 12
Apollonius of Perga 64, 94
Aprilis
  April 141
Arabs 51
  and the astrolabe 64
Archimedes 88
Archimedis Syracusani Arenarius. See The Sand
  Reckoner

Area velocity 184
Arenarius. *See* The Sand Reckoner
Aristarchus 88, 89, 90
Aristophanes 107
Aristotle 48, 91, 130, 158, 159, 160, 162, 165, 181, 199, 200
  and the geocentric theory 91
  idea of inertia 159, 199
  idea of space 201
  idea of time 129
Arrhenius, Svante
  ideas about Venus 23
Aryabhatta 90
Asteroid belt 33
Asteroids 32
  origin 34
Astræa
  asteroid 33
Astrolabe 63, 64
Astronomiae Pars Optica
  (The Optical Part of Astronomy) 119
Astronomia Nova 121
Astronomical Unit (AU) 20
Atlas
  of Greek mythology 85, 86
Atomic clock 129, 137
AU. Astronomical Unit
Aurora
  of Greek mythology 86
Ayscough, Hannah 171
Ayscough, William 172

B

Balance wheel 151
Bank
  of oars 56
Barberini
  Cardinal 126
Barbosa, Diogo 78
Baron
  title 162
Barrow, Isaac 172
Barycentre 30, 185
Basilios Bessarion, Regiomontanus
  Cardinal 95
Battle of Diu 78
Battle of Lepanto 72
Beijing's Drum Tower 146
Bellarmine
  Cardinal 126, 128
Benátky nad Jizerou
  castle 118
  city 115
Benededitti, Giambattista 165

Bernard's star 18
Bessel, Friedrich Wilhelm 102, 128
Besso, Michele 194, 195
Big bang 9, 210
Bille, Beatte 109
Binary stars 181
Binary system 30
Binomial theorem 172
Biological clock 135, 136, 137
Biremes 56
Black Death 161
Board of Longitude 152
Bode, Johann Elert 32
Bode's Law 16
Boltzmann, Ludwig 196
Borelli 166
Borobodur
  ancient city 51
Bose Einstein condensation 222
Bose, Jagadish Chandra 221
Bose, Satyandra Nath 222
Boson 222
Brahe, Jörgen 109, 110, 111
Brahe, Otto 109
Brahe, Tycho 104, 109, 115, 118, 119, 120, 121, 122, 125, 128, 181, 185
  accurate observations 111
  and the crystal spheres 114
  builds Uraniborg 112
  death of his uncle 110
  early education 109
  early life 109
  his data passes to Kepler 120
  his death 120
  his duel 110
  his first interest in astronomy 110
  his first meeting with Kepler 118
  his life style 112
  his pursuit of excellence 114
  his second meeting with Kepler 119
  in Germany 111
  later education 110
  observes a new star 111
  returns to Denmark 111
  self banishment 115
  the accuracy of his observations 112
  the relevance of his work 114
Brocklesby Park 153
Brownian motion 195, 196, 197
Brown, Robert 195
Brudzewaski, Wojciech 106
Bruno, Giordano 96, 115, 127
Bubble 19
Byzantium 72, 90, 159

C

Cabo Deseado
  Cape Forward 80
Cabot, John 66
Caccini, Tomasso 126
Caesar, Julius 77, 142, 143
Calcium carbonate 60
Calendarium and Prognosticum 117
Calendars 84, 136, 141, 143, 144
Callisto
  Jupiter's satellite 35
Canali 31
Canals
  Martian canals 31
Candle clock 147
Canis Major
  constellation 12
Canis Minor
  constellation 12
Cantoni 196
Capillary forces
  Einstein's paper on 194
Capra, Balthazar 168
Caravel 75
  description 75
Carrack 75
Carriage clock 151
Cassini, Giovanni Domenic 36
Cassiopeia
  constellation of 111
Cathay
  China 75
Caulking 57
Celestial clock 58
Celestial navigation 59
Celichius, Andreas 39
Centrifugal force 186
  explained 186
Ceres 20, 33
Cesium atom 129
Chaloner, William 188
Champollion, Jean François 178
Charlemagne 160
Charles II
  King of Great Britain 151
Charles I of Spain 78
Charles Martel
  The Hammerer 161
Charon
  Pluto's companion 39
Childeric III 160
Chilias Logarithmorum 125
Chip log 68

Chlorophyll 26
Chola
  kings 51
Christian Church 160, 199
Christian IV
  King of Denmark 115
Chromosomes 84
Chronobiology 133
Church 79, 91, 92, 94, 96, 97, 98, 103, 115, 123, 127,
    143, 160, 161, 162, 163, 191
Cicada 134
Circa 133
Circadian 133
Circadian rhythm 133, 134
Circinus (Pair of compasses)
  constellation 13
Cladius Ptolemacus 91
Clarkia pulchella 195
Clausius, R. J. E. 196
Cleanthes 88
Clement, William 150
Clinker 57
Clock
  Seven day 150
Clovis I 160
Clubmoss 27
Cluster 6
Coin reform 107
Columbus, Christopher 66
Comet, long-range 20
Comets 187
  description 40
  ideas about comets in the past 39
  superstitions 39
Comma
  of a comet 40
Commenta Riolus
  (Little Commentary) 108
Compass rose 69
Compensated gridiron pendulum 150
Concave mirror 174
Conifers 27
Conservation of energy 208
Conservation of mass 209
Conservation of momentum 209
Constantinople. See Byzantium
Constellations
  introduction 11
Content of space-time continuum 218
Copernicus, Nicolas 88, 89, 90, 91, 95, 96, 97, 98, 99,
    102, 104, 106, 115, 117, 118, 121, 126, 128,
    163, 170, 184
  an appraisal of his achievement 109
  and the heliocentric theory 107

birth and ancestry 105
goes to study law 106
his connections on his maternal side 106
his education 106
peace negotiations and advice on monetary reforms 107
the story of publication of his book 108
Coriolis force 201
Corpuscular theory of light 176
Cosimo II
Grand Duke of Tuscany 167
Cosmic rays 20, 22
Council of Nicaea 191
Council of Trent
of 1545 98
Covalent bond 211
Cross- staff 63
Crown glass 174
Crusades 161
Cryatal spheres
and Tycho Brahe 115
Crystal spheres 86, 128
ancient beliefs 86
Cycads 27

D

d'Abreu, Antonio 78
da Ferrara, Domenico Maria Novarra 106
da Gama, Vasco 66, 71, 76, 77
Dante, Alighieri 95
d'Arrest, Heinrich 16
Darwin, Charles 96
da Vinci, Leonardo 82
Dead reckoning 59, 66
de Albuquerque, Afonso 78
de Almeida, Fransisco 77
de Broglie, Louis Victor 211
December 141
de Cervantes, Saavedra Miguel 144
Deferent 93
Deferents 92, 94, 108
de Fermat, Pierre 176
de Fonseca, Juan Rodriguez 78
De Harmonice Mundi
(Harmony of the World) 123
de Haro, Christopher 78
Deimos 31
del Cano, Juan Sebastian 81
Delsaulx 196
de Magalhães, Ferdinand. *See* Magellan, Ferdinand
de Magalhães, Pedro 77
De Revolutionibus Orbium Coelestium
(On the Revolutions of Celestial Orbs) 108
Descartes, René 166, 177

Dessertatio cum Nuncio Sidereo
(Conversation with the Starry Messenger) 122
De vero Anno 123
Dial face. *See* Dial plate
Dial furniture
features of a sundial 146
Dialogue Concerning the Two Chief World Systems 126
Dial plate
parts of a sundial 146
Diana 12
Dicotyledon 27
Diet 106
Differential calculus 172
Diffraction 177
Dioptrice 122
Discourse on Method. 177
Discourses and Mathematical Demonstrations Concerning the Two New Sciences 127
Donato, Leonardo
Dodge of Venice 168
Don John of Austria 72
d'ortoust Mairan, Jean-Jacques 132
Drosophila melanogaster
fruit fly 135
Dwarf planet 33
Dwarf planets 20, 39
Dysnomia
Eris' moon 39

E

Earth
and asteroid orbits 34
cues to the shape of the Earth 45
description 25
early atmosphere 25
early beliefs about its shape 44
evolving atmosphere 26
ideas of the Earth's place in the universe 85
lunar eclipse 47
measurement in light years 4
measuring the 47
place in the order of the universe 3
relative speeds of stars 14
shape of constellations 13
sidereal year 14
solar eclipse 47
true shape 81
various features in relation to evolution of life 28
Earth-Moon system 30, 181
East Indies 51
East Prussian Diet 107
Eclogae Chronicae 123

Eduard
    Einstein's second son 195
Egyptian
    fleet 78
Einstein, Albert 129, 130, 180, 181, 183, 187, 193, 194, 208
    a grand offer from Berlin 212
    and general relativity - the significance of his contribution 220
    and German anti-Semitism 222
    and Mach's principle 194
    and paradoxes 210
    and Professor Stumpf 213
    and the country house 222
    and the mathematics of general relativity 217
    as a boy 193
    as a teacher 211
    as viewed by his peers 213
    at Berlin 213
    during the Great War 214
    enstrangement with his first wife 213
    extends his idea of special relativity 212
    extends his work on general relativity 218
    formality a burden 212
    his early education 193
    his early life 193
    his explanation 206
    his first marriage 195
    his first paper 193
    his golden year 195
    his intellectual pursuits 194, 195
    his last illness 223
    his lecture tours 220
    his letter 223
    his real blunder 220
    in Belgium 222
    in England 221
    in France 221
    in Japan 221
    in persuit of the unified field theory 222
    in Zurich 212
    marries Elsa 220
    on Brownian motion 197
    on equivalence of mass and energy 208
    paper on Brownian motion 197
    paper on special relativity 207
    paper on the photo electric effect 209
    rescues Newton 209
    settles in America 222
    sums up the discovery of special relativity 208
    the concept of general relativity 214
    the consequences of fame 222
    the Nobel Prize 221
    theories on contribution to his work by Mileva 211
    theory of special relativity 198
    years that followed the golden year 211
Einstein, Rudolf 220
Eleanor of Aquitaine
    queen 47
Embolon 56
Embolos 56
Embolus 56
$E = mc^2$ 209
Emmer 84
Emperor Huang-ti
    Chinese legend of the compass 70
Emperor Montezuma 39
Energy-momentum tensor 218
Enzymes 135
epicycles 94
Epicycles 92, 93
Epitome Astronomia Copernicanae
    (Epitome of Copernician Astronomy) 125
Equant 94, 95
Equants 92, 108
Equivalence of mass and energy 195
Eratosthenes 48, 50, 51, 192
    cirticism 48
    measurement of the Earth 48
    sources of error in measuring the Earth 50
Eris 20
    dwarf planet 39
Eros
    asteroid 34
Escapement mechanism 149
Ether 180, 181, 203
Eudoxus of Cnidus 91
    and the geocentric theory 91
Europa
    Jupiter's satellite 35
Evening Star 23
Experiments and Calculations Relative to Physical Optics 179
Extra ordinary ray
    polarization 61
Eyepiece 175

F

Face. *See* Dial plate
Falerio, Ruy 78
February 142
Februus 142
Federal Office for Intellectual Property
    Swiss patent office 195
Felony 162
Ferdinand II
    Archduke of Stria 119
Fermat's principle 176

Ferns 27
Feudal system
　structure 162
Fictitious forces 201
Fidalgo escudeiro 78
Fief 161, 162
Field equations 217, 218
First law of planetary motion 121
Fitzgerald, George 198, 205, 206
Fixed star 14
Flamsteed, John 190
Flint glass 174
Floki the Raven. See Floki Vilgerdarson
Floki Vilgerdarson 60
Florentine Academy 164
Flower clock 132
Fluxions 172
Foliot 149
Force 181, 183
Forecastles 74
Fore-mast 74
Foucault, Jean Bernard Léon 180
Frame–dragging 219
Frankish tribes 160
Franklin 162
Frauenburg Cathedral 106
Frederick II
　Holy Roman Emperor 95
　King of Denmark 95, 110
Fresnel, Augustin-Jean 178, 180
Freud 39
Fruit fly. See Drosophila melanogaster

G

Galactic magnetic field 20
Galactic year 20
Galilean moons 35, 122
Galilean transformation 202
Galilei, Vincenzo 125, 164
Galileo Galilei 9, 15, 88, 96, 99, 103, 115, 118, 122,
　　128, 163, 164, 168, 170, 181, 199, 200, 202
　also looks at the Moon, phases of Venus and the
　　Milky Way 169
　an apprasial 170
　and Saturn's rings 37
　and the telescope 165
　birth and his parents 164
　discovery of Jupiter's moons substantiated 168
　discovery of the idea of the pendulum 148
　draft letter to Leonardo Donato 168
　his education 164
　his last days 171
　his recommendation about Kepler 123
　honoured by the Venetian senate 169
　insight about force 165
　insight into motion 165
　last phase of his life 169
　makes a telescope 166
　names Jupiter's moons 167
　opens a new era in astronomy 167
　studies on motion 165
　the story of his fall 126
　the story of the coin and the cannon ball 164
　turns his telescope towards Jupiter 36
　why he missed out on gravitation 165
Galle, Johann, Gottfried 16
Galleys 56, 73
Gamba, Marina 171
Ganymede
　Jupiter's satellite 35
Gas giants 34
Ge 30
Gear train 149
General relativity
　discussion 214
　Einstein's paper on 214
　experimental verifications 218
Genes 135
Genome 135
Geocentric hypothesis 91, 92, 99, 100, 106, 126, 127,
　　128
Geocentric theory 90, 91, 92, 93, 96, 98, 100, 102,
　　128
Geodesy 48
Geoid 81
　definition 81
George II
　King of Great Britain 207
George III
　King of Great Britain 156
Gerard of Cremona 95
Germanic tribes 162
Ghiraldi, Luigi Lilio. See Lilius, Aloysius
Global positioning system
　(GPS) 157
Gnomon
　parts of a sundial 146
Goat grass 84
Göttingen University 207
Graham, George 153
Grandfather clock 150
Grassi, Orazio 170
Gravitation 130, 165, 170, 172, 181, 183, 184, 185,
　　186, 212, 214, 216, 217
Gravitational time dilation 219
Great Book. See Almagest
Great Comet
　of 1577 116

Great Plague 187
Greco
   name of wind 69
Greek idea
   of how the sky was held up 85
Greek Orthodox Church 144
Greeks 44, 47, 170
   and their metaphysical idea of ether 181
Greenwich meridian 152
Gregorian calendar 123, 143, 171, 201
Gresham's Law 107
Grimaldi, Francesco Maria 177
Grossmann, Marcel 212, 217
Guilds 163
Gulf Stream 54
Guttenberg 161
Gymnosperms 27

H

H1 154
H2 154
H3 154
H4 155, 156
H5 156
Hahn, Otto 222
Halberg, Franz 133
Half-minute sand glass 68
Halley, Edmond 39, 153, 155, 190
Halley's Comet 39
Hannibal 77
Hans Albert
   Einstein's first son 195
Harlan Graduate Library
   Special collection 169
Harriot, Thomas 177
Harrison, John 150, 152, 153, 154, 155, 156, 157, 190
Harun ar-Rashid 50
Head
   of a comet 40
Heimholtz, Hermann 194
Heliacal rising of Sirius 138
Heliocentric theory 88, 89, 90, 91, 96, 98, 100, 102, 103, 108, 117, 118, 127, 128
Heliochronometers 147
Heliopause 22, 41
Heliosheath 22, 41
Heliosphere 41
   Oort cloud 42
Helm 56
Hè Megalè Syntaxis. *See* Almagest
Hencke 33
Henry II
   King of England 47
Henry the Navigator
   Prince 75, 162
Hera 11
Heracles 11
Hermes
   asteroid 34
Hero of Alexandria 176
Herschel, William Fredrick 10, 15, 101, 127
Hertz, Gustav 213
Hertz, Heinrich 194
Hesiod 85, 140
Hidalgo
   asteroid 34
Hieroglyphics 178
Hilbert, David 207, 217, 220
Hillf 56
Hindus 44, 45, 89
Hipparchus 48, 64, 94, 95, 101
Hitler, Adolf 222
Hlios 22
HMS Centurion 154
Holdfast 26
Holy Land 106
Holy Roman Emperor 161
Holy War 106
Homer 46, 85, 86, 87
Hooke, Robert 36
Hooke. Robert 176, 177, 189, 190
Horologium 69
Horologium (Clock)
   constellation 13
Horsetails 27
Hourglass 147
Hour lines
   features of a sundial 146
Hugyens, Christiaan 180
Hume, David 39, 195
Hus, Jan 98
Hussey, Rev. T. J. 15
Hutton, Charles 165
Huygens, Christiaan 36, 148, 150, 175, 176, 203
   shape of the Earth 81
Hven 115
Hven in the Sont 111
Hydra
   Pluto's moon 39
Hydrostatic balance
   Galileo's 164
Hydrostatic equilibrium 82
Hylotheism 191

I

Ibn al-Shatir 90

Ibn Sahl 177
Icarus
    asteroid 34
Ice Age 84
Icelandic feldspar 60
Ices 34, 35, 39
Incense clock 147
Indica 90
Indu
    River Indus 44
Indulgences 97
Industrial revolution 157
Industrial Revolution 136
    and guild workers 163
Inertia 9, 99, 128, 159, 160, 181, 183, 194, 199, 200, 215, 216
Inertia forces 200
Inertial 195, 200, 202, 214, 215
Inertial frames 200, 208, 214, 215
Inferior planet 24
Infradian rhythm 134
Inquisition 126, 127
Integral calculus 172
Interferometer 204
Internal energy 209
In Terra inest virtus, quae Lunam ciet
    (There is a force in the Earth which causes the Moon to move) 119
Interstellar cloud 19
Interstellar medium 19, 20, 22
Interstellar winds 41
Inverse square law 102
Investigation of the State of Aether in Magnetic Fields. 193
Io
    Jupiter's satellite 35
Island universes 128
Isochronism 149
Isostatic equilibrium 82

J

Jahangir
    Mogul Emperor 151
James I
    King of Great Britain 171
James II
    King of Great Britain 187
Jansen, Sacharias. *See* Jansen Zacharias
Jansen, Zacharias 166
Janssen, Zacharias 166
January 142
Janus 142
Jeffrys, John 154

Jesenius, Jan 119
João de Barros 71
John II the Perfect
    King of Portugal 77
Journeyman 163
Jovian planets 34
Junius
    June 141
Juno
    asteroid 33
Jupiter
    description 35

K

Kalends
    calendar 141
Kamal 63
Karlsen, Leif K 62
Katharina
    Kepler's mother 116, 125
    Kepler's mother accused of whichcraft 124
Kelvin, William Thompson 204
Kendal, Larcum 156
Kepler, Heinrich 116
Kepler, Johannes 14, 15, 88, 96, 99, 104, 109, 114, 115, 116, 117, 118, 119, 120, 122, 123, 124, 125, 126, 128, 165, 166, 168, 170, 177, 181, 184, 185, 187
    and his idea of force (gravity) in the Earth 119
    and Reimarus Ursus 118
    and religion 117
    and Tycho Brahe's data 119
    an outline of Astronomia Nova 121
    asked to fill Tycho's place as Imperial Mathematician 120
    asks for help from Jesenius 119
    as Tycho Brahe's guest 118
    at Linz 123
    banished from Graz 119
    birth and childhood 116
    caught up in uncertain times 118
    contribution to optics 119
    death of first wife 123
    declines offer to go to Italy 123
    discovers the first two laws of planetary motion 121
    espouses the heliocentric theory 116
    finally obtains Tycho's data 120
    his first book 117
    his first marriage 118
    ideas 118
    impediment as a teacher 116
    negotiations with Tycho fails 119

position reconfirmed as Imperial Mathematician 123
premonition of gravity 122
reconciliation with Tycho 119
replies to Galileo 122
second marriage 123
story of his mother's arrest 124
the connection between his third law and Newton's laws of gravitation 184
the third law of planetary motion 123
uncertainty 123
understanding of gravity 122
uses a telescope and suggests a modification 122
views on astrology 117
writes his most influential work Epitome Astronomia Copernicanae 125

Ketu
  of Hindu mythology 46
Keynes, John Maynard 191
King 162
King Harold 39
King of Cebu 80
King of Mactan 80
King of Siam 166
King Rajendra I 52
Kirchhoff, Gustav 194
Kirkwood gaps 33
Knight 162
Knots 56
  in navigation 63
Knotted line 66
  description 68
Krakowac Academy 106
Kronborg Library 110
Kuiper belt 33, 38
  description 38
Kuiper belt object 20

L

Laccadive Islands 52
Ladrones
  island 80
Lande 21185 18
Langevin, Paul 197
Lantern 147
Lanthorn
  see lantern 147
Larmor, Joseph 206
Lateen sail 74
Lateran Council
  of 1514 107
Latitude
  determining the 59
Latitude hook 63

Law of uniform acceleration 165
Laws of gravitation 114, 128
Laws of motion 128, 165
Laws of planetary motion 170, 181
Leaning Tower of Pisa 164
Leap year 143
Leibniz, Gottfreid Wilhelm 129, 130, 189, 190, 191
Leif Ericsson 57
Letters on Sunspots 170
Leventer
  name of wind 69
Le Verrier, Urbain Jean Joseph 16
Levi-Civita, Tullio 212
Lewis, Gilbert Newton 211
Libeccio. See Africo
Lieserl
  Einstein's daughter before marriage 195
Light waves 179
Light year 20
  definition 4
Lilius, Aloysius 143
Linnaeus, Carl 132
Lippershay, Hans. See Lippersheim, hans
Lippersheim, Hans 166
Liverworts 27
Lobachevski, Nikolai Ivanovich 207
Local Bubble 19
Local Fluff 19
Local group 5, 6
Local spur 19
Longitude
  knowing the 58
Longitude and latitude 58
Longitude Prize 154
Long period comets 41
Long ships 57
Lord Vishnu 46
Lorentz, Hendrik 198, 204, 205, 206, 208, 210
Lorenzitti, Ambrogio 148
Louis XIV
  King of France 201
Lowell, Percival 32
Lucasian Professor 172, 191
Luciferase 135
Luminiferous æther 180
Lunar calendar 97, 138, 141, 142
Lunar Distance Method 151
Lunar revolution 142
Lunar rhythm 134
Lunation 142
Luni-solar calendar 97
Luyten 726 and 728 18

M

Mach, Ernst 194
Mactan
  island 81
Maestlin, Michael 116, 117
Maestro
  name of wind 69
Magellan, Ferdinand 66, 77, 78, 79, 81
  Magellanic Clouds 5
  proof of circumnavigation 81
  significance of his journey to Banda 78
  voyage 79
Magellanic Clouds 5
Magicicada septendecim 134
Magnetic compass 66, 70
Magnetite 71
Main-mast 74
Maius
  May 141
Malacca 78
Maldive
  islands 52
Manifold
  what is a 217
Manuel I the Fortunate
  King of Portugal 77
Mari (Marity), Mileva 194
marine chronometer 66, 152, 153, 154
Marine chronometer 153
  history of development of the 151
Marius, Simon 168
  and his claim of priority over Galileo 168
Mars 110
  description 31
Mars' orbit
  the significance of 119
Martin Luther 89, 98, 108, 163
Martius
  March 141
Marx, Karl 220
Maskelyne, Nevil 155, 156, 190
Master craftsman 163
Mastering the waters 53
Masts
  ship's 73
Mater
  parts of an astrolabe 64
Mathematical and Philosophical Dictionary 165
Mathematike Syntaxis. *See* Almagest
Matthias
  Emperor 122
Mausim 59
Maxwell, James Clarke 194, 196, 203, 222

Mayans 9, 89
Mayors of the palace 160
McGregor, Tracy W. 169
Mechanica 115
Medicean Stars 167
Medici's
  of Florence 162
Meitner, Lise 213, 222
Melanopsin ganglion cells 133, 134
Mercedinus 142
Mercury
  description 22
  result of close proximity to Sun 23
Mercury poisoning
  and Tycho Brahe's death 120
Merovingian Kingdom 160
Metius, Jacob (or James) 166
Metric tensor 218
Metric theory 217
Michelson, Albert Abraham 193, 203, 204, 205, 220
Michelson and Morley's experiment. 205
Michelson-Morley experiment 204, 205
Michelson-Morley's experiment 207
Microscopium (Microscope)
  constellation 13
Middelburg
  town in Netherlands 166
Middle Ages 161
Mileva 195, 212, 213
  Einstein's divorce with 220
Military compass
  Galileo's 168
Milky Way 10
  its identity 11
  the story of its creation 11
Milky Way Galaxy 3, 16, 19, 128
  distinction from Milky Way 10
  size 5
Mimosa pudica 132
Minkowski, Hermann 207, 208, 220
Minute hand 151
Mister Time 131
Mizzen-mast 74
Moluccas 79
  proof of Magellan's circumnavigation 81
Monetae Cudendae Ratio 107
Monocotyledon 27
Montague, Charles 188
Moon 30
  distance in light years 4
  place in the order of the universe 3
Morley, Edward Williams 193, 204, 205, 220
Morning Star 23
Mosses 27

Motto
  features of a sundial 146
Mount Olympus 31
Moveable type 161
Müller, Johannes 95
Mundus Jovialis 168
Muslin
  cloth 52

N

Napoleon Bonaparte 178
Narritio Prima
  (Primary Narration) 108
Natural History 142
Navigation 54
  concept of 58
Neckham, Alexander 72
Nelson, Horatio 78
Neptune 15, 33
  description 38
  of Greek mythology 12
Nernst, Walther 213
Neutrinos 22
New Style (NS)
  calendar 144
Newton
  applies the concept of gravity 183
Newton, Isaac 9, 15, 45, 99, 102, 114, 117, 119, 120, 122, 127, 128, 129, 130, 144, 152, 165, 171, 172, 174, 175, 176, 177, 181, 186, 187, 188, 189, 190, 191, 199, 200, 201, 207, 209, 215
  achievements 187
  and John Flansteed 190
  and Kepler's laws 184
  and Ralph Cudworth 190
  and Robert Hooke's papers 190
  and the corpuscular theory 176
  and the riddle of the colours 173
  and William Chaloner 188
  at Woolsthorpe during the plague 172
  becomes the warden and then master of the mint 188
  birth and parents 171
  brought back from school to help at the farm 172
  early education 171
  enters politics 187
  first insight into gravitation 172
  his contribution to Kepler's work 181
  his corpuscular theory eclipsed 180
  his darker side 189
  his death 189
  his disappointment 176
  his dualistic theory 177
  his first law of motion 181
  his ideas on how scientific theories should be arrived at 176
  his interest in alchemy 191
  his law of universal gravitation 185
  his laws of motion 181
  his second law of motion 182
  his third law of motion 183
  in perspective 190
  re-elected as President of the Royal Society 189
  returns to the university followed by his mental breakdown 188
  return to Cambridge 172
  shape of the Earth 81
  stabilizes Britain's economy 188
  story of how he came to like school 171
  study of light 172
  takes up the challenge against counterfeiting 188
  the connection between his laws of gravitation and Kepler's third law 184
  the love of his youth 189
  the man 189
  the religious side of his nature 191
  the story of his conflict with Leibniz 189
  the story of the falling apple 183
Nix
  Pluto's moon 39
Nodus
  parts of a sundial 146
Nomadic way
  of life 84
Non-inertial frame 200
Non-inertial frames 214
Nova Steriometria Doliorum vinariorum
  (New Stereometry of Wine Barrels) 123
November 141
Nucleus
  of a comet 40
Null geodesics 218
Numa Pompilius 142

O

Oakum 58
Oar 54
Objective 174
oblate spheroid 81
Observations upon the Prophecies of Daniel 191
October 141
Oehl 196
Old Style (OS)
  calendar 144
Olsztyn Castle 107
Olympian Games 48
On a Heuristic Viewpoint Concerning the Produc-

tion and Transformation of Light 209
On Burning Mirrors and Lens 177
On the Electrodynamics of Moving Bodies 207
Oort cloud 20, 41
  description 42
Optical calcite 60
Optic chiasma 134
Opticks 177
Orcus
  dwarf planet 39
Ordinary ray
  polarization 61
Orion
  story from mythology 12
Orion arm 11, 19
Orion, of Greek mythology 12
Orion, the constellation 12
Osiander, Andreas 108, 109
Ostro
  name of wind 69
Ottoman
  fleet 78

P

Pacific
  Magellan's passage across the 80
  naming of the 80
Palembang
  ancient city 51
Pallas
  asteroid 33
Pallas Athena 11
Parabolic mirror 175
Parallax 14, 99, 100, 101, 102, 114, 119, 128
Parallax phenomenon 14
Parsburg, Manderup 110
Patagonians 79
Pendulum 148
  \ 150
  isochronism 149
Pendulum clock 148
  principle 148
Pepin III
  See Pepin the Short 161
Pepin of Héristal 161
Pepin the Short 160, 161
Perigee 30
Perihelion 22
Periodical cicada 134
Perseus arm 19
Philolaus
  of Kroton 88
Philosophiae Naturalis Principia Mathematica. *See*
  Principia

Phobos 31
Photon 21, 211
Photon theory of light 195
Piazzi, Giuseppe 33
Pick, Georg Alexander 217
P'ing-chou-k'o-t'an 71
Planck, Max 203, 209, 210, 213
Planet 15
Planetary Hypothesis 95
Plasma 22
Plato 91
Pliny the Elder 142
Plutarch 88
Pluto 15
  distance in light years 4
  dwarf planet 39
Pocket watch 151
Poincaré, Henri 130, 195, 207
Point to point navigation 58
Point-to-point navigation 151
Poisson, Simeon Denis 180
Polarization
  of light 61, 179
Pole Star 59, 99
  in navigation 62
Polish Teutonic War 107
Polo, Marco 75
Pomerania 106
Ponente
  name of wind 69
Pope Alexander VI 76
Pope Clement VII 108
Pope Gregory XIII 143
Pope Julius II 77
Pope Nicholas V 95
Pope Urban VIII 126
Porta, John Baptisa 166
Portolan charts 69
Port side 55
Poseidonius 50
Power source
  in the days of sail 54
Precessed 108
Pretzel 14
Prince-Bishop of Ermeland 106
Principia 128, 172, 187, 189, 190, 200
Principle
  of sailing 53
Prism 173
Privatdozent 211
Professor \"extraordinary\" 211
Proper motion 14
Protestant 143, 187
Proxima Centauri 5, 17

Ptolemy IV of Egypt 57
Ptolemy of Alexandria. *See* Cladius Ptolemacus
Ptolemy of Alexandria 72
Ptolemys of Egypt 87
Puthi 88
Pythagoras of Samos 47
Pythagoreans 88

Q

Quadrant 66
Quadreme 57
Quaestores 97
Quantum theory 209
Quaoar
  dwarf planet 39
Queen Leonor 77
Quib al-Din Shirazi 90
Quinqueremes 57
Quintilis
  Roman month 141

R

Radie 110
Radius vector 184
Rahu
  of Hindu mythology 46
Red dwarf 17
Reference frame 197
Regolith 27
  creation of soil 27
Regulators.
  clocks 136
Reimarus Ursus
  (Nicolaus Reimers Bär) 118
Reinel, Pedro 72
Relative motion 202
Relativistic 198
Renaissanc 161
Rest mass 209
Rete
  description 65
  parts of an astrolabe 64
Retrograde motion 93, 100, 108
Reuttinger, Susanna
  Kepler's second wife 123
Rhaticus, Georg Joachim 108
Rhodes 50
Ricci-Curbastro, Gregorio 217
Ricci, Ostilio 164
Riemannian manifold 217
Rings
  Saturn's and the story of their discovery 36
Robert Bellarmine
  Cardinal 103
Roche sphere 42
Roemer, Ole 201
Roe, Thomas
  of East India Company 151
Rolland, Romain 214
Roman calendar 141
Rome
  Roman empire 160
Rømer, Ole 169
Romulus 141, 142
Rosamund 47
Rosa ventorum. *See* Wind rose
Rosetta stone 178
Ross 154 18
Rostock University 110
Rother 56
Rower 56
Royal Observatory at Greenwich 151
Rudolf II
  Emperor 122
  King of Bohemia and Holy Roman Emperor 115
Rudolphine Tablets 96, 123

S

Sacy, Silvestre 178
Saggiatore
  (Assyer) 170
Sagittarius arm 19
Sagittarius, constellation of 16
Salviati 126
Satellite
  coining the term 122
Saturn
  description 36, 37
  satellites 37
Saturn's rings
  description 37
Scattered disc 38
  description 39
Sceptical Chymist 191
Schiaparelli, Giovanni 31
Scholer, Wolfgang 111
Schöner, Johannes 108
Schrödinger, Erwin 213
Second law of planetary motion 121
Seleucus of Seleucia 88
Selim II
  Sultan 151
Seljuk Turks 161
Seneca, Annaeus Lucius 40
September 141
SETI
  Search for Extra Terrestrial Intelligence 127

Seville 79, 81
Sextilis
 Roman month 141
Shailendra Kingdom 51
Shakespeare, William 144
Shatapatha Brahmana 88
Ship
 development 72
Ship's watch 67
Short period comets 41
 origin 39
Sidereal year 14
Sidereus Nuncius
 (Starry Messenger) 122, 167
Sigismund I
 King of Poland 107
Silk Route 53
Simplicius 126
Sir Gawain and the Green Knight
 story of 139
Sirius 13, 18
Sirius B 18
Sirocco
 name of wind 69
Sirturus 166
Smith, Barnabus 171
Snell, Willebrord 177
Sobral
 in Brazil 219
Socotra
 island 52
Sodium calcium aluminum silicate. *See* Sunstone
Sol. *See* Sun
Solar calendar 88, 97, 138, 140, 141, 143
Solar calendars 141
Solar system
 constituents of the 20
Solar wind 41
Solar wind. 22
Soli-lunar calendar 138
Solstices 84
Somnium
 (The Dream) 125
Sosigenes of Alexandria 142
Sound waves 179
Space-time 130, 208, 216, 217, 218
Space-time continuum 206
Spanish galleon 75
Special relativity 198, 208, 216, 217, 220
Special theory of relativity 195, 208, 209, 212, 215, 217
Spherical aberration 173
Spice Islands 51
Spice trade 51

Spinoza, Benedict 191
Srivijaya Kingdom 51
Stadia 48
 definition 48
Starboard 55
Star tables 72, 95, 96
St. Augustine 129
St. Bernard 98
Steerside
 see Starboard 55
Stern oar rudder 54
Sternpost rudder 56
Stjerneborg 112
Stonehenge 84, 137, 140
Storer, Anne 189
Straits of Magellan 80
Strassmann, Fritz 223
Stumpf
 Professor 213
St. Ursula
 story of 79
Style
 parts of a sundial 146
Subsizar 172
Substyle height
 fearures of a sundial 146
Sudarsan Chakra 46
Sultan of Gujarat 78
Sultan Soliman 110
Summer solstice 48, 49, 139
Sun 16, 18, 21, 88, 101, 128
 composition 21
 description 20
 distance in light years 4
 in navigation 62
 internal conditions 21
 Kepler's idea of the Sun having an orbit of its own 122
 place in the order of the universe 3
 proper movement 20
 relative motion 14
 source of energy 21
Sundials 146
Sun's
 chromosphere 21
 convection zone 21
 core 21
 photosphere 21
 radiation zone 21
Sunstone 62
Supercluster 6
Supernova 111
Suprachiasmatic nucleus 134
Syene 48, 49, 50

T

Tabulae Rudolphinae
  (Rudolphine Tablets) 125
Tail
  of a comet 40
Talmud, Max 193
Tamarind leaf 132
Tanner, Hans 211
Taqui al Din Muhammad ibn Ma'ruf al-Shami al-Asadi 151
Tau Ceti 18
Telescope
  history of invention of the 165
Telescopium (Telescope)
  constellation 13
Temple of Winds 69
Tensor
  what is 218
Tensor calculus 217, 218
Termination of shock 41
Territio verbalis 125
Teutonic Knights 106
Teutonic Order 105
The five perfect solids of the Greeks 117
The Frogs
  the play 107
The Great Treatise. See Almagest
Theory of gravitation 165
Theory of special relativity 195
The Prodromus Dissertationum Mathematacarum continens Mysterium Cosmographicum
  (Sacred Mystery of the Cosmos) 117
The Sand Reckoner 88
Third law of planetary motion 123
Thought experiment 194
Three Fates
  of Greek mythology 86
Tiberius
  Roman Emperor 52
Tierra del Fuego 79
Tiller 56
Time
  biological time - discussion 132
  deprivation 136
  man's innate sense of time - a discussion 131
  nature of time - discussion 129
Titans 30
Tombaugh, Clyde 16
Torricelli, Evangelista 171
Tramonta 70
  name of wind 69
Transformation 197
Trans-Neptunian region 38
Traverse board 67
  description 67
  description of bottom part 68
Treatise on Light 176
Treaty of Toresilla 77
Trinidad
  ship 81
Triremes 57
Trojans
  asteroids 34
Tychonian system 114, 128, 168
Tycho's nose
  the story of 110
Tyge Ottesen Brahe. See Tycho Brahe
Tympan
  description 64
  parts of an astrolabe 64
Tympans
  set of 66

U

Ultradian rhythm 134
Unified field theory 222
Unireme 56
Universal gravitational force. 183
Universe 127
  as a dynamic entity 8
  size 6
Universe is expanding
  a prediction of general relativity 219
University of Copenhagen 111
Uraniborg 112
Uranus 15
  description 37
  moons 38
  of Greek mythology 30, 112
Ursus, Reimarus 118, 119

V

Valles Marineris 31
van Roigen, Willebrord Snellius. See Snell, Willebrord
Varuna
  dwarf planet 39
Vassal 162
Vasuki
  the divine snake 44
Vedic texts 87
Venus 23
  description 23
Verge escapement 150
Vesta
  asteroid 33

Vikings
  and celestial navigation 59
  and the Sunstone 60
  navigation 57
Vineland 57
Viviani, Vincenzo 164, 171
Voigt, Woldemar 206
von Gleichen, F W 195
von Hohenberg, Henstart 118
von Laue, Max 213
von Muhleck, Barbra 118
von Pruskow, Oldrich Desiderius Pruskowsky 115
von Rosenbergs
  Baron 120
von Wittenburg, Johann Daniel Titus 32
von Zach, Baron Franz Xaver 33

W

Waczenrode, Lucas 106
Warmia 106, 107
War of Austrian Succession 154
Water clocks 144
Water currents
  in navigation 54
Water resistance
  in navigation 54
Wave theory of light 176, 178
White dwarf 18
Wild wheat 84
Wimanstadt, Johann Albrecht 108
Wind rose
  or rosa ventorum 68
Winteler, Jost.
  Professor 194
Winteler, Sofia Marie-Jeanne Amanda 194
Winter solstice 84
Winter solstices 139
Wittenburg Castle Church 98
Wolf 18
Wolfius 166
World line 208
Worshipful Company of Clockmakers'
  Guildhall, London 153
Wycliffe, John 98

Y

Yajnavalkya
  Hindu sage 87
Young, Thomas 178

Z

Zamorin of Calicut 78
Zenith distance 66

Zeus 11

About the Author

I asked him what he was going to write "about the author". He said, "It is not who I am, but what I have to say that is important." It sounded laconic, so I took over the task.

We were asked take a test after our schooling was completed, so we might have an idea of our future vocation. When the results came back, all of us had received at least one suggestion. Except Arun, who had none!

Looking back he would have fitted into in any profession. He is a man of ideas, but not a crank. I suspect he would like to have a new idea every day. Sometimes when I ask him what new ideas he has had today, he occasionally explains them. At other times he remains quite. One gets the feeling he is thinking about something new.

His ideas are always grand, but not grandiose. When he heard about the communication-gap between patients and doctors, which included nearly all aspects of patient management, resulting in misunderstandings, delays, even failure to get proper treatment, he became curious to know how they occurred. After long discussions, he came up with an idea of an all-embracing program, which was designed from the patients' point of view. It would allow the doctor more time to listen to patients, increase safety threshold, reduce malpractice, pinpoint medicolegal responsibility, reduce paper work and facilitate insurance claims. Being a surgeon of many years standing I was quite impressed. Unfortunately there were no takers.

His life has been interrupted by many illnesses, some life threatening. He has taken it all unflinchingly. Today with the sword of Damocles hanging over his head, he wants to give the finishing touches to the other two parts soon. The only thing that upsets him is his Cassandra syndrome.

Arun Mullick